Introductory Algebra

Introductory Algebra

Michael A. Gallo
Monroe Community College

Charles F. Kiehl
State University of New York, Brockport

WEST PUBLISHING COMPANY
St. Paul   New York   Los Angeles   San Francisco

Production Coordination: Editing, Design & Production, Inc.
Composition: Syntax International

Library of Congress Cataloging in Publication Data

Gallo, Michael A.
  Introductory algebra.

  Includes index.
  1. Algebra.   I. Kiehl, Charles F.   II. Title.
QA152.2.G35   1984   512.9   83-21825
ISBN 0-314-78001-7

To my parents, Joe and Rose Gallo,
and to two of the world's finest secondary mathematics teachers,
Jim Sudore and Dick Toole (MAG);
and
To my father, Charles Kiehl,
to the memory of my mother, Thelma,
and to my loyal home supporters:
my wife Margo, and my children,
Amy, Karen, and Christopher (CFK).

Contents

Preface

This text is written for college students who are taking a first course in algebra. In writing this text, we have assumed that students using the book have little or no algebraic background.

Specifically, the distinctive pedagogical features and organization of the text include the following:

Introductions: Every topic or concept of each section is first introduced through either a diagram or logic to motivate the student and present rationale for method.

Procedures: A step-by-step algorithm, developed naturally from the introduction and summarizing the necessary steps for the solution to a problem, is provided for nearly every section. All procedures are boxed off from the text and are highlighted in a second color for easy reference.

Examples: A large number of graded examples follow each procedure. The development of the examples are telescopic in nature. The first example, for instance, is completely worked out using the procedure as a directional guide. Successive examples, however, show less detail, encouraging students to gradually build their problem-solving skills.

Practice Exercises: At strategic points in the discussion practice exercises consisting of problems dealing with a specific concept are provided. These practice exercises afford students the opportunity to test their understanding of a given topic before doing the section exercises. Typically, practice exercises immediately follow the examples of a section, with the problems keyed directly to the previously completed examples. Answers to the practice exercises are provided in the right-hand margin for immediate reinforcement.

Section Exercises: A substantial number of graded exercises are provided at the end of each section. In addition to providing the necessary practice needed to master the material of each section, certain exercises also offer some thought-provoking problems designed to further enhance specific concepts. Answers to all odd-numbered problems for section exercises are provided in the back of the text.

Chapter Review Exercises: A separate set of chapter review exercises that provides a mix of problems discussed in the chapter can be found at the end of each chapter. Exercises in color represent typical problems that might be included on a chapter test. Answers to all chapter review exercises are provided in the back of the text.

Notes: Special notes in boxes provide added depth to a particular concept. In most cases the notes reinforce a previously learned idea. In other instances students are alerted to typical errors to be avoided.

Verbal Problems: Chapter 5 is devoted entirely to the development of problem-solving techniques for solving verbal problems. The problems presented in this chapter are of the traditional type (number, coin, mixture, investment, and motion), and are carefully developed with highly detailed examples. This technique is further applied in subsequent chapters (e.g., Chapter 8—work problems; Chapter 11—using systems of equations to solve verbal problems). Each type of problem is treated separately, with matching exercise sets. The chapter review exercises do, however, provide a mix of problem types.

Graphics: In addition to enhancing the overall physical attraction of the text, the use of a second color and shading have been thoughtfully and carefully incorporated within the examples to aid in visualizing the solution of a problem. For the most part, shaded items represent a quantity that is being manipulated in a particular step of the process, and a second color is used to represent the result or answer of this manipulation. Such use of graphics is both effective and complements the learning process.

Basic Arithmetic Review: Chapter 1 provides a complete and thorough review of basic arithmetic.

Sets: A separate discussion of sets is provided in the Appendix.

INSTRUCTIONAL RESOURCES

In addition to the main text, an instructor's manual, student study guide, and computer-assisted instruction disks are available.

The instructor's manual contains six different but parallel chapter tests for each chapter in the textbook, two comprehensive final examinations, answers to the chapter tests included in the manual, answers to the final examinations included in the manual, and answers to all exercises found in the text.

The student study guide briefly reviews key concepts and all procedures for students. It also contains examples taken from the odd-numbered problems in the text to illustrate procedures. Additional drill exercises with answers and additional chapter tests with answers are included in the study guide.

The computer-assisted instruction disks provide students with additional drill, practice, and review in areas of difficulty.

ACKNOWLEDGMENTS

As it is with any work of this magnitude, there are many behind-the-scene individuals who either have assisted us or have provided us with guidance, direction, and encouragement. It is here that we acknowledge their tangible, and, in some cases, intangible effect.

The following reviewers read all or part of the manuscript during its various stages. We thank you for sharing your suggestions and insights with us.

Benjamin Bockstege, Jr. (Broward Community College)
Ray Boersema (Community College of Denver)

Tommy Caldcleuth (North Harris County College)
Ellen Casey (Massachusetts Bay Community College)
Robert Christie (Miami-Dade Community College, South)
Eunice Everett (Broward Community College)
K. Elayn Gay (University of New Orleans, Lakefront)
Alleyne Hartnett (Corning Community College)
Paul Hutchens (St. Louis Community College at Florissant Valley)
Corinne Irwin (Austin Community College)
Richard Linder (Southwestern College)
Rose Novey (Saginaw Valley State College)
Jack Porter (Cuyoga Community College)
Susan Wagner (Northern Virginia Community College, Annandale)
Edward B. Wright (Oregon College of Education)

Typists Barbara Kays, Taeko Heiser, Linda McIntyre, LuAnn Merz, and Dorothy Reed were responsible for typing various parts of the manuscript. We appreciate your time and skill and are grateful for the quick turn-around time that you provided us.

We also thank Marta Fahrenz, Production Editor, West Publishing Company, and Mark Walter, Project Editor, Editing, Design & Production, Inc., for their harmonious efforts in producing this text, and to Kristen McCarthy, Marketing Coordinator. To Peter Marshall, editor and friend, we extend a very warm thank you for having faith in this project and helping us through our "dark period." Also, special acknowledgments are deserving to Jack Pritchard, Larry Gilligan (Mattatuck Community College), Jane Edgar (Brevard Community College), and Nancy Keymel (Steno Services, Monroe Community College), who all helped in his or her own way.

Finally, we thank our wives, Mary Ellen and Margo, for their patience, love, and understanding during our periods of writing and frustration.

MAG
CFK

Introductory Algebra

CHAPTER 1

The Structure of Arithmetic

1-1 THE NUMBER LINE AND THE NUMBERS OF ARITHMETIC

In studying arithmetic, we operate with many different types of numbers—counting numbers, whole numbers, decimal numbers, fractional numbers, and so on. In this section, we will review these *numbers of arithmetic*.

A. The Number Line

Before we begin our review of the numbers of arithmetic, we introduce the *number line*. A number line is simply a line whose points are named by numbers. By way of an example, a number line with some of the numbers of arithmetic is pictured here.

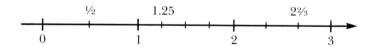

To construct a number line:

1. Draw a straight line.

2. Place an arrowhead at the right end of the line. (This means the line may be extended without ever ending.) The number line also extends to the left, but this will not be discussed now.

3. Pick a point on the line and label it "0."

4. To the right of 0, pick another point and label it "1."

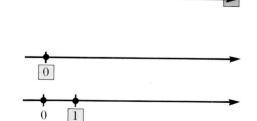

1

5. Mark off more points to the right of 1, using the length between 0 and 1 as a unit of measure. Label these points 2, 3, 4, and so on.

We now use the number line as an aid in our review of the numbers of arithmetic.

B.   Counting Numbers or Natural Numbers

Whenever we count anything we begin with a first number [which we call *one* (1)], followed by a second number [which we call *two* (2)], followed by a third number [which we call *three* (3)], and so on. Each of these numbers is one more than the number it follows (e.g., 2 is *one more* than 1; 3 is *one more* than 2; etc.).

These *counting numbers* are also consecutive in nature. That is, each number follows one after the other, in order, without any interruptions. As a result, the method of counting has no end; that is, there is no last counting number. (The counting numbers are also commonly referred to as the *natural numbers.*)

The counting numbers are easily pictured on a number line.

The counting numbers may also be represented as

$$1, 2, 3, 4, 5, 6, 7, 8, 9, 10, \ldots$$

NOTE:

> The three dots after the 10 have the same meaning as the arrowhead on the number line; these numbers continue in the exact same pattern without ever ending.

C.   Whole Numbers

As previously mentioned, whenever we count something we start counting with the number 1, followed by 2, 3, and so on. We never begin counting with zero (0). Thus, zero is not a counting number. However, if we include zero with the counting numbers, we have a new collection of numbers which we call the *whole numbers.*

The whole numbers are easily pictured on a number line.

The whole numbers may also be represented as

$$0, 1, 2, 3, 4, 5, 6, 7, 8, 9, 10, \ldots$$

D.   Decimal Numbers

Decimal numbers (or just *decimals*) represent a subdivision of whole numbers. This subdivision is regarded as a special subdivision because decimal values are always expressed in powers of ten.

Decimal numbers contain three parts: whole number(s), decimal point, and decimal digit(s). For example, in the decimal 12.06:

The whole number part is 12.	12 .06
The decimal digits are 0 and 6.	12.0 6
The decimal point separates the whole number part and decimal number part.	12 . 06

The term *decimal places* means the number of decimal digits (or the number of places to the right of the decimal point). Thus,

12.06 contains two decimal places   12. 06
                                             ⌄
                                      2 PLACES

Decimals may also be represented on a number line. For example, to represent the decimal three tenths (0.3):

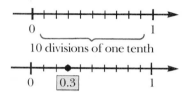

Divide the interval between the whole numbers into ten equal parts—each part represents one tenth (0.1) of the whole.

The decimal, 0.3, is the point that represents three tenths of the whole.

If we want to represent the point three hundredths (0.03) or three thousandths (0.003), we would divide the interval between the whole numbers into one hundred parts (where each part represents one hundredth [0.01] of the whole) or into one thousand parts (where each part represents one thousandth [0.001] of the whole).

E.   Fractional Numbers

Fractional numbers (or just *fractions*) also represent a part of a whole. Fractions are written in the form of a/b where a is a whole number (0, 1, 2, 3, . . .) and b is a natural number (1, 2, 3, . . .).

NOTE:

> The top number, a, is called the *numerator*, and the bottom number, b, is called the *denominator*. The line separating the top and bottom numbers is called a *fraction bar* and represents a division.
>
> Fraction bar ↝ $\dfrac{a}{b}$ ← numerator
>   ← denominator

Some examples of fractions are

$$\frac{0}{5}, \frac{1}{3}, \frac{2}{8}, \frac{3}{7}, \frac{5}{9}, \frac{12}{16}, \frac{99}{100}, \frac{8}{5}, \text{ etc.}$$

In arithmetic, we learned that there are many different classifications of fractions. We briefly review them here.

■ *Proper fractions* are fractions in which the numerator is smaller in value than the denominator.

$\frac{1}{5}, \frac{2}{3}, \frac{7}{10}, \frac{99}{100}$, etc.
proper fractions

■ *Improper fractions* are fractions in which the numerator is equal to or greater in value than the denominator.

$\frac{4}{4}, \frac{5}{3}, \frac{7}{2}, \frac{99}{80}$, etc.
improper fractions

■ *Like fractions* are fractions that contain the same denominator.

$\frac{1}{7}, \frac{2}{7}, \frac{3}{7}, \frac{4}{7}, \frac{5}{7}$, etc.
like fractions

■ *Unlike fractions* are fractions that contain different denominators.

$\frac{1}{7}, \frac{3}{6}, \frac{5}{4}, \frac{7}{8}, \frac{9}{10}$, etc.
unlike fractions

Fractions may easily be illustrated by using a number line. We simply divide the interval between the whole numbers into the number of equal parts indicated by the denominator.

For example, we can divide the interval into

■ *Halves*:

■ *Thirds*:

■ *Fourths*:

■ etc.

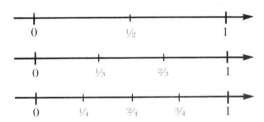

F. Mixed Numbers

A mixed number is considered a fractional number. However, it is a special fractional number. Mixed numbers represent fractional parts greater than one.

Mixed numbers are usually written in the form of: whole number + fraction . Some examples follow.

$$1\frac{2}{3}, \; 2\frac{1}{5}, \; 3\frac{4}{8}, \; 5\frac{6}{7}, \; \text{etc.}$$

NOTE:

> Mixed numbers represent the *sum* of a whole number and a proper fraction. Thus, $1\frac{2}{3}$ means $1 + \frac{2}{3}$.

Mixed numbers are also easily illustrated by using a number line. For example, $1\frac{2}{3}$ means one (1) whole interval plus an additional $\frac{2}{3}$ of an interval.

$2\frac{1}{5}$ means two (2) whole intervals plus an additional $\frac{1}{5}$ of an interval.

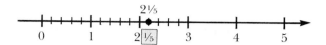

$3\frac{4}{8}$ means three (3) whole intervals plus an additional $\frac{4}{8}$ of an interval.

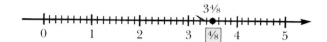

$5\frac{6}{7}$ means five (5) whole intervals plus an additional $\frac{6}{7}$ of an interval.

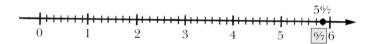

etc.

   We offer the following examples involving the number line and the numbers of arithmetic.

EXAMPLE 1:   State the number that represents the points A, B, and C on the number line shown here.

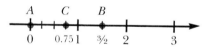

SOLUTIONS:
- ■ Point A is represented by 0.
- ■ Point B is represented by $\frac{3}{2}$.
- ■ Point C is represented by 0.75.

NOTE:

> Whenever a number represents a point on the number line the number is commonly referred to as the *coordinate* of that point. Thus, in example 1:
>
> - ■ The *coordinate* of point A is 0.
> - ■ The *coordinate* of point B is $\frac{3}{2}$.
> - ■ The *coordinate* of point C is 0.75.

EXAMPLE 2: *Graph* the following numbers on the number line: 0.5; $1\frac{3}{4}$; $\frac{12}{12}$.

To graph a number on the number line means to locate the point that can be represented by that number.

SOLUTIONS: ■ 0.5 is located halfway between 0 and 1.

■ $1\frac{3}{4}$ is located three fourths of the interval between 1 and 2.

■ $\frac{12}{12}$ is simply the number 1.

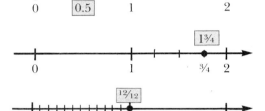

EXAMPLE 3: Using a number line, determine how many *whole numbers* there are between 2 and 6.

There are three whole numbers between 2 and 6. They are 3, 4, and 5.

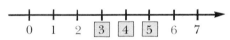

NOTE:

> If we want to *include* 2 and 6, we would use the word inclusive. For example, "How many whole numbers are there between 2 and 6, *inclusive*?" The answer is now five: 2, 3, 4, 5, and 6.
>
>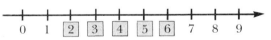

Do the following practice set. Check your answers with the answers in the right-hand margin.

PRACTICE SET 1-1

Using the number line given, state the coordinates of each of the following points (see example 1).

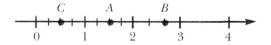

1. Point A 2. Point B 3. Point C

1. $1\frac{1}{2}$

2. $2\frac{2}{3}$

3. $\frac{2}{4}$ or $\frac{1}{2}$

Graph the following numbers on the number line (see example 2).

4. $\frac{3}{4}$

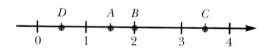
4.

5. 1.25

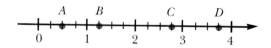
5.

6. $\frac{6}{2}$

6.

Determine how many whole numbers there are between each of the following intervals. List these numbers (see example 3).

7. Between 0 and 5

8. Between 2 and 3

9. Between 0 and 6 inclusive

7. four $(1, 2, 3, 4)$

8. none

9. seven $(0, 1, 2, 3, 4, 5, 6)$

EXERCISE 1-1

In each of the following problems, state the coordinate of points A, B, C, and D.

1. a. coordinate of point $A = $ _____
    b. coordinate of point $B = $ _____
    c. coordinate of point $C = $ _____
    d. coordinate of point $D = $ _____

2. a. coordinate of point $A = $ _____
    b. coordinate of point $B = $ _____
    c. coordinate of point $C = $ _____
    d. coordinate of point $D = $ _____

3. a. coordinate of point $A = $ _____
    b. coordinate of point $B = $ _____
    c. coordinate of point $C = $ _____
    d. coordinate of point $D = $ _____

For each of the following problems, graph the numbers on a number line.

4. The first four counting numbers.

5. The first four whole numbers.

6. $\dfrac{1}{3}, \dfrac{3}{3}, \dfrac{5}{3}, \dfrac{7}{3}, \dfrac{12}{3}$

7. $\dfrac{1}{4}, \dfrac{5}{4}, \dfrac{7}{4}, \dfrac{8}{4}, 2\dfrac{1}{4}, 2\dfrac{3}{4}$

8. $\dfrac{3}{10}, \dfrac{6}{10}, \dfrac{9}{10}, \dfrac{30}{10}, \dfrac{35}{10}$

9. $\dfrac{1}{10}, \dfrac{1}{5}, \dfrac{1}{4}, \dfrac{1}{3}, \dfrac{1}{2}$

10. $\dfrac{2}{4}, 0.75, 1.5, 2\dfrac{1}{8}$

Using the number line shown here, answer the following questions.

11. List all the points that represent natural numbers.

12. List all the points that represent whole numbers but *not* natural numbers.

13. List all the points that represent fractions.

14. List all the points that represent mixed numbers.

1-2 THE RATIONAL NUMBERS

A. Fractions

In arithmetic, we learned how to raise a fraction to higher terms and how to reduce a fraction to lower terms. We briefly review these now.

RAISING A FRACTION TO HIGHER TERMS

To raise a fraction to higher terms, we simply multiply both the numerator and denominator by the same number. For example, to raise $\frac{3}{5}$ to a higher term:

Multiply the numerator and denominator by the same number (2).

$$\frac{3}{5} = \frac{3 \times 2}{5 \times 2} = \frac{6}{10}$$

Thus,

$$\frac{3}{5} = \frac{6}{10}$$

NOTE:

> When raising a fraction to higher terms, we find that the two fractions are still equal in value. Thus, $\frac{3}{5}$ and $\frac{6}{10}$ are *equivalent fractions*.

REDUCING A FRACTION TO LOWER TERMS

To reduce a fraction to lower terms, we simply divide both the numerator and denominator by the same number that is also a *factor* of both numbers. The concept of dividing by a common factor is called *canceling a common factor*. For example, to reduce $\frac{8}{12}$ to lower terms:

Simply cancel the common factor, 2, in both the numerator and denominator
(2 goes into 8, 4 times; 2 goes into 12, 6 times).

$$\frac{8}{12} = \frac{\overset{4}{\cancel{8}}}{\underset{6}{\cancel{12}}} = \frac{4}{6}$$

Thus,

$$\frac{8}{12} = \frac{4}{6}$$

NOTE:

When the only common factor to both the numerator and denominator is 1, then the fraction is said to be *reduced to lowest terms*. For example:

- The common factor to 8 and 12 is 2 (2 goes into 8, 4 times; 2 goes into 12, 6 times).

$$\frac{\overset{4}{\cancel{8}}}{\underset{6}{\cancel{12}}} = \frac{4}{6}$$

- The common factor to 4 and 6 is 2 (2 goes into 4, 2 times; 2 goes into 6, 3 times).

$$\frac{\overset{2}{\cancel{4}}}{\underset{3}{\cancel{6}}} = \frac{2}{3}$$

- The common factor to 2 and 3 is 1.

$$\frac{2}{3}$$

Thus,

$$\frac{8}{12} \text{ reduced to lowest terms is } \frac{2}{3}.$$

This process may be done in one step if you recognize the largest common factor. For example:

- The largest common factor to 8 and 12 is 4 (4 goes into 8, 2 times; 4 goes into 12, 3 times).

$$\frac{\overset{2}{\cancel{8}}}{\underset{3}{\cancel{12}}} = \frac{2}{3}$$

- The common factor to 2 and 3 is 1. Thus, $\frac{8}{12}$ reduced to lowest terms is $\frac{2}{3}$.

$$\frac{2}{3}$$

Do the following practice set. Check your answers with the answers in the right-hand margin.

PRACTICE SET 1-2A

Reduce to lowest terms.

1. $\frac{4}{6}$

2. $\frac{9}{12}$

3. $\frac{18}{24}$

1. $\frac{2}{3}$

2. $\frac{3}{4}$

3. $\frac{3}{4}$

B. Decimal–Fraction–Mixed Number Conversions

In arithmetic, we learned that decimals may be expressed as fractions and fractions as decimals. We review this here.

CHANGING A DECIMAL TO A FRACTION

To express the decimal, 0.7, as a fraction:

We write a fraction whose numerator is the given decimal without the decimal point,

$$0.\,7 \qquad \dfrac{7}{\,}$$

And whose denominator is a power of ten (10, 100, 1000, etc.), depending on the number of decimal places in the given decimal. That is:

$$0.7 = \dfrac{7}{10}$$

1 decimal place

■ For one decimal place, the denominator is 10.

■ For two decimal places, the denominator is 100.

■ For three decimal places, the denominator is 1000.

    etc.

Thus,

$$0.7 = \frac{7}{10}$$

NOTE:

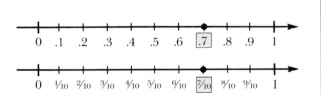

On the number line, the graph of 0.7 and $\frac{7}{10}$ represent the same point.

Another example follows. To express 0.065 as a fraction, we note:

The numerator is the given decimal without the decimal point.

$$0.\,065 \qquad \underline{65}$$

The denominator is 1000 since there are three decimal places in the given decimal.

$$0.065 = \dfrac{65}{1000}$$

3 decimal places

Thus,

$$0.065 = \frac{65}{1000}$$

CHANGING A FRACTION TO A DECIMAL

To express the fraction, $\frac{3}{5}$, as a decimal,

Divide the denominator into the numerator.

$$\frac{3}{5} = \begin{array}{r} 0.6 \\ 5\overline{)3.00} \\ -0 \\ \hline 30 \\ -30 \\ \hline 0 \end{array}$$

Thus,

$$\frac{3}{5} = 0.6$$

NOTE:

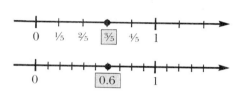

On the number line, the graph of $\frac{3}{5}$ and 0.6 represent the same point.

In arithmetic, we also learned that all mixed numbers may be expressed as improper fractions and improper fractions may be expressed as mixed numbers. We review this here.

CHANGING A MIXED NUMBER TO AN IMPROPER FRACTION

The mixed number, $1\frac{2}{3}$, may be quickly and easily converted to an improper fraction by:

Multiplying the denominator of the fractional part (of the mixed number) by the whole number. $1\frac{2}{3}$ $3 \times 1 = 3$

Adding this product to the numerator of the fractional part (of the mixed number). $1\frac{2}{3}$ $3 + 2 = 5$

Then, putting this sum over the denominator. $1\frac{2}{3} = \frac{5}{3}$

Thus,

$$1\frac{2}{3} = \frac{5}{3}$$

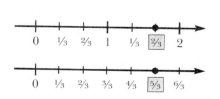

On the number line, the graphs of $1\frac{2}{3}$ and $\frac{5}{3}$ represent the same point.

CHANGING AN IMPROPER FRACTION TO A MIXED NUMBER

The improper fraction, $\frac{7}{3}$, may be quickly and easily converted to a mixed number by:

Dividing the denominator into the numerator. (The answer, 1, is the whole number part of the mixed number.)

$$\frac{7}{5} = 5\overline{\smash{\big)}7}$$
$$\frac{-5}{2} = \text{remainder}$$

Placing the remainder over the denominator (this fraction now is the fractional part of the mixed number).

$$\frac{7}{5} = 1\frac{2}{5} \quad \begin{array}{l}\text{remainder}\\ \text{denominator}\end{array}$$

Thus,

$$\frac{7}{5} = 1\frac{2}{5}$$

NOTE:

On a number line, the graphs of $\frac{7}{5}$ and $1\frac{2}{5}$ represent the same point.

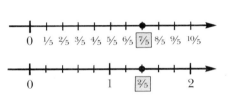

Do the following practice set. Check your answers with the answers in the right-hand margin.

PRACTICE SET 1-2B

Change each decimal to a fraction (reduce if possible).

1.   0.8

2.   0.09

3.   1.53

Change each fraction to a decimal.

4. $\frac{1}{4}$

5. $\frac{3}{5}$

6. $\frac{2}{50}$

Change each mixed number to an improper fraction (reduce if possible).

7. $3\frac{2}{3}$

8. $5\frac{3}{7}$

9. $12\frac{1}{9}$

1. $\frac{8}{10} = \frac{4}{5}$
2. $\frac{9}{100}$
3. $1\frac{53}{100}$

4. 0.25
5. 0.6
6. 0.04

7. $\frac{11}{3}$
8. $\frac{38}{7}$
9. $\frac{109}{9}$

Change each improper fraction to a mixed number (reduce if possible).

10. $\dfrac{15}{2}$ 11. $\dfrac{7}{7}$ 12. $\dfrac{24}{5}$ | 10. $7\frac{1}{2}$
          11. 1
          12. $4\frac{4}{5}$

C. Comparing the Numbers of Arithmetic

In arithmetic, we learned that if we are given two numbers, then one of the following will occur.

1. The numbers will be equal to each other.   *8 is equal to 8.*

2. The first number will be larger than the second number.   *16 is greater than 5.*

3. The first number will be smaller than the second number.   *12 is less than 23.*

We also learned that we use special symbols to compare the values of numbers.

Symbol	Meaning	Example	Read
$=$	is equal to	$8 = 8$	8 is equal to 8
$>$	is greater than	$16 > 5$	16 is greater than 5
$<$	is less than	$12 < 23$	12 is less than 23

We now discuss each of these symbols.

1. THE EQUAL SIGN ($=$)

To show that two numbers are equal in value, we place the equal sign ($=$) between them. For example, to show that the sum of $4 + 3$ is equal in value to the sum of $6 + 1$ (i.e., they both represent the same number),

Place an equal sign between them. $4 + 3 = 6 + 1$

The mathematical statement, $4 + 3 = 6 + 1$ is read

"Four plus three *is equal to* (or *equals*) six plus one."

This type of statement is commonly referred to as a *statement of equality*. These statements may be either true or false. For example:

$4 + 3 = 6 + 1$ is a true statement (both $4 + 3$ and $6 + 1$ name the same number, 7).

$4 - 3 = 6 - 1$ is a false statement ($4 - 3$ names the number 1, and $6 - 1$ names the number 5).

Placing a line through the equal sign changes its meaning from "is equal to" ($=$) to "is *not* equal to" ($\neq$).

Thus, we can make the false statement, $4 - 3 = 6 - 1$, into a true statement by replacing the equal sign ($=$) with the not equal sign ($\neq$).

$$4 - 3 = 6 - 1 \quad \text{(false)}$$
$$4 - 3 \neq 6 - 1 \quad \text{(true)}$$

NOTE:

> The properties of equality of numerals are: reflexive, symmetric, and transitive. These properties are sometimes called the RST relations.
>
> 1. *Reflexive property of equality (R)*
>   Any number is equal to itself. For example:
> $$2 = 2 \qquad 10 = 10 \qquad 999 = 999$$
>
> 2. *Symmetric property of equality (S)*
>   Any equality may be reversed. For example:
>   If
> $$2 + 6 = 8$$
>   then
> $$8 = 2 + 6$$
>
> 3. *Transitive property of equality (T)*
>   If one number is equal to a second number, And, in turn, this second number is equal to a third number, Then, the first number is also equal to the third number.
> $$\underline{5 + 5} = \underline{6 + 4}$$
> $$\underline{6 + 4} = \underline{7 + 3}$$
> $$\underline{5 + 5} = \underline{7 + 3}$$

2. THE GREATER THAN AND LESS THAN SYMBOLS ($>$, $<$)

If we locate two numbers on a number line, the *smaller* of the two numbers is always to the *left* of the larger number and the *larger* of the two numbers is always to the *right* of the smaller number. For example:

2 is smaller than 5 since the graph of the number 2 is to the left of the graph of the number 5.

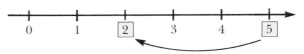

Also,

5 is greater than 2 since the graph of the number 5 is to the right of the graph of the number 2.

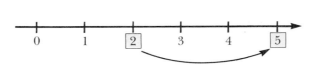

To show that one number is greater than or less than another number, we place either the *greater than* symbol ($>$) or the *less than* symbol ($<$) between the two numbers. For example:

To state that 2 is less than 5, we write $2 < 5$
"2 is less than 5."

To state that 5 is greater than 2, we write $5 > 2$
"5 is greater than 2."

These types of statements are commonly referred to as *statements of inequality*. Just like statements of equality, these statements are either true or false. Also, we can write the negative of the statement by placing a line through the symbol. For example:

$3 < 5$ is read "3 is less than 5" and is a true statement.

$5 < 3$ is read "5 is less than 3" and is a false statement.

$5 \not< 3$ is read "5 is not less than 3" and is a true statement.

$5 \not> 3$ is read "5 is not greater than 3" and is a false statement.

NOTE:

A good way to remember the meaning of the inequality symbols is to note that the "arrow" always points to the smaller number:

$7 < 15$ or $15 > 7$

7 is *less* than 15   15 is *greater* than 7

COMPARING DECIMALS

Before comparing decimals, we must be certain that all the decimals have the same number of decimal places. We do this by attaching (or annexing) zeros to the right-hand side of the decimal number. This is referred to as making the decimals *similar*.

NOTE:

Attaching zeros to a decimal does not alter its value. For example:

$0.5 = 0.50 = 0.500 = 0.5000$

Once the decimals are similar, we then compare them as if they were whole numbers. For example, to compare 0.18 and 0.0061:

First, make the decimals similar. (Since 0.0061 has four decimal places, we must attach two zeros to 0.18 to give *it* four decimal places.)

0.18	0.0061
0.18 00	0.0061
4 decimal places	4 decimal places

Now, ignore the decimal point and compare the numbers as if they were whole numbers.

| 0.1800 | 0.0061 |
| 1800 > | 61 |

Thus,

$$0.18 > 0.0061$$

COMPARING FRACTIONS

To compare fractions, we must be certain that the fractions are like fractions (i.e., same denominators), then we can simply compare their numerators. For example, to compare $\frac{2}{3}$

and $\frac{3}{5}$:

First, change them to like fractions.

$$\frac{2}{3} = \frac{10}{15} \qquad \frac{3}{5} = \frac{9}{15}$$

Now, compare the numerators of the like fractions.

$$\frac{2}{3} = \frac{10}{15} \qquad \frac{3}{15} = \frac{9}{15}$$

$$10 > 9$$

Thus,

$$\frac{2}{3} > \frac{3}{5}$$

Do the following practice set. Check your answers with the answers in the right-hand margin.

PRACTICE SET 1-2C

Compare the following whole numbers (use the symbols, $<$, $>$, or $=$).

1. 3, 2   2. 7, 10   3. 5, 5

Compare the following decimals (use the symbols, $<$, $>$, or $=$).

4. 0.653, 0.666   5. 0.006, 0.06   6. 0.87, 0.8

Compare the following fractions (use the symbols, $<$, $>$, or $=$).

7. $\frac{7}{8}, \frac{5}{8}$ 8. $\frac{5}{9}, \frac{2}{3}$ 9. $\frac{3}{4}, \frac{2}{3}$

1. $3 > 2$
2. $7 < 10$
3. $5 = 5$

4. $0.653 < 0.666$
5. $0.006 < 0.06$
6. $0.87 > 0.8$

7. $\frac{7}{8} > \frac{5}{8}$
8. $\frac{5}{9} < \frac{2}{3}$
9. $\frac{3}{4} > \frac{2}{3}$

D. The Rational Numbers

If a number can be represented by a fraction, that is, in the form a/b, where a is a whole number and b is a natural number, then the number is called a *rational number*. For example:

■ *Natural numbers* and *whole numbers* are rational numbers since they can be written in fractional form.

$$0 = \frac{0}{1} \qquad 1 = \frac{1}{1} \qquad 2 = \frac{2}{1} \qquad 3 = \frac{3}{1} \qquad 100 = \frac{100}{1} \qquad \text{etc.}$$

■ *Decimals* are rational numbers since they too can be expressed in fractional form.

$$0.3 = \frac{3}{10} \qquad 0.75 = \frac{75}{100} \qquad 2.7 = 2\frac{7}{10} = \frac{27}{10} \qquad \text{etc.}$$

■ *Fractions* are rational numbers.

$$\frac{1}{8} \quad \frac{1}{4} \quad \frac{1}{2} \quad \frac{3}{5} \quad \frac{2}{3} \quad \text{etc.}$$

■ *Mixed Numbers* are rational numbers.

$$1\frac{3}{5} = \frac{8}{5} \qquad 1\frac{2}{3} = \frac{5}{3} \qquad 5\frac{3}{9} = \frac{48}{9} \qquad \text{etc.}$$

It is obvious to see, then, that all the numbers of arithmetic that we have discussed thus far are rational numbers.

NOTE:

In Section 2-1B, we will provide a more complete definition of a rational number.

EXERCISE 1-2

Reduce to lowest terms.

1. $\frac{3}{9}$ 2. $\frac{12}{24}$ 3. $\frac{2}{8}$

4. $\frac{12}{20}$ 5. $\frac{18}{30}$ 6. $\frac{20}{24}$

7. $\frac{9}{6}$ 8. $\frac{12}{8}$ 9. $\frac{15}{3}$

Change each decimal to a fraction (reduce if possible).

10. 0.4   11. 0.825   12. 0.008

13. 0.75   14. 0.012   15. 0.85

16. 1.4   17. 5.12   18. 6.48

Change each fraction to a decimal.

19. $\frac{3}{4}$ 20. $\frac{1}{5}$ 21. $\frac{1}{10}$

22. $\frac{1}{2}$ 23. $\frac{7}{20}$ 24. $\frac{4}{5}$

25. $\frac{3}{8}$ 26. $\frac{5}{6}$ 27. $\frac{2}{3}$

Change each mixed number to an improper fraction (reduce if possible).

28. $2\dfrac{1}{2}$

29. $5\dfrac{1}{6}$

30. $7\dfrac{2}{3}$

31. $4\dfrac{2}{3}$

32. $10\dfrac{11}{12}$

33. $11\dfrac{6}{7}$

34. $2\dfrac{1}{9}$

35. $24\dfrac{7}{10}$

36. $66\dfrac{2}{3}$

Change each improper fraction to a mixed number (reduce if possible).

37. $\dfrac{5}{2}$

38. $\dfrac{4}{1}$

39. $\dfrac{3}{3}$

40. $\dfrac{18}{2}$

41. $\dfrac{11}{5}$

42. $\dfrac{103}{16}$

43. $\dfrac{24}{7}$

44. $\dfrac{11}{3}$

45. $\dfrac{38}{7}$

Write each inequality in words.

46. $14 < 112$

47. $3.6 \not> 4.2$

48. $\dfrac{1}{2} > \dfrac{1}{8}$

49. $\dfrac{3}{8} \neq \dfrac{2}{3}$

50. $3 < 5 < 8$

51. $\dfrac{1}{2} < \dfrac{3}{4} < \dfrac{2}{3}$

Compare using the symbols, $>$, $<$, or $=$.

52. $4, 7$

53. $3, 0$

54. $6, 6$

55. $\dfrac{5}{6}, \dfrac{7}{6}$

56. $\dfrac{8}{9}, \dfrac{3}{6}$

57. $\dfrac{3}{8}, 1\dfrac{3}{4}$

58. $0.6, 7.2$

59. $0.9, 0.256$

60. $1.36, 1.4$

1-3   OPERATIONS ON RATIONAL NUMBERS

In arithmetic, we learned how to add, subtract, multiply, and divide rational numbers. In this section, we will briefly review these operations.

A.   Addition and Subtraction of Rational Numbers

Addition is the operation of combining two or more quantities to obtain a third quantity, whereas subtraction is the operation of finding the difference between two quantities. Addition is denoted by a "plus" sign ($+$), and subtraction is denoted by a "minus" sign ($-$). The answer to an addition problem is called the *sum* or *total* and the answer to a subtraction problem is called the *difference* or *remainder*. Thus:

In the addition problem $10 + 5$, the sum is 15. $10 + 5 = 15$

In the subtraction problem $5 - 3$, the difference is 2. $5 - 3 = 2$

ADDING AND SUBTRACTING DECIMALS

To add or subtract decimals, we simply "line up" the decimal points and add or subtract the numbers.

EXAMPLE 1:   Add $4.6 + 3.95 + 0.113 + 246$.

SOLUTION: Write the numbers under each other, with the decimal points in line, and add the numbers.

$$
\begin{array}{r}
4.6 \\
3.95 \\
0.113 \\
246. \\
\hline
254.663
\end{array}
$$

Final answer: 254.663

EXAMPLE 2:   Subtract 34.26 from 298.7.

SOLUTION: Write the numbers under each other, with the decimal points in line, and subtract the numbers.

$$
\begin{array}{r}
298.70 \\
-\ \ 34.26 \\
\hline
264.44
\end{array}
$$

Final answer: 264.44

NOTE:

It was necessary first to attach a zero to 298.7 in order to make the decimals similar.

ADDING AND SUBTRACTING FRACTIONS

When adding or subtracting fractions we must first be certain that the fractions we are adding or subtracting are like fractions (i.e., their denominators are the same). If this is the case, then we simply add or subtract the numerators of the fractions and place this answer over the common denominators. For example:

$$\frac{3}{7} + \frac{2}{7} = \frac{5}{7} \quad \text{and} \quad \frac{7}{9} - \frac{3}{9} = \frac{4}{9}$$

On the other hand, if the fractions being added or subtracted are not like, then we must first change them to like fractions by finding their *least common multiple* (LCM).

EXAMPLE 3: Add $\frac{3}{4} + \frac{5}{6}$.

SOLUTION: The LCM of 4 and 6 is found by listing multiples of 4 and 6 and selecting the least common multiple.

Multiples of 4: 4, 8, 12 , 16, 20
Multiples of 6: 6, 12 , 18, 24
The LCM of 4 and 6 is 12.

We now change $\frac{3}{4}$ and $\frac{5}{6}$ to like fractions and add.

$$\frac{3}{4} = \frac{9}{12} \quad \text{and} \quad \frac{5}{6} = \frac{10}{12}$$

$$\frac{9}{12} + \frac{10}{12}$$

$$= \frac{19}{12} \quad \text{or} \quad 1\frac{7}{12}$$

Final answer: $\frac{19}{12}$ or $1\frac{7}{12}$

EXAMPLE 4: Subtract $\frac{4}{5} - \frac{3}{7}$.

SOLUTION: Since 7 and 5 are prime numbers, the LCM is simply their product. Thus, the LCM of 7 and 5 is 35. We now change $\frac{4}{5}$ and $\frac{3}{7}$ to like fractions and subtract.

$$\frac{4}{5} = \frac{28}{35} \quad \text{and} \quad \frac{3}{7} = \frac{15}{35}$$

$$\frac{28}{35} - \frac{15}{35}$$

$$= \frac{13}{35}$$

Final answer: $\frac{13}{35}$

ADDING AND SUBTRACTING MIXED NUMBERS

To add or subtract mixed numbers, we first add or subtract the fractional parts and then we add or subtract the whole number parts.

EXAMPLE 5:   Add $1\frac{2}{3} + 3\frac{4}{6}$.

SOLUTION: Simply add the fractions together and add the whole numbers together. Note that the fractions are first expressed as like fractions.

$$1\frac{2}{3} + 3\frac{4}{6}$$

$$= 1\frac{4}{6} + 3\frac{4}{6}$$

$$= \quad 4\frac{8}{6}$$

$$= \quad 4\frac{4}{3}$$

$$= 4 + 1\frac{1}{3}$$

$$= \quad 5\frac{1}{3}$$

Final answer: $5\frac{1}{3}$

EXAMPLE 6:   Subtract $3\frac{5}{8} - 1\frac{3}{8}$.

SOLUTION: Simply subtract the fractions and subtract the whole numbers.

$$3\frac{5}{8} - 1\frac{3}{8}$$

$$= \quad 2\frac{2}{8}$$

$$= \quad 2\frac{1}{4}$$

Final answer: $2\frac{1}{4}$

Do the following practice set. Check your answers with the answers in the right-hand margin.

PRACTICE SET 1-3A

Add or subtract (see examples 1 and 2).

1. $0.563 + 0.101 + 0.248$ 2. $0.01 + 0.512 + 1.06$

3. $13.41 + 2.997 + 16$ 4. $0.76 - 0.54$

5. $17.69 - 1.2$ 6. $1.2044 - 0.16$

1. 0.912
2. 1.582
3. 32.407
4. 0.22
5. 16.49
6. 1.0444

Add or subtract (see examples 3 and 4).

7. $\dfrac{1}{4} + \dfrac{2}{4}$ 8. $\dfrac{1}{4} + \dfrac{3}{8}$ 9. $\dfrac{1}{4} + \dfrac{3}{6}$

10. $\dfrac{5}{6} - \dfrac{4}{6}$ 11. $\dfrac{3}{4} - \dfrac{1}{2}$ 12. $\dfrac{5}{9} - \dfrac{1}{6}$

7. $\frac{3}{4}$
8. $\frac{5}{8}$
9. $\frac{3}{4}$
10. $\frac{1}{6}$
11. $\frac{1}{4}$
12. $\frac{7}{18}$

Add or subtract (see examples 5 and 6).

13. $1\dfrac{1}{3} + 2\dfrac{1}{3}$ 14. $3\dfrac{3}{8} + 2\dfrac{3}{8}$ 15. $5\dfrac{3}{4} + 3\dfrac{2}{3}$

16. $3\dfrac{7}{8} - 2\dfrac{1}{8}$ 17. $4\dfrac{3}{6} - 2\dfrac{3}{12}$ 18. $9\dfrac{3}{4} - 1\dfrac{3}{6}$

13. $3\frac{2}{3}$
14. $5\frac{3}{4}$
15. $9\frac{5}{12}$
16. $1\frac{3}{4}$
17. $2\frac{1}{4}$
18. $8\frac{1}{4}$

B. Multiplication of Rational Numbers

Multiplication is the process of repeatedly adding a quantity to itself a certain number of times. In multiplication, the numbers being multiplied are called *factors*, and the result or answer to a multiplication is called a *product*. Thus:

In the multiplication problem,
$5 \times 3 = 15$,
5 and 3 are factors
and 15 is their product.

$$\underset{\text{factors}}{5 \times 3} = \underset{\text{product}}{15}$$

A multiplication problem may be expressed in three different ways.

1. We may place an " $\times$ " between the factors. 5×3

2. We may place a raised dot between the factors. $5 \cdot 3$

3.   We may separate the factors by parentheses.

$$(5)\ (3)$$
$$(5)3$$
$$5(3)$$

MULTIPLYING DECIMALS

To multiply decimal numbers, we simply multiply the numbers as if they were whole numbers. The product must then contain the same number of decimal places as the total number of decimal places that are contained in the numbers being multiplied. For example, to multiply 2.6 by 0.48:

First, multiply them as if they are whole numbers.

$$\begin{array}{r} 26 \\ \times\ 48 \\ \hline 208 \\ 1040 \\ \hline 1248 \end{array}$$

Since the original numbers to be multiplied contain a total of three decimal places,

2. 6   1 decimal place
.48   2 decimal places

The product must contain three decimal places.

$$\begin{array}{r} 2.6 \\ \times\ .48 \\ \hline 208 \\ 1040 \\ \hline 1248 = 1.\,248 \end{array}$$

3 decimal places

Thus,

$$2.6 \times 0.48 = 1.248$$

MULTIPLYING FRACTIONS

To multiply fractions, we sometimes find it helpful to cancel out common factors first before multiplying. Once this is done, we then simply multiply the numerators together and multiply the denominators together. For example, to multiply $\frac{5}{8}$ by $\frac{12}{25}$:

First, cancel out the common factor 5 in both 5 and 25 and the common factor 4 in both 8 and 12.

$$\frac{\overset{1}{5}}{\underset{2}{8}} \cdot \frac{\overset{3}{12}}{\underset{5}{25}}$$

Now, simply multiply numerators together and denominators together.

$$\frac{\overset{1}{5}}{\underset{2}{8}} \cdot \frac{\overset{3}{12}}{\underset{5}{25}} = \frac{(1)(3)}{(2)(5)} = \frac{3}{10}$$

Thus,

$$\frac{5}{8} \times \frac{12}{25} = \frac{3}{10}$$

MULTIPLYING MIXED NUMBERS

To multiply mixed numbers, we must first change each mixed number into an improper fraction. After doing this, we then proceed as in multiplying fractions. For example, to multiply $2\frac{3}{4}$ by $5\frac{3}{11}$:

First, change each mixed number into an improper fraction.

$$2\frac{3}{4} \cdot 5\frac{3}{11} = \frac{11}{4} \cdot \frac{58}{11}$$

Next, cancel out the common factors.

$$\frac{{}^{1}\cancel{11}}{{}_{2}\cancel{4}} \cdot \frac{{}^{29}58}{{}_{1}\cancel{11}}$$

Now, multiply.

$$\frac{{}^{1}\cancel{11}}{{}_{2}\cancel{4}} \cdot \frac{{}^{29}58}{{}_{1}\cancel{11}} = \frac{(1)(29)}{(2)(1)} = \frac{29}{2}$$

Thus:

$$2\frac{3}{4} \times 5\frac{3}{11} = \frac{29}{2}\left(\text{or } 14\frac{1}{2}\right)$$

Do the following practice set. Check your answers with the answers in the right-hand margin.

PRACTICE SET 1-3B

Multiply the following decimals.

1.　　0.43
　　　× 0.2

2.　0.3 × 0.3

3.　6.4 · 0.26

1.　0.086
2.　0.09
3.　1.664

Multiply the following fractions (reduce if possible).

4.　$\frac{3}{8} \cdot \frac{4}{7}$

5.　$\frac{5}{7} \cdot \frac{21}{15}$

6.　$\frac{9}{10} \cdot \frac{15}{21}$

4.　$\frac{3}{14}$
5.　$\frac{3}{3} = 1$
6.　$\frac{9}{14}$

Multiply the following mixed numbers (reduce if possible).

7.　$\left(1\frac{2}{3}\right)\left(2\frac{4}{7}\right)$

8.　$\left(3\frac{1}{2}\right)\left(2\frac{4}{9}\right)$

9.　$\left(4\frac{3}{5}\right)\left(1\frac{1}{2}\right)$

7.　$\frac{30}{7}$ or $4\frac{2}{7}$
8.　$\frac{77}{9}$ or $8\frac{5}{9}$
9.　$\frac{69}{10}$ or $6\frac{9}{10}$

C.　Division of Rational Numbers

Division is the process of finding how many times one number or quantity is contained in another. In a division problem, the number being divided is called the *dividend*, the number we are dividing by is called the *divisor*, and the answer to a division is called the *quotient*. Thus:

In the problem of dividing 20 by 5, 20 is the dividend, 5 is the divisor, and the quotient is 4.

$$\begin{array}{r} 4 \longrightarrow \text{quotient} \\ 5\,\overline{)\,20} \end{array}$$

divisor　dividend

Division may be symbolically expressed in two ways other than the standard division box (as depicted here). One way makes use of the division symbol ÷, and the other uses the fraction bar –. Both symbols are read as "divided by." Using the preceding example with these symbols, we have

$$\underbrace{20}_{\text{dividend}} \div \underbrace{5}_{\text{divisor}} = \underbrace{4}_{\text{quotient}}$$

20 divided by 5 = 4

dividend

$$\frac{20}{5} = 4 \longleftarrow \text{quotient}$$

divisor

20 divided by 5 = 4
or 20 over 5 = 4

DIVIDING DECIMALS

To divide decimals, we simply make the divisor a whole number by moving the decimal point all the way to the right. We then move the decimal point in the dividend the same number of places we moved it in the divisor. The decimal point in the quotient is placed directly above the new decimal position in the dividend. After doing this, we divide as we would for whole numbers. For example, to divide 4.508 by 0.23:

1. Move the decimal point in the divisor two places to the right.

$$0.23\overline{)4.508}$$

2. Move the decimal point in the dividend two places to the right.

$$023.\overline{)4.508}$$

3. Place the decimal point in the quotient directly above the new decimal position in the dividend.

$$023.\overline{)450.8}$$

4. Divide as if we were dividing whole numbers.

$$
\begin{array}{r}
19.6 \\
23.\overline{)450.8} \\
-23 \\
\hline
220 \\
-207 \\
\hline
138 \\
-138 \\
\hline
0
\end{array}
$$

Thus,

$$4.508 \div 0.23 = 19.6$$

DIVIDING FRACTIONS

To divide fractions, we simply multiply the first fraction by the *reciprocal* of the second fraction. For example, to divide $\frac{3}{5}$ by $\frac{7}{10}$:

Multiply $\frac{3}{5}$ by the reciprocal of $\frac{7}{10}$.

$$\frac{3}{5} \div \frac{7}{10} = \frac{3}{5} \times \frac{10}{7}$$

$$= \frac{3}{\underset{1}{5}} \times \frac{\overset{2}{10}}{7} = \frac{(3)(2)}{(1)(7)} = \frac{6}{7}$$

NOTE:

$\frac{7}{10}$ and $\frac{10}{7}$ are reciprocals since $\frac{7}{10} \cdot \frac{10}{7} = 1$.

Thus,

$$\frac{3}{5} \div \frac{7}{10} = \frac{6}{7}$$

DIVIDING MIXED NUMBERS

To divide mixed numbers, we must first change each mixed number to an improper fraction *before* we divide. We then divide as before. For example, to divide $2\frac{3}{8}$ by $1\frac{4}{5}$:

First, change the mixed numbers to improper fractions.

$$2\frac{3}{8} \div 1\frac{4}{5} = \frac{19}{8} \div \frac{9}{5}$$

Now, divide by multiplying by the reciprocal of the second fraction.

$$\frac{19}{8} \div \frac{9}{5} = \frac{19}{8} \times \frac{5}{9} = \frac{95}{72} = 1\frac{23}{72}$$

Thus,

$$2\frac{3}{8} \div 1\frac{4}{5} = 1\frac{23}{72}$$

Do the following practice set. Check your answers with the answers in the right-hand margin.

PRACTICE SET 1-3C

Divide the following decimals.

1. $0.07\overline{)1.491}$ 2. $0.069 \div 0.3$ 3. $6.5 \div 0.05$

1. 21.3
2. 0.23
3. 130

Divide the following fractions (reduce if possible).

4. $\frac{5}{6} \div \frac{2}{3}$ 5. $\frac{3}{5} \div \frac{3}{4}$ 6. $\frac{3}{8} \div \frac{3}{4}$

4. $\frac{5}{4}$ or $1\frac{1}{4}$
5. $\frac{4}{5}$
6. $\frac{1}{2}$

Divide the following mixed numbers (reduce if possible).

7. $4\frac{4}{5} \div 2\frac{2}{3}$ 8. $3\frac{1}{5} \div 2\frac{2}{3}$ 9. $1\frac{1}{2} \div 1\frac{2}{3}$

7. $\frac{9}{5}$ or $1\frac{4}{5}$
8. $\frac{6}{5}$ or $1\frac{1}{5}$
9. $\frac{9}{10}$

EXERCISE 1-3

Perform the indicated operation.

1. $\dfrac{3}{5} + \dfrac{5}{7}$

2. $\dfrac{5}{6} + \dfrac{5}{18}$

3. $\dfrac{7}{8} + \dfrac{5}{6}$

4. $\dfrac{3}{4} - \dfrac{1}{6}$

5. $\dfrac{5}{12} - \dfrac{5}{18}$

6. $\dfrac{7}{6} - \dfrac{2}{3}$

7. $\dfrac{3}{4} \cdot \dfrac{5}{7}$

8. $\dfrac{3}{4} \cdot \dfrac{8}{3}$

9. $\dfrac{3}{8} \cdot \dfrac{2}{15}$

10. $\dfrac{2}{3} \div \dfrac{3}{4}$

11. $\dfrac{3}{4} \div \dfrac{1}{5}$

12. $\dfrac{9}{5} \div \dfrac{3}{5}$

13. $1\dfrac{3}{4} + 2\dfrac{1}{6}$

14. $7\dfrac{7}{9} + 8\dfrac{1}{3}$

15. $18\dfrac{5}{6} + 21\dfrac{3}{8}$

16. $4\dfrac{3}{7} - 3\dfrac{1}{7}$

17. $5\dfrac{2}{3} - 2\dfrac{3}{4}$

18. $6\dfrac{1}{4} - 2\dfrac{5}{8}$

19. $1\dfrac{1}{2} \cdot 4\dfrac{1}{4}$

20. $3\dfrac{3}{4} \cdot 1\dfrac{3}{5}$

21. $4\dfrac{1}{2} \cdot 2\dfrac{2}{3}$

22. $4\dfrac{4}{5} \div 2\dfrac{2}{3}$

23. $1\dfrac{1}{2} \div 1\dfrac{2}{3}$

24. $3\dfrac{1}{5} \div 2\dfrac{2}{3}$

25. $3.26 + 9.6 + 34 + 3.0938$

26. $9.98 + 43.2 + 4.301 + 18$

27. $6.2 - 3.9$

28. $601.4 - 8.33$

29. $(0.006)(66.5)$

30. $(0.09)(0.1008)$

31. $0.6 \div 15$

32. $27.2 \div 0.32$

1-4 EXPONENTS

A. Exponents

From our work in arithmetic, we learned that we can express the product of a number in which all the factors are the same by using exponential notation. For example, the product $2 \times 2 \times 2$ may be expressed in exponential form as 2^3. Furthermore:

The repeating factor, 2, is called the *base*.

$$\underbrace{2 \cdot 2 \cdot 2}_{\substack{\text{repeating} \\ \text{factors}}} = 2^{\underset{\text{base}}{3}}$$

The number of times the base is being used as a factor (i.e., the number of times 2 is repeated), is called the *exponent*.

$$\underbrace{2 \cdot 2 \cdot 2}_{\text{3 repeats}} = 2^{3} \text{——exponent}$$

Similarly:

■ 3^2 has a base of 3 and an exponent of 2. Also, $3^2 = 3 \cdot 3$.

■ 5^3 has a base of 5 and an exponent of 3. Also, $5^3 = 5 \cdot 5 \cdot 5$.

■ 4^1 has a base of 4 and an exponent of 1. Also, $4^1 = 4$.

NOTE:

> Any number that is written without an exponent is assumed to have a power of 1. Thus,
>
> $$2 = 2^1, \qquad 3 = 3^1, \qquad 5 = 5^1, \qquad \text{etc.}$$
>
> Additionally, any number raised to the second power is commonly read as "squared," and any number raised to the third power is commonly read as "cubed." So:
>
> ■ 3^2 is read as three *squared*.
>
> ■ 5^3 is read as five *cubed*.

Frequently, in mathematics, it is necessary either to express the product of repeating factors by using exponents or to express a product written in exponential notation without using exponents. In either case, the procedures are fairly simple. Consider the following examples.

EXAMPLE 1: Express the product $4 \cdot 4 \cdot 4$ by using exponents.

SOLUTION: Since 4 is being repeated three times, the base is 4 and the exponent is 3.

$$4 \cdot 4 \cdot 4$$
$$= 4^3$$

Final answer: 4^3

EXAMPLE 2: Express the product $2 \cdot 2 \cdot 2 \cdot 3 \cdot 3$ by using exponents.

SOLUTION: 2 is being repeated three times and 3 is being repeated twice.

$$\underbrace{2 \cdot 2 \cdot 2} \cdot \underbrace{3 \cdot 3}$$
$$= \quad 2^3 \quad \cdot \quad 3^2$$

Final answer: $2^3 \cdot 3^2$

EXAMPLE 3: Write 7^4 without using exponents.

SOLUTION: The base 7 must be multiplied by itself four times.

$$7^4$$
$$= 7 \cdot 7 \cdot 7 \cdot 7$$

Final answer: $7 \cdot 7 \cdot 7 \cdot 7$

EXAMPLE 4:   Write $7^3 \cdot 9^2$ without using exponents.

SOLUTION:

$$7^3 \quad \cdot \quad 9^2$$
$$= 7 \cdot 7 \cdot 7 \cdot 9 \cdot 9$$

Final answer: $7 \cdot 7 \cdot 7 \cdot 9 \cdot 9$

Do the following practice set. Check your answers with the answers in the right-hand margin. If you have any questions, refer back to the preceding examples.

PRACTICE SET 1-4A

Do the following for each of these problems: (a) identify the base, (b) identify the power, (c) read the number (see example 1).

1. 3^4 2. 5^6 3. 6

Express each product below in exponential form (see examples 2 or 3).

4. $2 \cdot 2 \cdot 2 \cdot 2$ 5. $2 \cdot 2 \cdot 3$ 6. $4 \cdot 4 \cdot 4 \cdot 5 \cdot 5$

Write each of the following without using exponents (see examples 4 or 5).

7. 7^2 8. $2^2 \cdot 3^3$ 9. $2 \cdot 5^3$

1. a. base is 3
    b. power is 4
    c. 3^4 is read "three to the fourth power"
2. a. base is 5
    b. power is 6
    c. 5^6 is read "five to the sixth power"
3. a. base is 6
    b. power is 1
    c. 6^1 is read "six to the first power" (or just "six")
4. 2^4
5. $2^2 \cdot 3^1$
6. $4^3 \cdot 5^2$
7. $7 \cdot 7$
8. $2 \cdot 2 \cdot 3 \cdot 3 \cdot 3$
9. $2 \cdot 5 \cdot 5 \cdot 5$

B.   Evaluating Expressions That Contain Exponents

An expression written in exponential form may be evaluated by first writing the expression without using exponents and then multiplying the factors together. For example, to evaluate 5^3:

First, write 5^3 without using exponents. $5^3 = 5 \cdot 5 \cdot 5$

Then, multiply the factors. $5 \cdot 5 \cdot 5 = 125$

We provide two more examples.

EXAMPLE 1: Evaluate 2^4.

SOLUTION:

$$
\begin{aligned}
2^4 &= \underbrace{2 \cdot 2} \cdot 2 \cdot 2 \\
&= \underbrace{4 \cdot 2} \cdot 2 \\
&= \underbrace{8 \cdot 2} \\
&= 16
\end{aligned}
$$

Final answer: 16

EXAMPLE 2: Evaluate $2^3 \cdot 3 \cdot 5^2$.

SOLUTION:

$$
\begin{aligned}
2^3 \cdot 3 \cdot 5^2 &= \underbrace{2 \cdot 2 \cdot 2} \cdot 3 \cdot \underbrace{5 \cdot 5} \\
&= \underbrace{8 \cdot 3 \cdot} 25 \\
&= \underbrace{24 \cdot 25} \\
&= 600
\end{aligned}
$$

Final answer: 600

Do the following practice set. Check your answers with the answers in the right-hand margin.

PRACTICE SET 1-4B

Evaluate (see example 1).

1. 2^3	2. 5^2	3. 1^6	1.	8
			2.	25
			3.	1

Evaluate (see example 2).

4. $2^4 \cdot 5^2$	5. $2^2 \cdot 3^3 \cdot 7$	6. $2^2 \cdot 3^2 \cdot 5^2$	4.	400
			5.	756
			6.	900

EXERCISE 1-4

Write each of the following by using exponential notation.

1. 10 squared
2. 4 cubed
3. 7 to the fifth power
4. 12 to the fourth power
5. 6 to the zero power
6. The first power of 2
7. The eighth power of 2
8. 15 raised to the third power
9. The zero power of 9
10. x cubed

Express each of the following products by using exponents.

11. $2 \cdot 2 \cdot 2$
12. $4 \cdot 4 \cdot 4 \cdot 4 \cdot 4$
13. $(10)(10)(10)$
14. $x \cdot x \cdot x \cdot x \cdot x$
15. $y \cdot y \cdot y$
16. $(3 \cdot x)(3 \cdot x)(3 \cdot x)$
17. $2 \cdot 2 \cdot 2 \cdot 3 \cdot 3 \cdot 3 \cdot 3$
18. $4 \cdot 4 \cdot 3 \cdot 3 \cdot 3 \cdot 5 \cdot 5 \cdot 5 \cdot 5 \cdot 5$
19. $9 \cdot x \cdot x \cdot x$
20. $10 \cdot 10 \cdot x \cdot x \cdot x \cdot y \cdot y \cdot y \cdot y$

Evaluate.

21. 2^4
22. 1^3
23. 10^4
24. 4^2
25. 5^3
26. 6^2
27. $2^3 \cdot 6$
28. $3^4 \cdot 8$
29. $5^2 \cdot 3$
30. $3 \cdot 3^2$
31. $5 \cdot 2^3$
32. $3 \cdot 8^2$
33. $2^3 \cdot 3 \cdot 5$
34. $4^3 \cdot 2 \cdot 7$
35. $3 \cdot 2^3 \cdot 7$
36. $3^2 \cdot 5^2$
37. $10^2 \cdot 3^2$
38. $3^3 \cdot 2^2$
39. $3^1 \cdot 2^1 \cdot 4^2$
40. $5^2 \cdot 3^2 \cdot 4^2$
41. $2^3 \cdot 3^4 \cdot 5^1$
42. $7^2 \cdot 3^2 \cdot 2$
43. $3^2 \cdot 2^5 \cdot 5$
44. $3^5 \cdot 2^3 \cdot 7^2$
45. $1^{10} \cdot 2^1 \cdot 3^1$
46. $9^2 \cdot 10^2 \cdot 100^1$
47. $1000^1 \cdot 1^{1000}$

48. Evaluate 2^1, 2^2, 2^3, 2^4, 2^5, and carefully look at the pattern. What do you think 2^0 is equal to?

1-5 ORDER OF OPERATIONS

A. Order of Operations

In mathematics, we are able to perform only one operation at a time. If a problem involves more than one operation, we must agree, then, on the order in which the operations should be performed.

For example, a problem such as $3 \times 2 - 1$ might be solved in two ways.

SOLUTION 1		**SOLUTION 2**	
Original problem:	$3 \times 2 - 1$	Original problem:	$3 \times 2 - 1$
Multiply first:	$3 \times 2 - 1$	Subtract first:	$3 \times 2 - 1$
	$6 - 1$		3×1
Subtract:	$6 - 1$	Multiply:	3×1
	5		3

Notice that solution 1 yields a result of 5, whereas solution 2 yields a result of 3—the same problem with two different answers.

To prevent situations such as those previously described, we must agree upon an order of operations. Mathematicians have agreed to use the following order of operations:

Working from left to right, perform all multiplication and division first; then starting again from left to right, perform all addition and subtraction.

Thus, in the preceding example, solution 1 is the correct method employed and 5 is the only acceptable answer to the problem.

However, if we had wanted to subtract 1 from the 2 before multiplying, we could indicate this change of operations by using parentheses or some other symbol of grouping.

NOTE:

> The three most commonly used symbols of grouping are: parentheses ( ); brackets [ ]; and braces { }.

By placing $(2 - 1)$ within parentheses, we have now changed our problem and must think of it as

3 times the value within parentheses.	$3 \times (2 - 1)$

Thus, we must

First, find the value within parentheses.	$3 \times (2 - 1)$
	3×1
Then, find the product of the factors.	3×1
	3

Thus,

$$3 \times (2 - 1) = 3$$

Parentheses (or other grouping symbols) are used to modify the order of operation. Whenever grouping symbols are used, we must perform the operations within the grouping symbols first.

In general, to evaluate problems that involve more than one operation, we proceed in the following manner:

1. Remove all grouping symbols by evaluating the expression(s) contained within the sym-

bols of grouping. (If more than one set of grouping symbols is used, always begin with the innermost set and work your way out.)

2. Perform all exponentiation, in order, working from left to right.

3. Perform all multiplication and division, in order, working from left to right.

4. Perform all addition and subtraction, in order, working from left to right.

EXAMPLE 1: Evaluate $8 \div 4 + 2 \times 3 + 1$.

SOLUTION: Since no grouping symbols are given, we first perform all multiplication and division, in order, working from left to right.

$$\underline{8 \div 4} + \underline{2 \times 3} + 1$$
$$= \quad 2 \quad + \quad 6 \quad + 1$$

Now, we simply perform the addition, working, in order, from left to right.

$$= \underline{2 + 6} + 1$$
$$= \quad \underline{8 \quad + 1}$$
$$= \quad\quad 9$$

Final answer: 9

EXAMPLE 2: Evaluate $9 \times 4 + 6 \div 3 \times 2 - 28$.

SOLUTION: Perform all multiplication and division, in order, working from left to right.

$$\underline{9 \times 4} + \underline{6 \div 3} \times 2 - 28$$
$$= \quad 36 \quad + \quad \underline{2 \quad \times 2} - 28$$
$$= \quad 36 \quad + \quad\quad 4 \quad - 28$$

Now, perform all the addition and subtraction working from left to right.

$$= \underline{36 + 4} - 28$$
$$= \quad \underline{40 \quad - 28}$$
$$= \quad\quad\quad 12$$

Final answer: 12

EXAMPLE 3: Evaluate $2 \times (3 + 5) - 5 \times 1 + 2$.

SOLUTION: First, remove parentheses by evaluating the expression within the parentheses.

$$2 \times \underbrace{(3 + 5)} - 5 \times 1 + 2$$
$$= 2 \times \quad 8 \quad - 5 \times 1 + 2$$

Next, perform all multiplication, working from left to right.

$$= \underbrace{2 \times 8} - \underbrace{5 \times 1} + 2$$
$$= \quad 16 \quad - \quad 5 \quad + 2$$

Finally, perform all addition and subtraction, in order, working from left to right.

$$= \underbrace{16 - 5} + 2$$
$$= \quad \underbrace{11 \quad + 2}$$
$$= \quad \quad 13$$

Final answer: 13

EXAMPLE 4: Evaluate $2 \times \{3 + (2 \times 4) - 5\} + 1$.

SOLUTION: Remove all grouping symbols, the innermost ones first.

$$2 \times \{3 + \underbrace{(2 \times 4)} - 5\} + 1$$
$$= 2 \times \{3 + \underbrace{\quad 8 \quad} - 5\} + 1$$
$$= 2 \times \{ \quad \underbrace{11 \quad \quad - 5} \} + 1$$
$$= 2 \times \{ \quad \quad 6 \quad \quad \} + 1$$
$$= 2 \times 6 + 1$$

Multiply.

$$= \underbrace{2 \times 6} + 1$$
$$= \quad 12 \quad + 1$$

Add.

$$= \underbrace{12 + 1}$$
$$= \quad 13$$

Final answer: 13

EXAMPLE 5: Evaluate $3 \times \{4 \times (7 - 2) + 8\} - 5$.

SOLUTION:

$$3 \times \{4 \times \underbrace{(7 - 2)}_{} + 8\} - 5$$
$$= 3 \times \{\underbrace{4 \times \quad 5}_{} + 8\} - 5$$
$$= 3 \times \{\underbrace{20 \quad + 8}_{}\} - 5$$
$$= \underbrace{3 \times \quad\quad 28}_{} - 5$$
$$= \underbrace{84 \quad\quad\quad - 5}_{}$$
$$= \quad\quad\quad 79$$

Final answer: 79

EXAMPLE 6: Evaluate $9 - 6 \times [15 + 4 - \{3 + (6 - 4) \times 8\} + 2] \div 4$.

SOLUTION:

$$9 - 6 \times [15 + 4 - \{3 + \underbrace{(6 - 4)}_{} \times 8\} + 2] \div 4$$
$$= 9 - 6 \times [15 + 4 - \{3 + \quad 2 \times 8\} + 2] \div 4$$
$$= 9 - 6 \times [15 + 4 - \{3 + 16\} + 2] \div 4$$
$$= 9 - 6 \times [15 + 4 - 19 + 2] \div 4$$
$$= 9 - 6 \times [19 - 19 + 2] \div 4$$
$$= 9 - 6 \times [0 + 2] \div 4$$
$$= 9 - \underbrace{6 \times [2]}_{} \div 4$$
$$= 9 - \quad 12 \div 4$$
$$= \underbrace{9 - 3}_{}$$
$$= \quad 6$$

Final answer: 6

Do the following practice set. Check your answers with the answers in the right-hand margin.

PRACTICE SET 1-5A

Evaluate each of the following (see example 1 or 2).

1. $5 \times 2 - 2 \times 3$ 1. 4

2. $6 \times 3 \div 9 + 6 - 1$ 2. 7

3. $17 + 3 \div 1 \times 4 - 6 \times 2$ 3. 17

$$\overset{3}{} \quad \overset{4}{} \quad \overset{12}{}$$

$$17 + 1 - 12$$

$$18 \quad 12$$

Evaluate (see examples 3, 4, 5, or 6).

4. $2 \times (8 + 7) - 3$ 4. 27

5. $[4 \times (8 - 2) - 20] \times 2 - 6$ 5. 2

6. $[1 + 3 - 6 \div 3] + [5 \times \{3 \times (1 + 4)\}]$ 6. 77

B.   Order of Operations with Exponents

In section 1-4, we learned that 4^2 means 4×4. Therefore, a number such as $20 - 4^2$ would mean $20 - (4 \times 4)$. Thus, to evaluate a number such as $20 - 4^2$:

First, evaluate the power. $20 - 4^2 = 20 - (4 \cdot 4)$
$$= 20 - 16$$

Now, subtract. $20 - 16 = 4$

Thus,
$$20 - 4^2 = 4$$

EXAMPLE 1: Evaluate $4^2 - 2^3 + 6$.

SOLUTION: Since there are no grouping symbols, we first evaluate the powers.

$$\underbrace{4^2} - \underbrace{2^3} + 6$$
$$(4 \times 4) - (2 \times 2 \times 2) + 6$$
$$= \quad 16 \quad - \quad 8 \quad + 6$$

Now, simply subtract and add.

$$= \underbrace{16 - 8} + 6$$
$$= \quad \underbrace{8 \quad + 6}$$
$$= \quad 14$$

Final answer: 14

EXAMPLE 2: Evaluate $3 \times (1 + 6 \div 3)^3 - 8$.

SOLUTION: Remove parentheses.

$$3 \times (1 + \underbrace{6 \div 3})^3 - 8$$
$$= 3 \times \underbrace{(1 + \quad 2 \quad)^3} - 8$$
$$= 3 \times \quad (3)^3 \quad - 8$$
$$= 3 \times \quad 3^3 \quad - 8$$

Evaluate the power.

$$= 3 \times \underbrace{3^3}_{} - 8$$
$$= 3 \times \underbrace{(3 \times 3 \times 3)}_{} - 8$$
$$= 3 \times \underset{27}{} - 8$$

Multiply.

$$= \underbrace{3 \times 27}_{} - 8$$
$$= \underset{81}{} - 8$$

Subtract.

$$= \underbrace{81 - 8}_{}$$
$$= 73$$

Final answer: 73

Do the following practice set. Check your answers with the answers in the right-hand margin.

PRACTICE SET 1-5B

Evaluate (see examples 1 and 2)

1. $3^2 + 2^3 \times 6$
2. $8 + (3 + 6)^2 \times 4$
3. $6 \times 2^4 - (4 - 1)^3 + 9$

1.   57
2.   332
3.   78

EXERCISE 1-5

Perform the indicated operations.

1. $20 + 4 \cdot 2$
2. $3 + 7 \cdot 5$
3. $12 - 3 \cdot 2$
4. $9 + 6 \cdot 5$
5. $12 \div 4 - 3$
6. $18 \div 6 + 2$
7. $3 \cdot 8 + 4 \cdot 6$
8. $28 - 10 \div 2 + 4$
9. $119 - 0.5 \cdot 100$
10. $13 + 6 \div 2 - 0 \div 14 + 0.6$

Perform the indicated operations.

11. $5 + (6 + 3)$
12. $19 - (4 + 6)$
13. $27 - (8 - 5)$
14. $10 \cdot (4 + 6)$

15. $15 \div 5 \cdot (3 + 9)$

16. $15 \div [5 \cdot (3 + 9)]$

17. $12 \cdot (6 - 5) - 4$

18. $6(3 + 5)(12 - 5)$

19. $5[6 + (8 - 2) \cdot 3]$

20. $28 \div (2 \cdot 2) - (8 + 2) \div 5$

21. $40 - \{(99 - 6) \cdot (6 - 6)\}$

22. $6[\{3 + (10 - 4) \cdot (8 + 6)\} - 5]$

23. $\{3 \cdot (8 - 5) + 4\} \div \{6 \cdot 3 - 5\}$

24. $6[2 + \{(3 + 4) \cdot 5\} \div \{(7 - 5) \cdot 8 - 9\}]$

Perform the indicated operations.

25. $3^2 - 2^3 + 5$

26. $5^3 \cdot (3^2 + 4)$

27. $5^1 \cdot (2 + 3)^2 - 16$

28. $3^3 \div [(14 - 8)^2 \cdot 0.25]$

29. $(1 + 5 \div 2)^3$

30. $[2 \cdot (6 + 4) - 2]^4$

31. $(6 - 3)^2 \cdot (3 + 6)^2$

32. $(12 - 3 \cdot 2) \div [12 + (3 \cdot 2)^2]$

33. $5^3 + (6 + 3)^2 \div 10^2$

34. $100 - 4 (6 - 2)^2$

35. $9 \times (5^2 + 4^2)$

36. $(3^2 + 2^2) \cdot (2^3 + 1^{10})$

37. $3 + \left(\dfrac{2}{3}\right)^2$

38. $\left(\dfrac{2}{7}\right)^2 + \left(\dfrac{4}{7}\right)^2$

39. $\left(\dfrac{4}{5} - \dfrac{3}{5}\right)^2 + \left(\dfrac{1}{3} + \dfrac{1}{4}\right)^2$

40. $\left(\dfrac{5}{2}\right)^3 - \dfrac{7}{8}$

REVIEW EXERCISES

Do the following problems.

1. Graph the first five counting numbers on a number line.

2. What number is a whole number but not a natural number?

3. Write three fractions that are equal in value for each of the following numbers.
   a. 2   b. 6   c. 0

4. Write three equivalent fractions for each of the following.
   a. $\dfrac{2}{3}$ b. $\dfrac{3}{5}$ c. $\dfrac{2}{4}$

5. Convert each of the following decimals to a fraction. Reduce to lowest terms.
   a. 0.3   b. 0.70   c. 2.6

6. Convert each of the following mixed numbers to an improper fraction. Reduce to lowest terms.
   a. $1\dfrac{2}{3}$ b. $3\dfrac{4}{8}$ c. $4\dfrac{2}{10}$

7. Using the number line shown here, name the point that is the graph of the number.

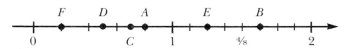

   a. $\dfrac{1}{5}$ b. $0.\overset{\bullet}{7}$ c. $1\dfrac{5}{8}$

In problems 8–25, add.

8. $3.26 + 9.7$

9. $9.98 + 43.23$

10. $102.81 + 62.7$

11. $3.0938 + 12.3016$

12. $212.87 + 46.25$

13. $326.08 + 207.11$

14. $2.6 + 0.114 + 3$

15. $97 + 2.47 + 0.1$

16. $8.0009 + 4 + 16.2$

17. $\dfrac{13}{24} + \dfrac{7}{24}$

18. $\dfrac{3}{7} + \dfrac{2}{7}$

19. $\dfrac{7}{11} + \dfrac{4}{11}$

20. $\dfrac{3}{8} + \dfrac{1}{4}$

21. $\dfrac{1}{3} + \dfrac{1}{4}$

22. $\dfrac{2}{9} + \dfrac{5}{12}$

23. $5\dfrac{7}{8} + 3\dfrac{6}{8}$

24. $2\dfrac{3}{4} + 6\dfrac{2}{3}$

25. $3 + 8\dfrac{2}{5}$

In problems 26–43, subtract.

26. $3.13 - 1.6$

27. $4.8 - 0.9$

28. $200.18 - 0.116$

29. $63.2 - 61.28$

30. $14.09 - 8.7$

31. $4.301 - 1.86$

32. $7.6 - 1.47$

33. $725.4 - 16.04$

34. $6 - 1.513$

35. $\dfrac{13}{24} - \dfrac{5}{24}$

36. $\dfrac{7}{18} - \dfrac{5}{18}$

37. $\dfrac{9}{13} - \dfrac{7}{13}$

38. $\dfrac{2}{3} - \dfrac{1}{6}$

39. $\dfrac{5}{6} - \dfrac{5}{8}$

40. $\dfrac{5}{6} - \dfrac{2}{9}$

41. $5\dfrac{7}{8} - 4\dfrac{3}{8}$

42. $5\dfrac{1}{3} - 3\dfrac{7}{12}$

43. $9 - 6\dfrac{5}{6}$

In problems 44–61, multiply.

44. 2.6×1.4

45. 1.32×0.06

46. 0.372×0.01

47. 4.208×1.25

48. 59.36×0.99

49. 2.006×5.6

50. 0.0001×9

51. 4.07×16

52. 300×0.17

53. $\dfrac{3}{4} \times \dfrac{5}{7}$

54. $\dfrac{1}{2} \times \dfrac{3}{5}$

55. $\dfrac{3}{4} \times \dfrac{1}{2}$

56. $\dfrac{7}{12} \times \dfrac{3}{4}$

57. $\dfrac{3}{10} \times \dfrac{5}{6}$

58. $\dfrac{6}{15} \times \dfrac{5}{9}$

59. $8\dfrac{1}{3} \times 3\dfrac{1}{5}$

60. $3\dfrac{3}{4} \times 1\dfrac{3}{5}$

61. $2\dfrac{2}{3} \times 8$

In problems 62–79, divide (round answers to the nearest tenth).

62. $18.4 \div 0.32$

63. $9.6 \div 0.16$

64. $59.04 \div 4.8$

65. $0.23 \div 0.16$

66. $24.852 \div 0.38$

67. $1.07 \div 0.3$

68. $16 \div 0.05$

69. $28 \div 2.6$

70. $5 \div 9$

71. $\dfrac{1}{2} \div \dfrac{1}{4}$

72. $\dfrac{3}{2} \div \dfrac{1}{3}$

73. $\dfrac{5}{6} \div \dfrac{5}{7}$

74. $\dfrac{35}{8} \div \dfrac{5}{6}$

75. $\dfrac{3}{4} \div \dfrac{7}{12}$

76. $10 \div \dfrac{9}{16}$

77. $1\dfrac{3}{4} \div 2\dfrac{1}{4}$

78. $2\dfrac{5}{6} \div 5\dfrac{1}{10}$

79. $1\dfrac{1}{2} \div 2\dfrac{3}{4}$

80. Compare the following numbers by using the symbols $<$, $>$, or $=$.

    a. 13, 5   b. 5.6, 2.007   c. $\dfrac{3}{4}, \dfrac{2}{3}$ d. 0.2, $\dfrac{3}{5}$

81. Evaluate each of the following.
    a. $3^2 \cdot 2^4$ b. $2^3 \cdot 3^2 \cdot 5^2$ c. $1^{10} \cdot 1^{100} \cdot 1^{1000}$

82. Perform the indicated operations.
    a. $20 + 4 \cdot 3$ b. $27 \div 3 \cdot 5$
    c. $15 \div 3 \cdot (5 + 10)$ d. $3(4 + 5)(5 - 3)$
    e. $3[5 + (7 - 4) \cdot 6]$ f. $14 \div (3 \cdot 2 + 1) \div (8 + 4)$
    g. $30 - [(33 \div 11) \cdot (6 - 6)] \div 30$ h. $[2 \cdot (7 + 4) - 6] \div [6 \cdot 3 - 2]$
    i. $2^2 + 4^3 + 5$ j. $5^3 \cdot (2^3 + 4^2)$
    k. $[3^3 + (2 + 4)^2] \div [2 - 1]$ l. $[30 - \{(33 - 11) \cdot (6 - 6)\} \div 30]^2$

CHAPTER 2

Signed Numbers and Their Operations

2-1 INTRODUCTION

A. The Integers

In Chapter 1, we showed how whole numbers could be represented by using a number line. We start at 0 and extend the line to the right.

We can also extend the number line to the left of zero. To do this, we simply:

1. Choose any point on the line as our zero point.

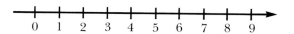

2. Mark off equal intervals to the left and right of the zero position.

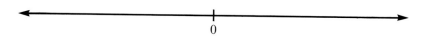

3. Label all the points to the right of zero with a number and a "+" sign.

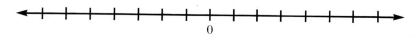

4. Label all the points to the left of zero with a number and a "−" sign.

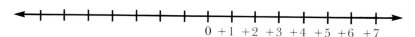

NOTE:

A sign attached to a number indicates a position to the left or right of zero. The positive sign (+) indicates a position to the right and the negative sign indicates a position to the left. For example:

A number +1 represents a position 1 unit to the *right* of 0.
A number −1 represents a position 1 unit to the *left* of 0.

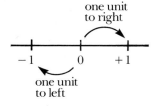

By extending the number line to the left of zero, we introduce another group of numbers called the *negative numbers*.

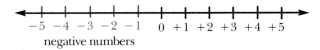

Note the following.

1. Numbers to the right of zero are called positive numbers and are written with a positive sign (+) placed before them, or have no sign at all.

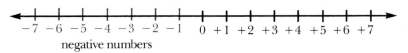

2. Numbers to the left of zero are called negative numbers and are written with a negative sign (−) placed before them.

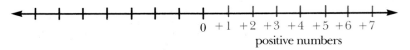

3. Zero is neither negative nor positive.

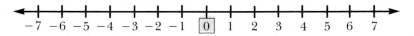

Numbers that are used to name the positive numbers, the negative numbers, and zero are called *integers*.

The integers are easily graphed on a number line.

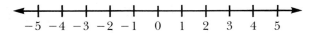

The integers may also be represented as

$$\ldots -5, -4, -3, -2, -1, 0, 1, 2, 3, 4, 5, \ldots$$

NOTE:

> The arrowheads at both ends of the number line have the same meaning as the three dots at both ends of the preceding listing, implying that the integers continue without end in both the positive direction and the negative direction.

READING INTEGERS

When reading integers, we always read the sign first, followed by the number. If a number does not have a sign in front of it, we assume the number to be positive. For example:

$+15$ is read as "*positive* fifteen"

-12 is read as "*negative* twelve"

6 is read as "*positive* six" (or just "six")

NOTE:

> The $+$ and $-$ signs have two different meanings in mathematics.
>
> ■ The $+$ sign refers to the operation of addition (plus), and it also indicates a positive number (a number greater than zero).
>
> ■ The $-$ sign refers to the operation of subtraction (minus), and it also indicates a negative number (a number less than zero).
>
> (Also, a number with no sign means positive).

Follow through the examples given here.

EXAMPLE 1: Write -32 in words.

SOLUTION: Since there is a "$-$" sign in front of the number, we write

$$-32$$

negative   thirty-two

Final answer: Negative thirty-two

EXAMPLE 2: Write negative two hundred seven by using integers.

SOLUTION:

negative   two hundred seven
$-$ 207

Final answer: -207

EXAMPLE 3: Compare -3 and -7.

SOLUTION: Since -3 is to the right of -7, -3 is greater than -7.

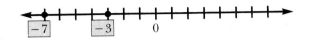

Final answer: $-3 > -7$

EXAMPLE 4: Represent a change of temperature from $2°$ below zero to $8°$ below zero.

SOLUTION: Since the temperature dropped from $2°$ below zero to $8°$ below zero, the change of temperature is $6°$. However, we indicate this downward change by using a negative sign.

Final answer: $-6°$

Do the following practice set. Check your answers with the answers in the right-hand margin.

PRACTICE SET 2-1A

Write the following integers in words (see example 1).

1. -34 2.   6   3. $+5$

Write the following using integers (see example 2).

4.   positive fourteen

5.   zero

6.   negative five thousand

1.   negative thirty-four
2.   six
3.   positive five

4.   14 (or $+14$)
5.   0
6. -5000

Using the symbols > and <, compare the following numbers (see example 3).

7. 0, −5   8. −4, −8   9. −7, −1

7. $0 > -5$
8. $-4 > -8$
9. $-7 < -1$

Answer the following questions (see example 4).

10. If sea level is represented by 0, represent 100 feet *below* sea level.

10. -100

11. If the temperature changes from *5° below zero to 10° above zero*, represent the change by using integers.

11. $+15$

12. In a football game, a quarterback is sacked for a *loss* of 12 yards. Represent this loss by using integers.

12. -12

B. The Rational Number System

In Chapter 1, we learned that all natural numbers, whole numbers, repeating or terminating decimals, fractions, and mixed numbers are rational numbers. We now expand this list to include the negative of these numbers.

On a number line, every number has a *reflection* or *opposite point*. That is, every number can be paired with another number that is the same distance from 0 and on the *opposite* side of 0. In fact, every negative number is really the opposite of its respective natural number. For example:

The numbers 1 and −1 are 1 unit away from zero and on the opposite sides of zero. (Likewise for 2 and −2, 3 and −3, etc.)

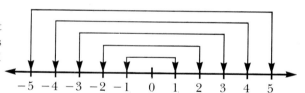

A number and its opposite are referred to as additive inverses of each other. Thus, 1 and −1 are additive inverses of each other; 2 and −2 are additive inverses; 3 and −3 are additive inverses, and so on.

Every natural number, whole number, decimal, fraction, and mixed number has an additive inverse. For example:

■ The additive inverse of 3 is −3.

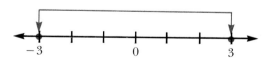

■ The additive inverse of 0 is 0.

■ The additive inverse of 0.5 is −0.5.

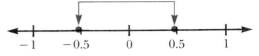

■ The additive inverse of $\frac{3}{4}$ is $-\frac{3}{4}$.

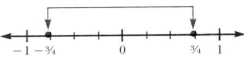

■ The additive inverse of $1\frac{2}{3}$ is $-1\frac{2}{3}$.

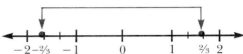

The rational number system consists of all the classifications of numbers we have discussed—natural numbers, whole numbers, decimals, fractions, mixed numbers, integers, and negative numbers.

We now give the following definition of a rational number.

RATIONAL NUMBERS

> Any number that can be written in the form a/b where both a and b are integers (and $b \neq 0$) is called a *rational number*.

Do the following practice set. Check your answers with the answers in the right-hand margin.

PRACTICE SET 2-1B

State the additive inverse for each of the following numbers.

1. 10

2. −5

3. $-\dfrac{2}{3}$

4. $3\dfrac{7}{8}$

5. 3.94

6. −2.6

1. −10
2. +5
3. $+\frac{2}{3}$
4. $-3\frac{7}{8}$
5. −3.94
6. +2.6

EXERCISE 2-1

Draw a number line for each of the following problems and graph the numbers given.

1. +5, 0, −5, 3, −2

2. $+\dfrac{3}{4}, -\dfrac{1}{4}, -3, +2\dfrac{1}{4}, -1\dfrac{1}{2}$

3. $+0.5, -0.5, +\dfrac{3}{10}, 0, -\dfrac{9}{10}$

4. $-\dfrac{1}{3}, -\dfrac{3}{3}, -\dfrac{5}{3}, 0, +\dfrac{8}{3}$

5. $-1, -2.3, +0.3, +0.9, +2.6$

Compare the following pairs of signed numbers by using the $<$ and $>$ symbols.

6. $-3, +5$

7. $+0, +6$

8. $-4, -5$

9. $+1\dfrac{1}{2}, -\dfrac{3}{2}$

10. $+5, -5$

11. $+1.2, -\dfrac{8}{4}$

12. $-3.1, -\dfrac{32}{10}$

13. $-9, +0.9$

14. $+5, -100$

15. $-11\dfrac{3}{4}, -\dfrac{15}{8}$

16. $+2\dfrac{3}{10}, -2.4$

17. $-2\dfrac{7}{8}, -2\dfrac{3}{4}$

18. $-5, -4, -7$

19. $-9.1, +8\dfrac{1}{8}, -0.6$

20. $+2.5, -3, -\dfrac{1}{2}$

Represent each of the following by using a signed number.

21. A $300 *profit*

22. One hundred feet *below sea level*

23. A *credit* of $10

24. A *rise* of 2 points

25. Twenty points *in the hole*

26. A *loss* of 12 pounds

27. A *debit* of $100

28. Twelve degrees *below zero*

29. A *loss* of 12 yards

30. A *negative amortization* of $206

State the additive inverse of each number.

31. -6

32. -3.4

33. $+0.5$

34. $\dfrac{9}{12}$

35. 0

36. $-\dfrac{7}{3}$

Read each statement carefully and then answer each question by selecting the correct choice given.

37. If the opposite of a number is positive, then the number itself must be ___?___. (positive/negative)

38. If a certain number is positive, then its opposite must be ___?___. (positive/negative)

39. If the opposite of a number is negative, then the number itself must be ___?___. (positive/negative).

40. If we can represent the opposite of a number by simply placing a negative sign in front of it, how would we represent:
    a. The opposite of $(+3)$? What is the opposite of $+3$?
    b. The opposite of (-3)? What is the opposite of -3?

2-2 THE ABSOLUTE VALUE OF A NUMBER

In the last section, we learned that every rational number has an additive inverse. That is, every number can be paired with another number that is the same distance from 0, and on the opposite side of 0. For example:

5 and −5 are additive inverses.

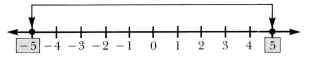

We also learned that the positive number is always to the right of its additive inverse, the negative number. Thus, the positive number is greater in value than its negative opposite. For example:

5 is to the right of its opposite, −5. Thus, 5 is greater in value than −5.

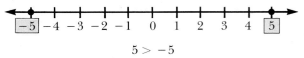

$$5 > -5$$

In fact, when dealing with any number and its opposite (except zero), the positive number will always be greater in value than its opposite. The absolute value of a number is the distance from zero to the graph of the number without regard to which direction it is from zero. For example:

5 is 5 units from 0.

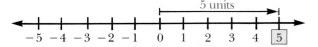

and

−5 is 5 units from 0.

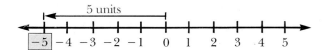

Therefore, the absolute value of 5 is 5, and the absolute value of −5 is 5.

Since distance is never negative, the absolute value of a number is never negative. This leads to the following definition.

The absolute value of a number is:

1. The number itself if the number is either zero or positive.

2. The additive inverse (or opposite) of the number if the number is negative.

We denote the absolute value of a number by writing vertical bars on either side of the number. Thus:

■ The absolute value of 2 is written $|2|$.

■ The absolute value of −3 is written $|-3|$.

■ The absolute value of 0 is written $|0|$.

etc.

In mathematics, it will be necessary to evaluate (or simplify) absolute value expressions. To do this, we must first write the expression without using the absolute value bars. Since the absolute value of a number is never negative, we may evaluate an absolute value expression by simply removing the bars and writing the nonnegative value of the expression.

Follow through the examples given here.

EXAMPLE 1:   Evaluate $|-20|$.

SOLUTION: Since the expression between the absolute value bars is negative, we remove the bars and write the value as a nonnegative expression.

$$|-20|$$
$$= (20)$$
$$= 20$$

Final answer: 20

EXAMPLE 2:   Evaluate $-|-6|$.

SOLUTION: Note that the expression between the bars is negative, and its absolute value will be nonnegative. However, we must still keep the negative sign that is outside the bars since it is part of the original problem.

$$-|-6|$$
$$= -(6)$$
$$= -6$$

Final answer: -6

EXAMPLE 3:   Evaluate $|20| - |-6|$.

SOLUTION:

$$|20| - |-6|$$
$$= 20 - (6)$$
$$= 14$$

Final answer: 14

EXAMPLE 4: Evaluate $|7 - 2| + |-3| - |-8|$.

SOLUTION:

$$
\begin{aligned}
&|7 - 2| + |-3| - |-8| \\
&= \quad |5| \quad + |-3| - |-8| \\
&= \quad 5 \quad + \;(3)\; - \;(8) \\
&= \qquad 5 + 3 - 8 \\
&= \qquad\quad 8 - 8 \\
&= \qquad\qquad 0
\end{aligned}
$$

Final answer: 0

Do the following practice set. Check your answers with the answers in the right-hand margin.

PRACTICE SET 2-2

Evaluate (see examples 1 or 2).

1. $|2|$ 2. $|-3|$ 3. $-|-6|$ 4. $-\left|-1\dfrac{1}{2}\right|$

1.	2
2.	3
3.	-6
4.	$-1\dfrac{1}{2}$

Evaluate (see examples 3 or 4).

5. $|4| + |6|$ 6. $|-5| - |3|$

7. $|-9| - |-8|$ 8. $|8 + 2| - |-7|$

5.	10
6.	2
7.	1
8.	3

EXERCISE 2-2

Write the absolute value for each of the following.

1. $+6$ 2. -3 3. $+8.5$ 4. $-7\dfrac{1}{2}$ 5. -9.42

6. $3\dfrac{2}{3}$ 7. $-4\dfrac{1}{4}$ 8. $-3\dfrac{2}{3}$ 9. 5   10. -0.9

Using the symbols, $<$, $>$, and $=$, compare the following pairs of numbers.

11. $|6|, -6$

12. $|-5|, 5$

13. $|-5|, -5$

14. $|-3|, |-3|$

15. $|-3|, |3|$

16. $|9|, |-9|$

17. $|-5|, 2$

18. $|-8|, |-12|$

19. $|14|, -36$

Evaluate each of the following.

20. $|3| + |5|$

21. $|16| - |-5|$

22. $|-9| + |18|$

23. $|-19| - |-18|$

24. $|-9| \cdot |-3|$

25. $|10| \cdot |-5|$

26. $\dfrac{|22|}{|-11|}$

27. $\dfrac{|-3| \cdot |-6|}{|-2|}$

28. $|(9 - 5) + 12| - |-16|$

29. $|3| + |8 - 6| + |3 + (4 - 3)|$

2-3 ADDITION AND SUBTRACTION OF SIGNED NUMBERS

In this section, we will discuss the operations of addition and subtraction as they are related to signed numbers. In our development, we will concentrate on adding or subtracting integers. However, you should keep in mind that any rule or procedure we develop for integers can be extended to all rational numbers that have been discussed thus far.

We begin our discussion by first pictorally considering addition, using a number line.

A. Adding Signed Numbers Using a Number Line

Since a positive number indicates direction to the right on a number line, and a negative number indicates direction to the left on a number line, we can add signed numbers by using a sequence of directed moves. For example, to add $2 + (-4)$ on a number line:

First, start at 0 and move 2 units in a positive direction.

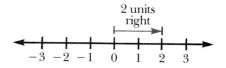

From this point, next, move 4 units in a negative direction.

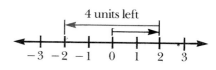

The sum is the number at which we end.

Thus,

$$2 + (-4) = -2$$

Follow through the examples given here.

EXAMPLE 1: Add $5 + (-7)$ on a number line.

SOLUTION: Start at 0 and move 5 units to the right (since 5 is positive).

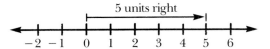

From this point move 7 units to the left (since 7 is negative).

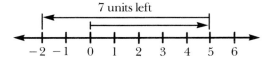

The sum is the final number reached.

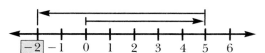

Thus,

$$5 + (-7) = -2$$

Final answer: -2

EXAMPLE 2: Add $(-3) + 4$ on a number line.

SOLUTION: Start at 0 and move 3 units left.

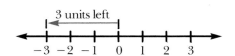

Now move 4 units to the right.

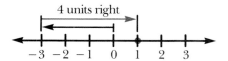

Thus,

$$(-3) + 4 = 1$$

Final answer: 1

EXAMPLE 3: Add $(-3) + (-5)$ on a number line.

SOLUTION:

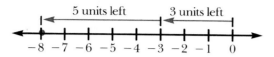

Final answer: -8

Do the following practice set. Check your answers with the answers in the right-hand margin.

PRACTICE SET 2-3A

Add the following signed numbers by using a number line (see example 1 or 2).

1. $4 + (-1)$
2. $3 + (-3)$
3. $(-8) + 2$

1.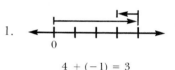

$$4 + (-1) = 3$$

2.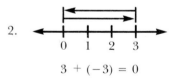

$$3 + (-3) = 0$$

3.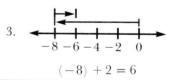

$$(-8) + 2 = 6$$

B. Expressing Subtraction as Addition

Earlier in our work, we learned that an opposite value, or additive inverse, is associated with every number. (e.g., the additive inverse of 3 is -3 and the additive inverse of -8 is 8).

In algebra, it is possible to express a typical subtraction problem as an equivalent addition problem by using additive inverses. To do this, instead of subtracting a number, we add its additive inverse. For example, the typical subtraction problem

$$5 - 3$$

may be expressed as the addition problem

$$5 + (-3)$$

where instead of subtracting 3 from 5, we add the additive inverse of 3 (namely, -3), to 5. By using a number line, we confirm the fact that $5 + (-3)$ does indeed equal $5 - 3$, which equals 2.

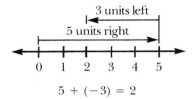

$$5 + (-3) = 2$$

As a result, we can conclude that subtracting a number is the same as adding its additive inverse (see the chart that follows).

Subtraction	Addition	Explanation
$4 - 4 = 0$	$4 + (-4) = 0$	Subtracting 4 is the same as adding the opposite value of 4
$4 - 3 = 1$	$4 + (-3) = 1$	Subtracting 3 is the same as adding the opposite value of 3
$4 - 2 = 2$	$4 + (-2) = 2$	Subtracting 2 is the same as adding the opposite value of 2
$4 - 1 = 3$	$4 + (-1) = 3$	Subtracting 1 is the same as adding the opposite value of 1.

What all this means, in algebra, is that every subtraction problem can be written as an equivalent addition problem. Thus, to subtract one signed number from another, we simply change the subtraction sign to an addition sign and replace the number being subtracted with its additive inverse. For example:

$$6 - 4 \quad \text{becomes} \quad 6 + (-4)$$
$$6 - (-4) \quad \text{becomes} \quad 6 + (+4)$$
$$(-6) - 4 \quad \text{becomes} \quad -6 + (-4)$$
$$(-6) - (-4) \quad \text{becomes} \quad -6 + (+4)$$

Subtraction problems, then, can be treated as addition problems that, in turn, can be evaluated by using a number line. Thus:

$6 - 4 \quad \text{equals} \quad 6 + (-4) = 2$

$6 - (-4) \quad \text{equals} \quad 6 + (4) = 10$

$(-6) - 4 \quad \text{equals} \quad (-6) + (-4) = -10$

$(-6) - (-4) \quad \text{equals} \quad (-6) + (4) = -2$

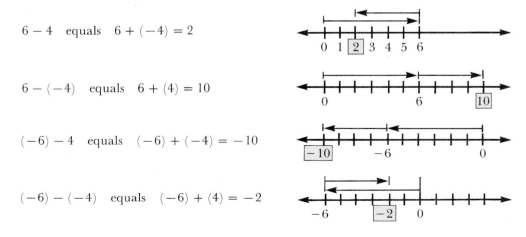

Do the following practice set. Check your answers with the answers in the right-hand margin.

PRACTICE SET 2-3B

Solve the following subtraction problems by first expressing them as equivalent addition problems. Then, using a number line, add the numbers.

1. $9 - (-2)$
2. $(-1) - 4$
3. $(-6) - (-2)$
4. $8 - (2)$

1. $9 - (-2) = 9 + (2)$
    $\qquad = 11$

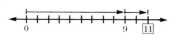

2. $\quad (-1) - 4$
    $\quad = (-1) + (-4)$
    $\quad = -5$

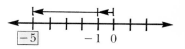

3. $\quad (-6) - (-2)$
    $\quad = (-6) + (2)$
    $\quad = -4$

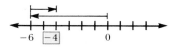

4. $8 - 2 = 8 + (-2)$
    $\qquad = 6$

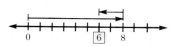

C.   Adding and Subtracting Signed Numbers

PART 1:   ELIMINATING DOUBLE SIGNS

In the last section, we discovered that a subtraction problem can be expressed as an equivalent addition problem. It would seem logical, then, that an addition problem can be expressed as an equivalent subtraction problem. This is indeed the case. For example:

$$3 - (-6) = 3 + (+6)$$
$$= 3 + 6$$

subtraction expressed as an
equivalent addition

and similarly,

$$3 + 6 = 3 + (+6) \left.\right\} \quad \text{addition expressed as an}$$
$$= 3 - (-6) \left.\right\} \quad \text{equivalent subtraction}$$

Thus, we can "flip/flop" between addition and subtraction by simply changing both the sign of the operation and the sign of the number being added or subtracted to its opposite. In general terms, we have

$$a + b = a - (-b)$$
$$a + (-b) = a - (+b)$$
$$a - (+b) = a + (-b)$$
$$a - (-b) = a + (+b)$$

When flip/flopping between addition and subtraction, we frequently encounter *double signs*, where the first sign indicates the operation (addition or subtraction) and the second sign indicates the sign of the number (positive or negative). For example:

■ **2 MINUS POSITIVE 3** or **2 MINUS NEGATIVE 3**

$$2 - (+3) \qquad\qquad 2 - (-3)$$
$$\diagdown\diagup \qquad\qquad\quad \diagdown\diagup$$
$$\text{double} \qquad\qquad\quad \text{double}$$
$$\text{signs} \qquad\qquad\quad\;\; \text{signs}$$

■ **2 PLUS POSITIVE 3** or **2 PLUS NEGATIVE 3**

$$2 + (+3) \qquad\qquad 2 + (-3)$$
$$\diagdown\diagup \qquad\qquad\quad \diagdown\diagup$$
$$\text{double} \qquad\qquad\quad \text{double}$$
$$\text{signs} \qquad\qquad\quad\;\; \text{signs}$$

When we deal with such problems, our goal will be to eliminate all double signs. Since we can flip/flop between addition and subtraction, the elimination of double signs can be easily accomplished. For example:

■ In the problem $2 - (+3)$, since $+3$ can be written without the $+$ sign, we have $2 - 3$.

$$2 - (+3)$$
$$\diagdown\diagup$$
$$= \;\; 2 - 3$$

■ In the problem $2 + (-3)$, we can switch from adding a negative number to subtracting a positive number. Now, we have the same problem as before.

$$2 + (-3)$$
$$= 2 - (+3)$$
$$\diagdown\diagup$$
$$= \;\; 2 - 3$$

■ In the problem $2 + (+3)$, $+3$ can be written without the $+$ sign.

$$2 + (+3)$$
$$\diagdown\diagup$$
$$2 + 3$$

■ In the problem $2 - (-3)$, we can switch from subtracting a negative number to adding a positive number. Now, we have the same problem as before.

$$2 - (-3)$$
$$2 + (+3)$$
$$\diagdown\diagup$$
$$2 + 3$$

This leads to the following rule of thumb.

---

1.   If the double signs are opposite, that is,

$$-(+)   \text{ or }   +(-)$$

replace them with $-$.

2.   If the double signs are the same, that is,

$$+(+)   \text{ or }   -(-)$$

replace them with $+$.

In more general terms, we have

$$a + (+b) = a + b$$
$$a + (-b) = a - b$$
$$a - (+b) = a - b$$
$$a - (-b) = a + b$$

and conversely.

Do the following practice set. Check your answers with the answers in the right-hand margin.

PRACTICE SET 2-3C (PART 1)

Rewrite each of the following problems without double signs. Do *not* evaluate.

1. $3 + (-6)$	2. $5 - (+9)$	1. $3 - 6$
3. $4 - (-8)$	4. $6 + (+6)$	2. $5 - 9$
5. $-7 - (-2)$	6. $-8 + (-6)$	3. $4 + 8$
7. $-3 + (+4)$	8. $-2 - (+3)$	4. $6 + 6$
		5. $-7 + 2$
		6. $-8 - 6$
		7. $-3 + 4$
		8. $-2 - 3$

PART 2:   COMBINING ACCORDING TO SIGNS

In part 1 of this section, we discovered that we can flip/flop between addition and subtraction of signed numbers. In algebra, rather than add or subtract per se, we *combine according to signs*. To do this, we follow the procedure outlined on the next page.

PROCEDURE:

> To add or subtract signed numbers:
> 1. Eliminate all double signs.
> 2. If the sign in front of each respective number is the same, that is,
> $$(+) \text{ and } (+) \qquad \text{or} \qquad (-) \text{ and } (-)$$
> add the absolute value of the numbers and keep the common sign for the answer.
> 3. If the sign in front of each respective number is different, that is,
> $$(+) \text{ and } (-) \qquad \text{or} \qquad (-) \text{ and } (+)$$
> subtract the absolute value of the numbers (larger number minus smaller number) and keep the sign of the larger number for the answer.

EXAMPLE 1: Evaluate $-4 + (-3)$.

SOLUTION: 1. First eliminate double signs.
$$-4 + (-3)$$
$$= \quad -4 - 3$$

2. Since the signs are the same, add the numbers and keep the common sign.
$$-4 - 3$$
$$= \quad -7$$

Final answer: -7

EXAMPLE 2: Evaluate $5 - (-8)$.

SOLUTION: 1. Eliminate double signs.
$$5 - (-8)$$
$$= \quad 5 + 8$$

2. The signs are both $(+)$, so add the numbers and keep the common sign.
$$5 + 8$$
$$= \quad +13$$

Final answer: 13

EXAMPLE 3: Evaluate $-4 + 3$.

SOLUTION: (No double signs.) Since the signs are different, subtract the smaller absolute value from the larger absolute value $(4 - 3)$ and attach the sign of the number with the larger absolute value.

$$-4 + 3$$
$$= \quad -1$$

Final answer: -1

EXAMPLE 4: Evaluate $-3 - (-4)$.

SOLUTION: 1. Eliminate double signs.

$$-3 - (-4)$$
$$= \quad -3 + 4$$

2. Subtract the numbers and keep the sign of the larger number.

$$-3 + 4$$
$$= \quad +1 \qquad [4 - 3 = 1]$$

Final answer: 1

EXAMPLE 5: Evaluate each problem that follows.

a. $-17 - (+45)$

b. $-\dfrac{2}{7} + \left(-\dfrac{3}{7}\right)$

c. $0.75 + (-4.2)$

SOLUTION:

$$-17 - (+45) = -17 - 45$$
$$= -62$$

Final answer: -62

SOLUTION:

$$-\frac{2}{7} + \left(-\frac{3}{7}\right) = -\frac{2}{7} - \frac{3}{7}$$
$$= -\frac{5}{7}$$

Final answer: $-\dfrac{5}{7}$

SOLUTION:

$$0.75 + (-4.2) = 0.75 - 4.2$$
$$= -3.45$$

Final answer: -3.45

Do the following practice set. Check your answers with the answers in the right-hand margin.

PRACTICE SET 2-3C (PART 2)

Evaluate.

1. $4 + 2$

2. $-1 + 4$

3. $6 - 7$

4. $-8 + (-2)$

5. $-8 + (+2)$

6. $4 - (-8)$

7. $-6 - (-10)$

8. $-3 - (+4)$

9. $0 - (-4)$

10. $-1.5 - 0.3$

11. $-0.25 - (-2.5)$

12. $-\dfrac{5}{7} + \left(-\dfrac{4}{7}\right)$

1. 6
2. 3
3. -1
4. -10
5. -6
6. 12
7. 4
8. -7
9. 4
10. -1.8
11. 2.25
12. $-\frac{9}{7}$ or $-1\frac{2}{7}$

EXERCISE 2-3

Evaluate.

1. $7 + 4$

2. $-4 + 3$

3. $9 + (-5)$

4. $18 - 3$

5. $-12 - 16$

6. $16 - (-12)$

7. $-2 + (-2)$

8. $5 + (-4)$

9. $-7 + 9$

10. $3 + (-5)$

11. $0 - (-2)$

12. $0 + (-2)$

13. $4 - (-4)$

14. $-4 - (-4)$

15. $0 - 9$

16. $-25 + (-18)$

17. $-146 + (-10)$

18. $29 - 106$

Evaluate.

19. $-0.113 + 2.4$

20. $5.31 + (-2.6)$

21. $-3.8 + (-0.12)$

22. $-4.2 - 5.6$

23. $-3.8 - (-2.1)$

24. $0.5 - 2.7$

25. $6.71 - (-2.14)$

26. $-1.28 + (-4.6)$

27. $-0.23 - (-1.47)$

28. $-\dfrac{1}{3} - \left(-\dfrac{2}{3}\right)$

29. $-\dfrac{5}{9} - \dfrac{3}{9}$

30. $\dfrac{2}{7} - \left(-\dfrac{5}{7}\right)$

31. $-\dfrac{1}{3} + \left(-\dfrac{2}{3}\right)$

32. $-\dfrac{3}{5} + \dfrac{2}{5}$

33. $\dfrac{5}{8} + \left(-\dfrac{3}{8}\right)$

34. $3\dfrac{1}{2} - \dfrac{5}{2}$

35. $-2\dfrac{1}{4} - \left(-\dfrac{7}{4}\right)$

36. $3\dfrac{2}{3} - \left(+\dfrac{20}{3}\right)$

37. $-2\dfrac{1}{3} + \dfrac{3}{2}$

38. $\dfrac{5}{9} + \left(-1\dfrac{2}{7}\right)$

39. $-3\dfrac{4}{5} + \left(-\dfrac{16}{15}\right)$

Evaluate.

40. $7 + 5 - (-19)$

41. $3 + (-8) - (-6)$

42. $13 + (-8) - 21$

43. $-23 - (-18) - 41$

44. $-4 - (-6) + (-15)$

45. $-21 + (-8) - (-6) - (-9)$

46. $12 + (-10) + 12$

47. $14 + 14 + (-28)$

48. $-4 + (-12) + 10$

49. $37 + (-18) + (-23)$

50. $-50 + 13 + (-8)$

51. $13 + 5 + (-6)$

52. $21 + (-3) + 14 + (-9)$

53. $20 + 13 + (-8) + 5$

54. $-6 + (-5) - (-6) - (+3)$

55. $-5 - 5 + 5 - 5 + 5 - 5$

Solve each of the following problems.

56. At Finger Lakes Racetrack, Ron started the day with $100. He lost $27 on the first race, won $16 on the second race, won $82 on the third race, and then lost $36 on the trifecta. Determine if Ron actually won or lost money and how much it was.

57. If Ronolog, Inc. opened trading on the MCC Exchange at $5\frac{3}{8}$ and closed at $4\frac{7}{8}$, what was the net change?

58. What is the change in temperature when the temperature drops from $-3°C$ to $-11°C$?

59. A football player made the following yardage on five plays: $(+17)$, (-3), $(+8)$, (-6), $(+4)$. What was his total gain or loss? What was his average per play?

60. On takeoff, an airplane rises to 18,500 feet. What would its new altitude be if the plane must drop 3000 feet?

61. Mary's checking account showed the following transactions.

Balance:	$502.67
Deposit $(+)$	210.00
Withdrawal $(-)$	390.00
Withdrawal $(-)$	116.57
Withdrawal $(-)$	22.18
Deposit $(+)$	25.00
Withdrawal $(-)$	108.93

What is Mary's balance? Is she overdrawn (yes/no)?

62. In a game of Scrabble, Jim had a total of 197 points but had to deduct 15 points, whereas Mike had a total of 192 points but had to deduct 8 points. Who won the game and by how many points?

63. A PFC freight truck weighed in at 83,216 pounds. If the total weight of the empty truck (tractor-trailer combination) is 28,920 pounds, what was the total weight of the freight the truck was carrying?

64. In New York State, the maximum gross weight (MGW) of a tractor-trailer combination allowed on the roads is 80,000 pounds. Would the truck in problem 63 be overweight? If so, by how much?

65. At the base of the Junfreau Mountain in Switzerland, the temperature is $82°F$. At the top of the mountain the temperature drops to $-5°F$. What is the total drop in temperature?

2-4   MULTIPLICATION OF SIGNED NUMBERS

As with the addition and subtraction of signed numbers, multiplication of signed numbers requires that we pay attention to the sign of the numbers being multiplied and the sign of the answer, as well as to the numbers being multiplied. Specifically, we encounter three distinct situations:

- The product of two positive numbers
- The product of a positive number and a negative number
- The product of two negative numbers

We will develop a general rule or procedure for each of these cases separately and then provide a summary at the end.

THE PRODUCT OF TWO POSITIVE NUMBERS

In arithmetic, we learned how to multiply positive numbers. For example:

$$(+2) \times (+4) = +8$$
$$2.6 \times 0.4 = +1.04$$
$$\frac{1}{5} \cdot \frac{3}{8} = \frac{3}{40}$$

Thus,

the product of two positive numbers is positive

THE PRODUCT OF A POSITIVE NUMBER AND A NEGATIVE NUMBER

Since multiplication is the process of repeatedly adding a quantity to itself a certain number of times, we may treat a multiplication problem as a repeated addition problem. For example,

$$3 \times 2 \quad \text{means} \quad 2 + 2 + 2 \quad \text{or} \quad 6.$$

Similarly,

$$3 \times (-2) \quad \text{means} \quad -2 + (-2) + (-2) \quad \text{or} \quad -6.$$

Consequently,

the product of a positive number and a negative number is negative

To see this more visually, we use a number line and follow the pattern that develops when one factor is consistently decreased by 1.

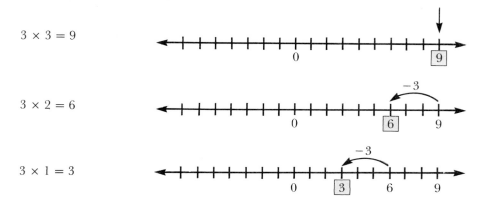

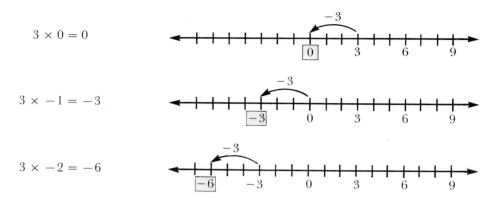

$$3 \times 0 = 0$$

$$3 \times -1 = -3$$

$$3 \times -2 = -6$$

Notice that the products are moving to the left in multiples of 3 and that by continuing in this manner, we find that the product of a positive number and a negative number will be a negative number.

THE PRODUCT OF TWO NEGATIVE NUMBERS

Using the idea of consistently decreasing one factor of a multiplication problem by 1, we can also develop a rule for multiplying two negative numbers. Again, follow the pattern that develops

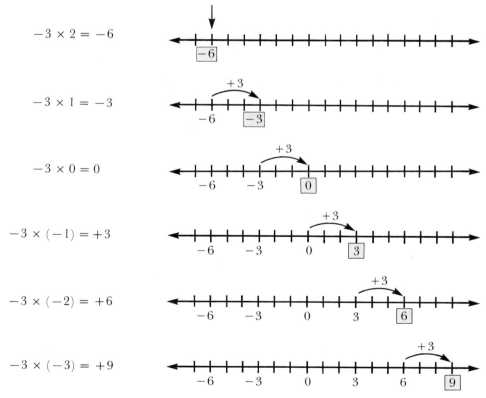

$$-3 \times 2 = -6$$

$$-3 \times 1 = -3$$

$$-3 \times 0 = 0$$

$$-3 \times (-1) = +3$$

$$-3 \times (-2) = +6$$

$$-3 \times (-3) = +9$$

Notice that the products are moving to the right in multiples of 3. By continuing this procedure, we see clearly that the product of two negative numbers is positive.

As a result of these three cases, we conclude the following:

> 1. If the signs of the numbers being multiplied are the same, that is,
>
> $$(+)(+) \qquad \text{or} \qquad (-)(-)$$
>
> then the product will be positive.
>
> 2. If the signs of the numbers being multiplied are different, that is,
>
> $$(-)(+) \qquad \text{or} \qquad (+)(-)$$
>
> then the product will be negative.

More generally,

$$(+a) \times (+b) = +ab$$
$$(-a) \times (+b) = -ab$$
$$(a) \times (-b) = -ab$$
$$(-a) \times (-b) = +ab$$

Consequently, to multiply signed numbers we simply multiply the numbers as if they were positive (i.e., we multiply their absolute values), and then we determine the sign of the product by examining the signs of the individual factors.

PROCEDURE:

To multiply signed numbers:

1. Multiply the absolute value of the numbers.

2. Determine the sign of the product by the following scheme.

$$(+)(+) = +$$
$$(-)(-) = +$$
$$(+)(-) = -$$
$$(-)(+) = -$$

EXAMPLE 1: Evaluate $2 \cdot (-8)$.

SOLUTION: First, multiply the absolute value of the numbers.

$$2 \cdot (-8)$$
$$= 2 \cdot (8)$$
$$= \quad 16$$

Now examine the signs.

$$+2 \cdot (-8)$$
$$(+) \quad (-) = -$$

Thus,

$$2 \cdot (-8) = -16$$

Final answer: -16

EXAMPLE 2: Evaluate $(-5) \times (-8)$.

SOLUTION: Multiply the absolute value of the numbers.

$$(-5) \times (-8)$$
$$= \quad 5 \quad \times \quad 8$$
$$= \quad\quad 40$$

Examine the signs.

$$(-5) \times (-8)$$
$$(-) \quad\quad (-) = +$$

Thus,

$$(-5) \times (-8) = +40$$

Final answer: $+40$

With practice, such multiplication can be performed in one step.

EXAMPLE 3: Evaluate each of the following.

a. $-6 \times (5)$ b. $-9 \times (-8)$ c. $3 \times (-5)$

SOLUTION:

$-6 \times 5 = 30$

$\begin{bmatrix} \text{Since } 6 \times 5 = 30 \\ \text{and since } (-)(+) = - \end{bmatrix}$

SOLUTION:

$-9 \times (-8) = +72$

$\begin{bmatrix} \text{Since } 9 \times 8 = 72 \\ \text{and since } (-)(-) = + \end{bmatrix}$

SOLUTION:

$3 + (-5) = -15$

$\begin{bmatrix} \text{Since } 3 \times 5 = 15 \\ \text{and since } (+)(-) = - \end{bmatrix}$

EXAMPLE 4: Evaluate $(-4) \times 8 \times (-3)$.

Example continued on next page.

SOLUTION: To evaluate this expression, we first multiply the first two factors.

$$(-4) \times 8 \times (-3)$$

$$= \quad -32 \quad \times (-3)$$

This product is now multiplied by the remaining factor.

$$-32 \times (-3)$$

$$= \quad +96$$

Final answer: $+96$

EXAMPLE 5: Evaluate $(-3) \times 5 \times (-2) \times (-7)$.

SOLUTION:

$$(-3) \times 5 \times (-2) \times (-7)$$

$$= \quad -15 \quad \times (-2) \times (-7)$$

$$= \quad +30 \quad \times (-7)$$

$$= \quad -210$$

Final answer: -210

NOTE:

If a product contains an odd number of negative factors, $(1, 3, 5,$ etc.$)$, the product is negative. If a product contains an even number of negative factors, $(0, 2, 4, 6,$ etc.$)$, the product is positive. For example:

TWO NEGATIVE FACTORS

$$(-1)(-1)$$

$$+1$$

Positive product

THREE NEGATIVE FACTORS

$$(-1)(-1)(-1)$$

$$(+1)(-1)$$

$$-1$$

Negative product

FOUR NEGATIVE FACTORS

$$(-1)(-1)(-1)(-1)$$

$$(+1)(-1)(-1)$$

$$(-1)(-1)$$

$$+1$$

Positive product

FIVE NEGATIVE FACTORS

$$(-1)(-1)(-1)(-1)(-1)$$

$$(+1)(-1)(-1)(-1)$$

$$(-1)(-1)(-1)$$

$$(+1)(-1)$$

$$-1$$

Negative product

EXAMPLE 6: Evaluate $(2)^2(-3)^3$.

SOLUTION: First, expand by writing out the factors.

$$(2)^2(-3)^3$$
$$= (2)(2)(-3)(-3)(-3)$$

Now, we can evaluate.

$$(2)(2)(-3)(-3)(-3)$$
$$= (4)(-3)(-3)(-3)$$
$$= (-12)(-3)(-3)$$
$$= (+36)(-3)$$
$$= -108$$

Final answer: -108

EXAMPLE 7: Evaluate $(-2)^2(-3)^2$.

SOLUTION:

$$(-2)^2(-3)^2$$
$$= (-2)(-2)(-3)(-3)$$
$$= (+4)(-3)(-3)$$
$$= (-12)(-3)$$
$$= (+36)$$

Final answer: $+36$

NOTE:

When evaluating signed numbers containing exponents, we must be careful to apply the exponent to the symbol that comes directly before it. For example:

■ $(-3)^2$ means $(-3)(-3) = +9$ since the exponent is being applied to the parentheses.

■ -3^2 means $-(3)(3) = -9$ since the exponent is being applied to the 3 only, not the entire quantity, -3.

Do the following practice set. Check your answers with the answers in the right-hand margin.

PRACTICE SET 2-4

Evaluate (see examples 1, 2, or 3).

1. $(3)(3)$	2. $(-1)(7)$	3. $(-3)(-7)$	1. 9
4. $(-8)(+3)$	5. $4 \times (-6)$	6. $-2 \times (-6)$	2. -7

Evaluate (see examples 4 and 5).

7. $(-2)(-3)(-1)$ 8. $(5)(-6)(-2)$ 9. $(6)(-3)(5)$

10. $(3)(-5)(-6)(-8)$ 11. $(4)(3)(8)(-1)$

Evaluate. (see examples 6 and 7).

12. $(3)(-2)^2$ 13. $(-3)^2(4^2)$ 14. $2^3(-4)^3$

15. $(-3)(-4)^2$ 16. $(-2)^3(-5)^2$ 17. $(2)(-2)^2(-5)^2$

1. 9
2. -7
3. $+21$
4. -24
5. -24
6. $+12$
7. -6
8. $+60$
9. -90
10. -720
11. -96
12. 12
13. 144
14. -512
15. -48
16. -200
17. 200

EXERCISE 2-4

Multiply.

1. $(9) \cdot (8)$ 2. $(-7) \cdot (4)$ 3. $(25) \cdot (-8)$

4. $(-12) \cdot (-4)$ 5. $(-6) \cdot (-5)$ 6. $(-10) \cdot (-10)$

7. $(-9) \cdot (8)$ 8. $4 \cdot (-4)$ 9. $10 \cdot (-20)$

10. $(-5.1) \cdot (10)$ 11. $3 \cdot (-1.8)$ 12. $12 \cdot (-1.20)$

13. $(-9.1) \cdot 13$ 14. $(-0.25) \cdot (-8)$ 15. $(-4.61) \cdot (-100)$

16. $0 \cdot (-6)$ 17. $(-100)(0)$ 18. $(-1)(-1)$

19. $\left(-\dfrac{1}{5}\right)\left(-\dfrac{3}{5}\right)$ 20. $\left(-\dfrac{2}{3}\right)(5)$ 21. $8\left(-\dfrac{12}{5}\right)$

22. $\left(-\dfrac{2}{7}\right)\left(-\dfrac{14}{21}\right)$ 23. $\left(2\dfrac{3}{5}\right)\left(-\dfrac{5}{13}\right)$ 24. $\left(-\dfrac{7}{8}\right)\left(\dfrac{23}{21}\right)$

Multiply.

25. $(-6)(-8)(-5)$ 26. $(4)(3)(-8)$ 27. $(-2)(5)(-10)$

28. $3[(-4)(-5)]$ 29. $-3[(-4)(-5)]$ 30. $-1[(3)(5)]$

31. $14(-3)(0)(-6)$ 32. $(2)(-5)(-1)(3)$ 33. $(-5)(-5)(-1)(1)$

34. $(|15|)(|-5|)(-1)$ 35. $(|3|)(|0|)(|-1|)$ 36. $(3)(|-3|)(1)$

Evaluate.

37. 2^3 38. $(-2)^3$ 39. -2^3

40. 0^2 41. $(-11)^1$ 42. 5^2

43. $(-2)^6$ 44. 25^2 45. $(-25)^2$

46. $4(-3)^2$ 47. $5(-6)^2$ 48. $2(-2)^3$

49. $(-3)(-3)^2$ 50. $(-3)^2(-3)^2$ 51. $-3^2(2)^2$

52. $-1(3)^2(2)^2$ 53. -1^{10} 54. $(-1)^{10}$

55. $5(-2)^2(-3)^3$ 56. $5^2(3)^2(-2)^3$ 57. $-3^2(-2^2)(-3)^2$

2-5   DIVISION OF SIGNED NUMBERS

In arithmetic, we learned that division is the opposite or inverse operation of multiplication. As an illustration of this relationship, consider the division problem $8 \div 4 = 2$.

To divide 8 by 4 means to find a number that when multiplied by 4 will yield a product of 8.

$\blacksquare \times 4 = 8$

$2 \times 4 = 8$

Thus,

$$8 \div 4 = 2 \quad \text{since} \quad 2 \times 4 = 8$$

As a result of this relationship between multiplication and division, we can extend and apply our rules for multiplication of signed numbers to the division of signed numbers. Thus:

■   A positive number divided by a positive number is positive since "plus times plus equals plus."

$\blacksquare \times (+) = (+)$

$(+) \times (+) = (+)$

■   A positive number divided by a negative number is negative since "minus times minus equals plus."

$\blacksquare \times (-) = (+)$

$(-) \times (-) = (+)$

■   A negative number divided by a positive number is negative since "minus times plus equals plus."

$\blacksquare \times (+) = (-)$

$(-) \times (+) = (-)$

■ A negative number divided by a negative number is positive since "plus times minus equals minus."

$$\blacksquare \times (-) = (-)$$
$$(+) \times (-) = (-)$$

The following statements can help you to remember these rules.

---

1. If the signs of the numbers being divided are the same, that is,

$$\frac{+}{+} \quad \text{or} \quad \frac{-}{-}$$

then the quotient will be positive.

2. If the signs of the numbers being divided are different, that is,

$$\frac{+}{-} \quad \text{or} \quad \frac{-}{+}$$

then the quotient will be negative.

More generally,

$$\frac{+a}{+b} = +c \qquad \frac{-a}{+b} = -c$$

$$\frac{+a}{-b} = -c \qquad \frac{-a}{-b} = +c$$

Consequently, to divide signed numbers, we simply divide the numbers as if they were positive (i.e., divide their absolute values), and then we determine the sign of the quotient by examining the signs of the divisor and dividend.

PROCEDURE:

To divide signed numbers:

1. Divide the absolute value of the numbers.

2. Determine the sign of the quotient by the following scheme.

$$\frac{+}{+} = +$$

$$\frac{-}{-} = +$$

$$\frac{+}{-} = -$$

$$\frac{-}{+} = -$$

EXAMPLE 1: Evaluate $\dfrac{-4}{2}$.

SOLUTION: First, divide the absolute value of the numbers.

$$\frac{|-4|}{|2|} = \frac{4}{2}$$
$$= 2$$

Now, examine the signs.

$$\frac{-4}{2} = \frac{-}{+}$$
$$= -$$

Thus,

$$\frac{-4}{2} = -2$$

Final answer: -2

EXAMPLE 2: Evaluate $\dfrac{-35}{-7}$.

SOLUTION: Divide the absolute value of the numbers.

$$\frac{|-35|}{|-7|} = \frac{35}{7}$$
$$= 5$$

Examine the signs.

$$\frac{-35}{-7} = \frac{-}{-}$$
$$= +$$

Thus,

$$\frac{-35}{-7} = +5$$

Final answer: $+5$

With practice, such division can be performed in one step.

EXAMPLE 3: Evaluate each of the following.

a. $\dfrac{-3}{9}$

b. $-\dfrac{1}{6} \div \left(-\dfrac{1}{8}\right)$

c. $\dfrac{-43.2}{2.4}$

SOLUTION:

$$\dfrac{-3}{9} = \dfrac{-1}{3}$$

Final answer: $-\dfrac{1}{3}$

SOLUTION:

$$-\dfrac{1}{6} \div -\dfrac{1}{8} = -\dfrac{1}{6} \times -\dfrac{8}{1}$$

$$= \dfrac{-8}{-6}$$

$$= \dfrac{4}{3}$$

Final answer: $+\dfrac{4}{3}$

SOLUTION:

$$\dfrac{-43.2}{2.4} = 2.4 \overline{\smash{\big)}\, 43.2}$$

$$\begin{array}{r} 18. \\ 2.4\overline{)43.2} \\ 24 \\ \hline 192 \\ 192 \\ \hline \end{array}$$

$$= -18$$

Final answer: -18

NOTE:

Frequently, in algebra, we write negative fractions by placing the negative sign in front of the fraction. Since a fraction can only be negative when either the numerator or denominator is negative (but not both), it should be observed that

$$-\dfrac{a}{b} = \dfrac{-a}{b} = \dfrac{a}{-b}$$

Do the following practice set. Check your answers with the answers in the right-hand margin.

PRACTICE SET 2-5

Evaluate (see examples 1, 2, 3, or 4).

1. $\dfrac{14}{2}$

2. $\dfrac{-28}{-4}$

3. $\dfrac{-5}{-1}$

4. $\dfrac{-18}{3}$

5. $\dfrac{16}{-4}$

6. $\dfrac{42}{-6}$

1. 7
2. 7
3. 5
4. (-6)
5. (-4)
6. (-7)

7. $\dfrac{4.41}{-2.1}$ 8. $(-6) \div \left(-\dfrac{1}{2}\right)$ 9. $\dfrac{-4}{0.5}$

7. (-2.1)
8. 12
9. (-8)

EXERCISE 2-5

Divide.

1. $(12) \div (6)$ 2. $(-9) \div 3$ 3. $100 \div (-5)$
4. $(-18) \div (-6)$ 5. $(-14) \div (2)$ 6. $(-6) \div (-3)$
7. $\dfrac{22}{11}$ 8. $\dfrac{-21}{-3}$ 9. $\dfrac{-9}{-9}$
10. $\dfrac{-25}{5}$ 11. $\dfrac{20}{-4}$ 12. $\dfrac{0}{-1}$
13. $\dfrac{-7}{-9}$ 14. $\dfrac{-6}{-9}$ 15. $\dfrac{5}{-10}$
16. $(-60) \div (20)$ 17. $(-28) \div (-10)$ 18. $(36) \div (-8)$
19. $\left(-\dfrac{2}{3}\right) \div \left(\dfrac{3}{4}\right)$ 20. $\left(\dfrac{5}{7}\right) \div \left(-\dfrac{15}{28}\right)$ 21. $\left(-\dfrac{7}{9}\right) \div (-14)$
22. $(-0.6) \div (0.3)$ 23. $(-5.8) \div (-0.2)$ 24. $(12.9) \div (-0.3)$
25. $\left(-2\dfrac{1}{2}\right) \div \left(\dfrac{5}{8}\right)$ 26. $\left(\dfrac{3}{7}\right) \div \left(-1\dfrac{2}{7}\right)$ 27. $\left(-1\dfrac{3}{4}\right) \div \left(-2\dfrac{5}{8}\right)$
28. $\dfrac{300}{-10}$ 29. $\dfrac{-567}{-3}$ 30. $\dfrac{-189}{7}$

2-6   MIXED OPERATIONS

In Section 1-5, we learned how to evaluate numerical expressions involving mixed operations. We now extend this evaluation to expressions containing signed numbers. We offer the same procedure as before.

PROCEDURE:

To evaluate problems involving more than one operation (containing signed numbers):

1. Remove all grouping symbols (by finding the value within the grouping symbols).
2. Evaluate all powers.
3. Perform all multiplication and division (working from left to right).
4. Perform all addition and subtraction (working from left to right).

EXAMPLE 1:   Evaluate $-8 \div 4 \times (-3) + (-1)$.

SOLUTION:

$$-8 \div 4 \times (-3) + (-1)$$
$$= \quad -8 \div 4 \times (-3) - 1$$
$$= \quad -2 \times (-3) - 1$$
$$= \quad +6 - 1$$
$$= \quad 5$$

Final answer: 5

EXAMPLE 2:   Evaluate $(-3)^2 + 4 \times (-5 + 3) \div (-2)$.

SOLUTION:

$$(-3)^2 + 4 \times (-5 + 3) \div (-2)$$
$$= \quad (-3)^2 + 4 \times (-2) \div (-2)$$
$$= \quad +9 + 4 \times (-2) \div (-2)$$
$$= \quad 9 - 8 \div (-2)$$
$$= \quad 9 + 4$$
$$= \quad 13$$

Final answer: 13

EXAMPLE 3:   Evaluate $[(-3 - (-2)^3) + (4 - 9)^2]^2$.

SOLUTION:

$$[(-3 - (-2)^3) + (4 - 9)^2]^2$$
$$= \quad [(-3 - (-8)) + (-5)^2]^2$$
$$= \quad [(-3 + 8) + (-5)^2]^2$$
$$= \quad [+5 + (+25)]^2$$
$$= \quad [+5 + 25]^2$$
$$= \quad [+30]^2$$
$$= \quad +900$$

Final answer: 900

EXAMPLE 4:   Evaluate $5 \div 5 - 5 \times 5 + 4 \div 4 - 4$.

SOLUTION:

$$5 \div 5 - 5 \times 5 + 4 \div 4 - 4$$
$$= \quad 1 - 5 \times 5 + 4 \div 4 - 4$$
$$= \quad 1 - 25 + 4 \div 4 - 4$$
$$= \quad 1 - 25 + 1 - 4$$
$$= \quad -24 + 1 - 4$$
$$= \quad -23 - 4$$
$$= \quad -27$$

Final answer: -27

Do the following practice set. Check your answers with the answers in the right-hand margin.

PRACTICE SET 2-6

Evaluate (see example 1 or 2).

1. $-4 \times (-6) \div 3$

2. $9 \times (-2) + (-3) \times (-5)$

3. $(-5)(-2) - (-2)(3)$

4. $2^3 \times (-8 + 7) - (-3)$

5. $4 \times [(8 - 3) - (-20)] \times (-6)^2$

6. $3 + (-4 + 2)^2$

Evaluate (see example 3 or 4).

7. $-(-1 - (-2)^2)^2 + (4 - 6)^3$

8. $[2 - (3 \times 4) \div 3] + (3 - (-6))^2$

9. $1 + 2 - 3 + 4 - 5 + 6 - 7 + 8 - 9$

10. $(3 - 3 + 3 \div 3 \times 3 - 3)^2$

11. $(-4 + 4 \div 4 \times 4 - 4)^3$

12. $(3 + 3 - 4)^2 + (5 - 5 + 5 \div 5)^2$

1. 8
2. -3
3. 16
4. -5
5. 3600
6. 7

7. -33
8. 79
9. -3
10. 0
11. -64
12. 5

EXERCISE 2-6

Evaluate.

1. $4 \div 2 + (-5) - 1$
2. $12 - 6(-5) + 2$
3. $(4)(-5) \div (-5) + 2$
4. $(-10) \div 2 - 4$
5. $(-6)(4) + (-12)$
6. $(-8) \cdot (-1) + 4$
7. $7(-10) + (12)(-9)$
8. $20 + (-6) \cdot (-3)$
9. $(-5)(6) \div (-15)$
10. $80 - 9(8) + (-6)(2)$
11. $28 \div 4 - 7 + (-1)(-8)$
12. $14(-2) - 5(-8) \div (-2)$
13. $3^2(-2) + 4$
14. $(-8)^2 \div 8 + (-9)$
15. $4(-3)^2 + (-4)(5)$
16. $16 \div (-2)^3 + (-3)(-4)$
17. $-8^2 + 18 \div (-9)$
18. $5 + 20 \div (-2)^2 - 8^2$
19. $2(5 - 6) + (-3)$
20. $(6 + 8) + 2(-3) + (2 - 4)$
21. $[2(3 - 2)] \div 2 + (-1)$
22. $28 \div [2(-2)] \div (-7)(-1)$
23. $3[3(-3) \div \{4 + (-1)\}]$
24. $25 - [(40 + 6) \cdot (-2)] \div (-23)$
25. $(4 + 6)^2 \div [3(-5)(-2) + 2(-10)]$
26. $(-5)(-3)^2 - 5 + (8 - 2)^2$
27. $\dfrac{(3 + 2)^2(5 - 4)}{5[5 + (-6)]}$
28. $12 + [6 - (3)(-4)] - \left[\dfrac{6^2}{(4 - 3)^3}\right]$
29. $\dfrac{6 + (-8)^2}{(-2)} + [8(-6) \div (-2)]^2$
30. $\dfrac{(5 + 2)^2 + 3(-2)^3 - 2(-6^2) + 5}{[3(-2) + \{24 \div (-8)\} + 4]^2}$

2-7 PROPERTIES OF SIGNED NUMBERS

A. Addition and Subtraction

COMMUTATIVE PROPERTY OF ADDITION

In arithmetic, we assumed that addition problems may be added in any order without changing the sum. For example:

$6 + 5 = 11$ $2.5 + 4.6 = 7.1$ $\dfrac{3}{8} + \dfrac{2}{8} = \dfrac{5}{8}$

$5 + 6 = 11$ $4.6 + 2.5 = 7.1$ $\dfrac{2}{8} + \dfrac{3}{8} = \dfrac{5}{8}$

Thus,   Thus,

$6 + 5 = 5 + 6$ $2.5 + 4.6 = 4.6 + 2.5$ Thus,

$$\dfrac{2}{8} + \dfrac{3}{8} = \dfrac{3}{8} + \dfrac{2}{8}$$

This is, in fact, true! We may change the order in which numbers are added without changing the sum.

The same holds true for adding signed numbers. Signed numbers may be added in any order without changing the sum. For example:

$(-5) + 2 = (-3)$ $(-4) + (-3) = (-7)$

$2 + (-5) = (-3)$ $(-3) + (-4) = (-7)$

Thus,   Thus,

$$(-5) + 2 = 2 + (-5)$$ $$(-4) + (-3) = (-3) + (-4)$$

This special property is called the *commutative property of addition* and may be stated in general terms as follows.

COMMUTATIVE PROPERTY OF ADDITION

> For any number a, and for any number b,
>
> $$a + b = b + a$$

ASSOCIATIVE PROPERTY OF ADDITION

When adding three or more numbers in an addition problem, we may group the numbers in any order without changing the sum. For example, we can find the sum of $3 + 4 + 2$ by using two different approaches.

APPROACH 1   **APPROACH 2**

Add $3 + 4$ first: 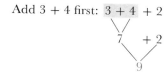   Add $4 + 2$ first: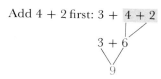

Thus,

$$(3 + 4) + 2 = 3 + (4 + 2)$$

Notice that we use parentheses to *group together* the numbers that are being added *first*.

The same holds true for adding signed numbers. When adding signed numbers, we may group the numbers in any manner without changing the sum. For example,

$$[8 + (-3)] + (-2) \qquad \text{and} \qquad 8 + [(-3) + (-2)]$$

Therefore,

$$[8 + (-3)] + (-2) = 8 + [(-3) + (-2)]$$

This example shows another special property of addition called the *associative property of addition*.

This property may be stated in general terms as follows.

ASSOCIATIVE PROPERTY OF ADDITION

> For any number a, any number b, and any number c:
>
> $$(a + b) + c = a + (b + c)$$

ADDITION PROPERTY OF ZERO

The sum of any number and zero is that number itself. For example:

$$5 + 0 = 5 \quad \text{and} \quad 0 + (-7) = -7$$

This is called the *addition property of zero*. The number zero is also called the identity element of addition (or the additive identity) because whatever number we add to zero, the sum is the number itself.

We state this property in general terms as follows.

ADDITION PROPERTY OF ZERO

For any signed number a,

$$a + 0 = a$$

and

$$0 + a = a.$$

THE ADDITIVE INVERSE

Every signed number has an opposite number such that the sum of the number and its opposite is zero. This opposite is called the additive inverse of the number. For example:

■ The additive inverse of 5 is (-5)

$$5 + (-5) = 0$$
number   opposite value

■ The additive inverse of (-3) is 3

$$(-3) + 3 = 0$$
number   opposite value

In general terms, we have

THE ADDITIVE INVERSE

For any signed number a, there exists an opposite of a, $(-a)$ such that

$$a + (-a) = 0$$

Unlike addition, subtraction does not have a commutative property. For example:

$$7 - 3 \neq 3 - 7$$

Unlike addition, subtraction is not associative. For example:

$$(12 - 6) - 3 \overset{?}{=} 12 - (6 - 3)$$
$$6 - 3 \overset{?}{=} 12 - 3$$
$$3 \neq 9$$

Do the following practice set. Check your answers with the answers in the right-hand margin.

PRACTICE SET 2-7A

Name the property being used in each of the following.

1. $3 + 4 = 4 + 3$
2. $(5 + 9) + 4 = 5 + (9 + 4)$
3. $9 + (x + 6) = (9 + x) + 6$
4. $45 + (-45) = 0$
5. $(-27) + 0 = (-27)$

1. Commutative property of addition
2. Associative property of addition
3. Associative property of addition
4. Additive inverse
5. Identity element of addition

B.   Multiplication

COMMUTATIVE PROPERTY OF MULTIPLICATION

In arithmetic, we assumed that when we multiply numbers, we may change the order of the factors without changing the product. For example:

$$3 \times 5 = 15 \qquad\qquad 3.1 \times 2.6 = 8.06 \qquad\qquad \frac{3}{4} \times \frac{1}{2} = \frac{3}{8}$$

$$5 \times 3 = 15 \qquad\qquad 2.6 \times 3.1 = 8.06$$

$$\frac{1}{2} \times \frac{3}{4} = \frac{3}{8}$$

Thus,   Thus,   Thus,

$$3 \times 5 = 5 \times 3 \qquad\qquad 3.1 \times 2.6 = 2.6 \times 3.1 \qquad\qquad \frac{3}{4} \times \frac{1}{2} = \frac{1}{2} \times \frac{3}{4}$$

This is, in fact, true! We may change the order of the factors of a multiplication problem, without changing the product.

The same holds true for multiplication of signed numbers. Signed numbers may be multiplied in any order without changing the product. For example:

$$3 \times (-2) = (-6) \qquad\qquad\qquad (-2) \times (-3) = +6$$

$$(-2) \times 3 = (-6) \qquad\qquad\qquad (-3) \times (-2) = +6$$

Thus,   Thus,

$$3 \times (-2) = (-2) \times 3 \qquad\qquad\qquad (-2) \times (-3) = (-3) \times (-2)$$

This special property is called the *commutative property of multiplication* and may be stated in general terms as follows.

COMMUTATIVE PROPERTY OF MULTIPLICATION

> For any number a and any number b,
>
> $$a \times b = b \times a$$

ASSOCIATIVE PROPERTY OF MULTIPLICATION

When multiplying three or more numbers, we may group the factors in any order without changing the product. For example, the product of $4 \times 6 \times 3$ may be found by using two different approaches.

APPROACH 1	**APPROACH 2**
Multiply 4×6:	Multiply 6×3:
$(4 \times 6) \times 3$	$4 \times (6 \times 3)$
$24 \quad \times 3$	$4 \times \quad 18$
Multiply 24×3:	Multiply 4×18:
24×3	4×18
72	72

Thus,

$$(4 \times 6) \times 3 = 4 \times (6 \times 3)$$

NOTE:

> We use parentheses to group the factors 4 and 6 together from approach 1 and 6 and 3 together from approach 2.

The same holds true for multiplication of signed numbers. When multiplying signed numbers, we may group them in any order without changing the product. For example,

$$[(2) \cdot (-4)] \cdot (-3) \qquad \text{and} \qquad (2) \cdot [(-4) \cdot (-3)]$$

$$(-8) \cdot (-3) \qquad\qquad (2) \cdot (+12)$$

$$(+24) \qquad\qquad\qquad (+24)$$

Therefore,

$$[(2) \cdot (-4)] \cdot (-3) = (2) \cdot [(-4) \cdot (-3)]$$

This example shows another special property of multiplication called the *associative property*. We state this property in general terms as follows.

ASSOCIATIVE PROPERTY OF MULTIPLICATION

> For any signed number a, any signed number b, and any signed number c,
>
> $$[(a) \cdot (b)] \cdot (c) = (a) \cdot [(b) \cdot (c)]$$

MULTIPLICATION PROPERTY OF ZERO

When multiplying two numbers, if one of the factors is zero, then the product is zero. For example:

$$5 \times 0 = 0 \quad \text{and} \quad 0 \times (-2) = 0$$

This is called the *multiplication property of zero* and may be stated in general terms as follows.

MULTIPLICATION PROPERTY OF ZERO

> For any number a,
>
> $$a \times 0 = 0 \quad \text{and} \quad 0 \times a = 0$$

MULTIPLICATION PROPERTY OF ONE

The product of any number and one is simply the number itself. For example:

$$3 \times 1 = 3 \quad \text{and} \quad 1 \times (-7) = -7$$

This is called the *multiplication property of one*. The number one is called the identity element of multiplication since the product of a number and one is the number itself. This property is stated in general terms as follows.

MULTIPLICATION PROPERTY OF ONE

> For any number a,
>
> $$a \times 1 = a \quad \text{and} \quad 1 \times a = a$$

Do the following practice set. Check your answers with the answers in the right-hand margin.

PRACTICE SET 2-7B

State the property being used.

1. $0 \cdot (-1) = (-1) \cdot 0$

2. $16 \cdot 1 = 16$

3. $(-4) \cdot [3 \cdot (-2)] = [(-4) \cdot 3] \cdot (-2)$

4. $-5 \cdot 0 = 0$

1. Commutative property of multiplication
2. Identity property of multiplication
3. Associative property of multiplication
4. Zero property of multiplication

C. The Distributive Property

In arithmetic, we are not able to multiply two numbers such as 14 and 12 in one operation. We must multiply in parts and then add the parts together. For example, to multiply 14 and 12, we:

First, multiply by the units (2 units times 14 are 28 units).

$$\begin{array}{r} 14 \\ \times\, 12 \\ \hline 28 \end{array}$$

Then, multiply by the tens (1 ten times 14 are 14 tens, or $10 \times 14 = 140$).

$$\begin{array}{r} 14 \\ \times\ 12 \\ \hline 28 \\ 140 \end{array}$$

Now, we add the parts together.

$$\begin{array}{r} 14 \\ \times\, 12 \\ \hline 28 \\ +\ 140 \\ \hline 168 \end{array}$$

Writing this problem horizontally, we work this problem out in the following manner.

Expand the 12 into two parts. 14×12
          $(12 = 1 \text{ ten} + 2 \text{ units})$ $14 \times (10 + 2)$

Now, multiply the 14 by each part in parentheses and add the parts together.

Multiply 14×10. $14 \times (10 + 2)$
                                          140

Multiply 14×2. $14 \times (10 + 2)$
                                          $140 \quad 28$

Add the products of the parts together. $140 + 28 = 168$

What we have done here is to distribute the factor 14 with each term of the sum, 10 and 2, and then add the two products.

$14 \times (10 + 2)$
$= (14 \times 10) + (14 \times 2)$
$= 140 + 28$
$= 168$

This example illustrates the *distributive property of multiplication over addition* (or, simply, the distributive property).

This property may also be applied to multiplication of signed numbers. For example, we can find the product of $-2(4 + 5)$ by:

First adding $4 + 5$ within parentheses

$$-2(4 + 5)$$
$$-2 \quad (9)$$

Then multiplying -2 times 9

$$-2(9)$$
$$-18$$

Thus,

$$-2(4 + 5) = -18$$

However, we may also find this product by using the distributive property. To do this, we simply *distribute* -2 with each term in parentheses.

Multiply -2×4.

$$-2(4 + 5)$$
$$-2(4)$$
$$-8$$

Multiply -2×5.

$$-2(4 + 5)$$
$$-2(4) + \;-2(5)$$
$$-8 \; + \; (-10)$$

Combine according to sign.

$$-8 + (-10)$$
$$= -8 - 10$$
$$= -18$$

This property may be stated in general terms as follows.

DISTRIBUTIVE PROPERTY

For any number a, any number b, and any number c,

$$a(b + c) = a(b) + a(c)$$

and

$$(b + c)a = b(a) + c(a)$$

EXAMPLE 1:   Evaluate $-5(-3 + 4)$ by using the distributive property.

SOLUTION: Multiply -3 by -5.

$$-5(-3 + 4)$$
$$(-5)(-3)$$
$$= \qquad +15$$

Example continued on next page.

Multiply 4 by -5.

$$-5(-3 + 4)$$
$$(-5)(-3) \quad (-5)(4)$$
$$= \quad +15 \qquad -20$$

Combine the two partial products.

$$-5(-3 + 4)$$
$$= (-5)(-3) \quad (-5)(4)$$
$$= \quad +15 \qquad -20$$
$$= \qquad -5$$

Final answer: -5

EXAMPLE 2: Evaluate $-3(5 - 4)$ by using the distributive property.

SOLUTION: Multiply 5 by -3.

$$-3(5 - 4)$$
$$(-3)(5)$$
$$= \quad -15$$

Multiply -4 by -3.

$$-3(5 - 4)$$
$$(-3)(5) \quad (-3)(-4)$$
$$= \quad -15 \qquad +12$$

Combine the two partial products.

$$-3(5 - 4)$$
$$= (-3)(5) \quad (-3)(-4)$$
$$= \quad -15 \qquad +12$$
$$= \qquad -3$$

Final answer: -3

We can also apply the distributive property in reverse by "pulling out" the common multiplier. For example, in the expression $2(5) + 2(6)$:

Since 2 is the common multiplier, $\qquad\qquad\qquad$ $2(5) + 2(6)$

We can pull 2 out and enclose the remaining sum $\qquad$ $2(5) + 2(6)$
within parentheses

$$2(5 + 6)$$

NOTE:

> The process of pulling out the common multiplier is called *factoring*.

Do the following practice set. Check your answers with the answers in the right-hand margin.

PRACTICE SET 2-7C

Fill in the blank to show the distributive property.

1. $2(3 + 8) = $ _____
2. _____ $\cdot (2 - 4) = -3(2) - 3(-4)$
3. _____ $= 5(-1) + 5(2)$

1. $2(3) + 2(8)$
2. (-3)
3. $5(-1 + 2)$

Evaluate by using the distributive property (see example 1 or 2).

4. $3(-5 + 9)$ 5. $7(2 - 5)$ 6. $-4(-3 - 6)$

4. 12
5. -21
6. 36

Use the distributive property in reverse for each of the following.

7. $3(8) + 3(6)$ 8. $5(4) - 6(4)$ 9. $2(x) + 2(y)$

7. $3(8 + 6)$
8. $4(5 - 6)$
9. $2(x + y)$

D. Division

Unlike multiplication, division of two numbers is not commutative. For example:

$$15 \div 3 \neq 3 \div 15$$

Also, division is not associative. For example:

$$(12 \div 6) \div 3 \neq 12 \div (6 \div 3)$$
$$2 \div 3 \neq 12 \div 2$$
$$\frac{2}{3} \neq 6$$

Although the operation of division is neither commutative nor associative, division does have certain properties with which you must be familiar. We now list these properties, along with an example to facilitate the reason they hold.

1. ANY NUMBER DIVIDED BY 1 IS EQUAL TO ITSELF

EXAMPLE: $5 \div 1 = 5$

EXPLANATION: To divide 5 by 1 means to find a number that when multiplied by 1 will yield a product of 5. This number can only be 5.

$$\blacksquare \times 1 = 5$$
$$5 \times 1 = 5$$

In general:

For any number a,

$$\frac{a}{1} = a$$

---

2. ANY NUMBER DIVIDED BY ZERO IS NOT DEFINED

EXAMPLE: $5 \div 0 = \text{(undefined)}$

EXPLANATION: To divide 5 by 0 means to find a number that when multiplied by 0 will yield a product of 5. There is no such number.

$$\blacksquare \times 0 = 5$$
$$\text{(not possible)}$$

In general:

For any number a,

$$\frac{a}{0} \text{ is undefined}$$

(except when $a = 0$)

---

3. ZERO DIVIDED BY ZERO CANNOT BE DETERMINED

EXAMPLE: $0 \div 0 = \text{(indeterminate)}$

EXPLANATION: To divide 0 by 0 means to find a number that when multiplied by 0 will yield a product of 0. There are many such numbers. Thus, we say indeterminate.

$\boxed{} \times 0 = 0$

(indeterminate)

In general:

$$\frac{0}{0} \text{ is indeterminate}$$

---

4. ZERO DIVIDED BY ANY NONZERO NUMBER IS EQUAL TO ZERO

EXAMPLE: $0 \div 13 = 0$

EXPLANATION: To divide 0 by 13 means to find a number that when multiplied by 13 will yield a product of 0. This number can only be zero.

$\boxed{} \times 13 = 0$

$0 \times 13 = 0$

In general:

For any nonzero a,

$$\frac{0}{a} = 0$$

---

5. ANY NONZERO NUMBER DIVIDED BY ITSELF IS EQUAL TO 1

EXAMPLE: $12 \div 12 = 1$

EXPLANATION: To divide 12 by 12 means to find a number that when multiplied by 12 will yield a product of 12. This number can only be 1.

$\boxed{} \times 12 = 12$

$1 \times 12 = 12$

Example continued on next page.

In general:

For any nonzero a,

$$\frac{a}{a} = 1$$

EXERCISE 2-7

Answer true or false. If the statement is true, indicate the property that makes it true.

1. $5 \cdot 8 = 8 \cdot 5$

2. $\frac{3}{4} - (2 - x) = \left(\frac{3}{4} - 2\right) - x$

3. $\frac{0}{0} = 1$

4. $x + (5 + 6) = x(5) + x(6)$

5. $3(8 + y) = 3(8) + 3(y)$

6. $4 \cdot 1 = 4$

7. $5 \cdot (x \cdot y) = (5 \cdot x) \cdot y$

8. $\frac{6}{0} = 0$

9. $6 + 0 = 6$

10. $3(x) + 3(y) = (x + y)3$

11. $4 - 6 = 6 - 4$

12. $4 + 0 = 0 + 4$

13. $(4 \cdot 9)x = x(4) \cdot x(9)$

14. $(x + y) \cdot z = z(x) + z(y)$

15. $2(6 \div 3) = 2(6) \div 2(3)$

16. $18 \div 6 = 6 \div 18$

17. $3 \cdot (x + y) = 3(x) + 3(y)$

18. $1 \cdot x = x$

19. $(10 \div 2) \div 1 = 10 \div (2 \div 1)$

20. $x + (y + z) = (x + y) + z$

State the property that makes each statement true.

21. $(-4) + (-3) = (-3) + (-4)$

22. $(0) + (-6) = (-6) + 0$

23. $(-3) + (3) = (0)$

24. $(-6) + (0) = (-6)$

25. $(-4) + [(-8) + (3)] = [(-4) + (-8)] + 3$

26. $(-3)(-5) = (-5)(-3)$

27. $(-6)(0) = 0$

28. $[(-2)(-4)] \cdot 8 = (-2)[(-4) \cdot 8]$

29. $0(-3) = (-3)0$

30. $-8[5 + (-6)] = (-8)(5) + (-8)(-6)$

Use the distributive property to write an equivalent statement.

31. $8(x) + 2(x)$

32. $6(3) - 6(9)$

33. $4(y) + (y)$

34. $z(5) - (z)$

35. $5(3 + 4)$

36. $9(12 + 2)$

37. $12(x - y)$

38. $(3 - 5)4$

39. $(8 + 1)6$

40. $(x - y)z$

Evaluate by using the distributive property.

41. $3(6 + 4)$ 42. $-3(6 + 4)$ 43. $-2[(-6) + 5]$ 44. $-5[(-6) - 3]$
45. $3(3 - 5)$ 46. $-3(3 - 5)$ 47. $-1[(-6) - 2]$ 48. $4[(-8) - (-6)]$

REVIEW EXERCISES

Match the number in column I with the letter in column II.

I

1. Additive inverse
2. Distributive property of multiplication
3. Absolute value of a number
4. Associative property of addition
5. Commutative property of addition
6. Addition property of zero
7. Multiplication property of one
8. Multiplication property of zero
9. Commutative property of multiplication

II

a. $-9 \cdot 1 = -9$
b. $(-3) + 4 = 4 + (-3)$
c. $6 \cdot 0 = 0$
d. $(-6)(-5) = (-5)(-6)$
e. $-3 + 0 = -3$
f. $|-3| = 3$
g. $4 + (-4) = 0$
h. $[8 + (-3)] + (-2) = 8 + [(-3) + (-2)]$
i. $-5(6 + 8) = -5(6) - 5(8)$

Perform the indicated operations.

10. $7 + (-5)$ 11. $-8 - 6$ 12. $4 - (-6)$
13. $27 - 43$ 14. $8 - (+10)$ 15. $49 - 32$
16. $18 + 9 - 15$ 17. $31 - 4 + 6$ 18. $12 - (-18) - 8$
19. $-48 + (-23) - 41$ 20. $7 + (+8) - (-9)$ 21. $(-12) - (7) + (-16)$
22. $|8| + |-7|$ 23. $|-15| + |4|$ 24. $|-12| - |-16|$
25. $6(-3)$ 26. $-3(-18)$ 27. $(-2)(-5)$
28. $(-6)^3$ 29. $(-8)^2 \cdot (-3)^3$ 30. $(-4 - 8)^2$
31. $\frac{-36}{-4}$ 32. $\frac{-8}{8}$ 33. $\frac{20}{-4}$
34. $(-12) \div (6)$ 35. $(48) \div (-3)$ 36. $\left(\frac{-8}{2}\right)^3$
37. $(-8)^2 + 18 \div (-9)$ 38. $2(5 - 6) + (-3)$ 39. $5 + 20 \div (-2)^2 - 8^2$
40. $|8| \cdot |-7|$ 41. $|-15| \cdot |4|$ 42. $|-12| \cdot |-16|$
43. $|10| \cdot |-3| \cdot (-5)$ 44. $\frac{|-12| \cdot |-9|}{3 - |-6|}$ 45. $\left(\frac{-8 + |-8|}{|2| + |-2|}\right)^2$

CHAPTER 3
Introduction to Algebra

3-1 THE LANGUAGE OF ALGEBRA

Although algebra can be regarded as an extension of arithmetic, it still has its own jargon and notation. As a result, to meet with any degree of success in algebra, you must become fluent in the language of algebra. In this section, you are provided with a listing of words (and appropriate definitions) and phrases that you will frequently encounter in your study of algebra.

A. Variables, Constants, and Coefficients

An important characteristic of algebra is that it uses letters to represent numbers or unknown quantities. These letters are called variables and can be virtually anything you wish—from capital letters to lower case letters to Greek letters, and so on.

Constants, on the other hand, are fixed numbers. That is, the value of a constant is known and does not change in a particular problem. Thus, in the problem, $x + 2$,

x is the variable that represents an unknown quantity, and 2 is the constant whose value will always remain 2.

$$x \quad + \quad 2$$
$$\uparrow \qquad \uparrow$$
$$\text{variable} \quad \text{constant}$$

Whenever a variable and a number are multiplied together, the number part is referred to as the numerical coefficient—or simply, coefficient. Thus, if we multiplied the number 2 by the variable x,

2 would be the coefficient of the variable x.

$$2 \cdot x$$
$$\uparrow$$
$$\text{coefficient}$$

In algebra, we commonly write the product of a specific number and a variable (e.g., $2 \cdot x$), or the product of two variables (e.g. $x \cdot y$), with nothing separating the quantities. For example:

$$6 \cdot x \quad \text{is written as} \quad 6x$$
$$3 \cdot y \quad \text{is written as} \quad 3y$$

$$3 \cdot x \cdot y \quad \text{is written as} \quad 3xy$$
$$3 \cdot x + 7 \cdot y \quad \text{is written as} \quad 3x + 7y$$

Get the idea? We will be using this notation throughout our study of algebra. Also, whenever a variable is written without a numerical coefficient, it is understood that the coefficient is 1. Thus,

$$x \quad \text{means} \quad (1)x \text{ and its coefficient is 1}$$

EXAMPLE 1: Identify the variables, constants, and coefficients for each of the following.

a. $2x + 3$

- The variable is x. $2\boxed{x} + 3$
- The coefficient is 2. $\boxed{2}x + 3$
- The constant is 3. $2x + \boxed{3}$

b. $5y^2 - x$

- The variables are y and x. $5\boxed{y}^2 - \boxed{x}$
- The coefficients are 5 (for the variable y) and -1 (for the variable x). $\boxed{5}y^2 \boxed{-1}x$
- There is no constant.

NOTE:

The coefficient of a term will always include the sign that comes immediately before it.

Do the following practice set. Check your answers with the answers in the right-hand margin.

PRACTICE SET 3-1A

In each of the following, identify (a) the variables, (b) the coefficients, and (c) the constant.

1. $x + 5$ 2. $2x + 8y + 4$

1. (a) x, (b) 1, (c) 5
2. (a) x and y
   (b) 2 and 8
   (c) 4

3. $\dfrac{3}{2}y^2 - \dfrac{6}{5}x$ 4. $-y^2$

3. (a) y and x
    (b) $\frac{3}{2}$ and $-\frac{6}{5}$
    (c) none
4. (a) y
    (b) -1
    (c) none

B.   Expressions and Terms

In arithmetic, whenever we have numbers connected by the signs of arithmetic $(+, -, \times, \div)$, we refer to them as numerical expressions. Thus,

$$3 + 6, \qquad 9 - 4, \qquad 16 \times 8, \qquad 28 \div 7$$

are all numerical expressions.

In algebra, since we use variables to represent numbers, whenever we have variables connected to variables or variables connected to numbers by the signs of arithmetic, we refer to them as algebraic expressions. For example:

$$x + 6, \qquad x + 2y, \qquad 3ab - b, \qquad 5(2x + 13), \qquad \dfrac{a + b}{2c}$$

are all algebraic expressions.

All algebraic expressions contain terms. Generally speaking, a *term* consists of a variable or a number, or both variables and numbers, connected by multiplication or division. For example, each of the following is a term.

- A number: 6

- A variable: x

- A number and a variable connected by multiplication: $6x$

- A number and two (or more) variables connected by multiplication: $3xy$

- A number and a variable connected by division: $\dfrac{x}{10}$

In any algebraic expression, terms are separated by $+$ or $-$ signs. Further, we always include the sign that comes immediately before the term as part of that term. For example:

- The expression, $9y + 5x$ contains two terms, namely, $9y$ and $5x$.

  $\quad 9y \;\; + \;\; 5x$
  $\quad$ 1st $\qquad$ 2nd
  $\quad$ term $\qquad$ term

- The expression, $9y + 5x + 6$ contains three terms, namely, $9y$, $5x$, and 6.

  $\quad 9y \;\; + \;\; 5x \;\; + \;\; 6$
  $\quad$ 1st $\qquad$ 2nd $\qquad$ 3rd
  $\quad$ term $\qquad$ term $\qquad$ term

- The expression $9y + 5x - 6$ contains three terms, namely, $9y$, $5x$, and -6.

  $\quad 9y \;\; + \;\; 5x \;\; + -6$
  $\quad$ 1st $\qquad$ 2nd $\qquad$ 3rd
  $\quad$ term $\qquad$ term $\qquad$ term

NOTE:

In this last expression, the last term is -6 rather than 6. We always include the sign that comes immediately before a term as part of that term. However, -6 can be thought of as $+(-6)$. Thus, $9x + 5y - 6$ is equivalent to $9x + 5y + (-6)$.

■ The expression, $(x - 3)/2 + (2x + 13)$ contains two terms, namely, $(x - 3)/2$ and $(2x + 13)$.

$$\underset{\text{1st term}}{\underline{\frac{x - 3}{2}}} + \underset{\text{2nd term}}{(2x + 13)}$$

Any expression within grouping symbols (as in the second term), is considered a single term.

In summary, the terms of an algebraic expression can be easily identified as products or quotients of numbers and/or variables, separated by $+$ or $-$ signs.

Do the following practice set. Check your answers with the answers in the right-hand margin.

PRACTICE SET 3-1B

Identify the terms of each expression.

1. $5x + 4y$	2. $3z - 2x$	1. $5x$ and $4y$
3. $3xy$	4. $5x + (2x + 3y)$	2. $3z$ and $-2x$
5. $\left(\dfrac{2 - x}{3}\right) + 3(y^2 + x^2)$	6. $(x + y - z)$	3. $3xy$
		4. $5x$ and $(2x+3y)$
		5. $(2-x)/3$ and $3(y^2+x^2)$
		6. $(x + y - z)$

C.   Factors

In arithmetic, whenever two or more numbers are multiplied together, we call the numbers being multiplied factors, and the result of the multiplication is called the product. For example, in the problem, $6 \times 3 = 18$,

6 and 3 are the factors and 18 is the product

$$\underset{\text{factors}}{6 \times 3} = \underset{\text{product}}{18}$$

In algebra, the same holds true. Whenever two or more variables, numbers, or expressions are multiplied together, the result is called a product and the variables, numbers, or expressions that are being multiplied together are called factors of the product. Consider the following examples.

■ In the product $2x$, 2 and the variable x are the factors.

$$\underset{\text{factors}}{(2)(x)} = \underset{\text{product}}{2x}$$

■ In the product xy, the variables x and y are the factors.

$$(x)(y) = \quad xy$$

factors   product

■ In the product $2(x + 4)$, 2 and the expression $(x + 4)$ are the factors.

$$(2)(x + 4) = 2(x + 4)$$

factors   product

■ In the product $10x$, 2 and $5x$ are the factors.

$$(2)(5x) = \quad 10x$$

factors   product

■ In the product $(x - 1)(x + 1)$, the expressions $(x - 1)$ and $(x + 1)$ are the factors.

$$(x - 1)(x + 1) = (x - 1)(x + 1)$$

factors   product

It is important to note the difference between factors and terms. A term is a number, a variable, or an expression that is written as a product of numbers and/or variables. Each of the numbers or variables being multiplied is a factor of the term. For example, the number 7 is a term and the variable y is a term. Also:

■ In the expression $8x$, $8x$ is a term, whereas 8 and x are factors of the term.

$8x$ term

$$(8)(x)$$

factors of term

■ In the expression $3x - 4y$, $3x$ and $-4y$ are terms, whereas 3 and x are factors of the first term; -4 and y are factors of the second term.

$3x$ $-4y$

1st   2nd

term   term

$$(3)(x) - (4)(y)$$

factors   factors

of 1st   of 2nd

term   term

Do the following practice set. Check your answers with the answers in the right-hand margin.

PRACTICE SET 3-1C

Identify the factors for each of the following products.

1. $-6x$

2. $(x - 6)(-3)$

3. xy

4. $x - 3y + 6$

1. -6 and x
2. $(x - 6)$ and -3
3. x and y
4. 1 and x;
    -3 and y;
    6 is a constant

D.   Exponential Notation

In Section 1-4, we learned to write the product of repeating factors by using exponential notation. We do this by simply writing the repeating factor as a base and the number of times the factor is repeated as an exponent. For example:

$3 \cdot 3$ may be written as 3^2 and
3^2 means $3 \cdot 3$

$3 \cdot 3$
3^2

$5 \cdot 5 \cdot 5$ may be written as 5^3
and 5^3 means $5 \cdot 5 \cdot 5$

$5 \cdot 5 \cdot 5$
5^3

$2 \cdot 2 \cdot 2 \cdot 3 \cdot 3$ may be written as $2^3 \cdot 3^2$
and $2^3 \cdot 3^2$ means $2 \cdot 2 \cdot 2 \cdot 3 \cdot 3$

$2 \cdot 2 \cdot 2 \cdot 3 \cdot 3$
$2^3 \quad \cdot \; 3^2$

We proceed in the same manner when using variables. For example,

$x \cdot x \cdot x$ may be written as x^3
and means $x \cdot x \cdot x$ (or xxx)

$x \cdot x \cdot x$
x^3

$x \cdot x \cdot y \cdot y \cdot y$ may be written as $x^2 \cdot y^3$
and means $x \cdot x \cdot y \cdot y \cdot y$ (or $xxyyy$)

$x \cdot x \cdot y \cdot y \cdot y$
$x^2 \cdot \quad y^3$

NOTE:

An exponent applies only to the base it directly follows. For example:

$5x^2$ means $5 \cdot x \cdot x$ or $(5xx)$

$(5x)^2$ means $5x \cdot 5x$

$a^2 b^3$ means $a \cdot a \cdot b \cdot b \cdot b$ or $(aabbb)$

$3a^2 b^3$ means $3 \cdot a \cdot a \cdot b \cdot b \cdot b$ or $(3aabbb)$

Our goal is to be able to express products of repeating factors by using exponents, or to express terms that are written in exponential form without using exponents. We do this in the same manner as that used with numerical expressions in Section 1-4.

PROCEDURE:

1. To express the product of repeating factors by using exponents:
   a. Write each different repeating factor as a base.
   b. Write the number of times each different repeating factor is repeated as an exponent.

2. To express a term written in exponential form without using exponents,
   a. Write each base as a factor as many times as indicated by its exponent.

EXAMPLE 1: Write the product $x \cdot x \cdot x$ by using exponents.

SOLUTION: Write the repeating factor x as the base.

$$x \cdot x \cdot x$$
$$x$$

Since x is being multiplied by itself three times, the exponent is 3.

$$x \cdot x \cdot x$$
$$x^3$$

Final answer: x^3

EXAMPLE 2: Write the product $c \cdot c \cdot c \cdot d \cdot d$ by using exponents.

SOLUTION:

$$\underbrace{c \cdot c \cdot c} \cdot \underbrace{d \cdot d}$$
$$= \quad c^3 \quad \cdot \quad d^2$$

Final answer: $c^3 \cdot d^2$

EXAMPLE 3: Write $3x^2$ without using exponents.

SOLUTION:

$$3x^2$$
$$= 3 \cdot x \cdot x$$

Final answer: $3xx$

EXAMPLE 4: Write $5x^3y^2$ without using exponents.

SOLUTION:

$$5x^3y^2$$
$$= 5 \cdot x \cdot x \cdot x \cdot y \cdot y$$

Final answer: $5xxxyy$

Do the following practice set. Check your answers with the answers in the right-hand margin.

PRACTICE SET 3-1D

Write each product by using exponents (see examples 1 or 2).

1. $a \cdot a \cdot a$ 2. $a \cdot a \cdot b \cdot b \cdot b$ 3. $8 \cdot x \cdot x \cdot y \cdot y \cdot y$

Write each term without using exponents (see examples 3 or 4).

4. x^4 5. $3y^3$ 6. $2x^2y^3$

1. a^3
2. $a^2 \cdot b^3$
3. $8x^2y^3$

4. $x \cdot x \cdot x \cdot x$
5. $3 \cdot y \cdot y \cdot y$
6. $2 \cdot x \cdot x \cdot y \cdot y \cdot y$

EXERCISE 3-1

In each of the following (a) identify the variables and (b) identify the constant term.

1. $x + 6$
2. $h - 8$
3. $4p + 15$
4. $5a^2 + 6b^2 + 8$
5. $9x - 7y - 16$
6. $4x^2 - 3x + 5$
7. $\dfrac{2x^2 + 4y}{3} - \dfrac{3}{4}$
8. $\dfrac{5x^3 - 16y}{4} + \dfrac{6}{17}$
9. $\dfrac{-3x - 6y^2}{9} - 8$

In each of the following expressions (a) determine the number of terms, (b) identify the terms, and (c) state the coefficient for each term.

10. $3x + 9y$
11. $4y - 6z$
12. $12xy$
13. $x - 4z + 16$
14. $5a - b - 8$
15. $x - y - z$
16. $3x^2 + 2x - 4$
17. $4x + (3y + 6z)$
18. $(x + y + z)$
19. $-4x(3y + 6z)$
20. $6x^3 + 3x^2 - 4(x + y)$
21. $3ab^2 + \left(\dfrac{2x - y}{3xy}\right) + 2(x^2 - 6y)$

Write each of the following products by using exponents.

22. $x \cdot x$
23. $x \cdot x \cdot x \cdot y$
24. $x \cdot x \cdot y \cdot y \cdot y \cdot y$

25. $5 \cdot x \cdot x \cdot y$ 26. $-3 \cdot x \cdot y \cdot y \cdot z$ 27. $-16xxyy$

28. $3 \cdot x \cdot x \cdot x \cdot y \cdot y \cdot y \cdot z$ 29. $3[(xxx) + yy]$ 30. $3[(2xxx) - (4yy)]$

Write each of the following terms without using exponents.

31. x^2 32. $2x^4$ 33. $-5y^3$

34. $3x^2y^4$ 35. $(-5x)^2$ 36. $15(-x)^3y^4$

3-2 EVALUATING ALGEBRAIC EXPRESSIONS AND FORMULAS

A. Evaluating Expressions

An algebraic expression such as $2n + 6$ represents an unknown value. When we replace (or substitute) the variable n by a specific number, we are then able to determine the value of the expression. For example, if we were to replace n with the number 5 in the expression $2n + 6$, the expression would be evaluated to 16.

$$\text{Replacing } n \text{ with 5 and evaluating} \qquad \begin{array}{c} 2n + 6 \\ \downarrow \\ 2(5) + 6 \\ \diagdown \\ 10 + 6 \\ \diagdown \\ 16 \end{array}$$

This method of finding the value of an expression by replacing the variable(s) of the expression with a specific number(s) is called evaluating algebraic expressions.

Thus, to evaluate an algebraic expression, we first substitute each variable with a number value, and then perform the indicated operations by using the correct order of operations.

PROCEDURE:

> To evaluate an algebraic expression:
>
> 1. Substitute all variables with their corresponding numerical values.
>
> 2. Evaluate the expression by using the correct order of operations.

EXAMPLE 1: Evaluate $2x + 5$ when $x = 8$.

SOLUTION: Substitute 8 for x and evaluate.

$$2x + 5$$
$$= 2(8) + 5$$
$$= 16 + 5$$
$$= 21$$

Final answer: 21

EXAMPLE 2:   Evaluate $3x/2 + 5y/8$ when $x = -4$ and $y = 8$.

SOLUTION: Substitute -4 for x and 8 for y and evaluate.

$$\frac{3x}{2} + \frac{5y}{8}$$
$$= \frac{3(-4)}{2} + \frac{5(8)}{8}$$
$$= \frac{-12}{2} + \frac{40}{8}$$
$$= -6 + 5$$
$$= -1$$

Final answer: -1

EXAMPLE 3:   Evaluate $4a - [b - (5x + 2y)]$ when $a = -1$, $b = 2$, $x = -4$, and $y = 8$.

SOLUTION: Substitute -1 for a, 2 for b, -4 for x, 8 for y, and evaluate.

$$4a - [b - (5x + 2y)]$$
$$= 4(-1) - [(2) - (5(-4) + 2(8))]$$
$$= -4 - [2 - (-20 + 16)]$$
$$= -4 - [2 - (-4)]$$
$$= -4 - [2 + 4]$$
$$= -4 - [6]$$
$$= -4 - 6$$
$$= -10$$

Final answer: -10

EXAMPLE 4:   Evaluate $-b - (b^2 - 4ac)$ when $a = 5$, $b = -4$, and $c = -2$.

SOLUTION: Substitute 5 for a, -4 for b, and -2 for c.

$$-b - (b^2 - 4ac)$$

$$= -(-4) - ((-4)^2 - 4(5)(-2))$$
$$= 4 - (16 - 4(-10))$$
$$= 4 - (16 + 40)$$
$$= 4 - (56)$$
$$= 4 - 56$$
$$= -52$$

Final answer: -52

Notice that in each of the preceding examples, whenever a number was substituted in place for a variable, we placed parentheses around the number being substituted. We strongly suggest that you follow this same practice.

Do the following practice set. Check your answers with the answers in the right-hand margin.

PRACTICE SET 3-2A

Evaluate each expression when $a = 3$, $b = -5$, $x = 8$, $y = -9$ (see examples 1 or 2).

1. $3a + b$ 2. $5xy - 6$ 3. $\dfrac{2x}{2} + \dfrac{y}{a}$

1.   4
2. -366
3. -5

Evaluate each expression when $a = 5$, $b = -20$, $x = -3$, $y = -6$ (see example 3).

4. $3(x + y)$ 5. $3a - (x - y)$

6. $-b + [4a - (2x - y)]$

4. -27
5.   12
6.   40

Evaluate each expression when $a = 5$, $b = -3$, $c = -2$, $x = -4$ (see example 4).

7. $x^2 + c^2$ 8. $b^2 - (4ac)$ 9. $-b + (b^2 - 4ac)$

7.   20
8.   49
9.   52

B. Evaluating Formulas

In many instances throughout our life we encounter formulas. Whether for a course that we are enrolled in (e.g., physics, electricity, mathematics, etc.), or our occupation (e.g., business or scientific work), we must be prepared to use formulas. Studying algebra helps us in this preparation.

   To evaluate a formula, we simply use the same procedures that are outlined for evaluating algebraic expressions. Follow through the examples given here, then do the subsequent practice set.

EXAMPLE 1: The formula $I = PRT$ is used in business, where $I =$ interest, $P =$ principal, $R =$ rate, and $T =$ time.

   Find I when $P = \$1000$, $R = 0.10$, $T = 2.5$ years.

SOLUTION:

$$I = PRT$$
$$= (1000)(0.10)(2.5)$$
$$= (100)(2.5)$$
$$= 250$$

   Final answer: $250

EXAMPLE 2: The formula $I = E/R$ is known as Ohm's law and is used in electricity, where $I =$ amperes, $E =$ voltage, and $R =$ resistance.

   Find I when $E = 12$ volts and $R = 0.048$ ohms.

SOLUTION:

$$I = \frac{E}{R}$$
$$= \frac{(12)}{0.048}$$
$$= 250$$

   Final answer: $I = 250$ amperes

EXAMPLE 3: The formula $A = \pi r^2$ is used in mathematics to find the area of a circle, where $A =$ area, π (pi) is a constant approximately equal to 3.14, and $r =$ radius of the circle.

   Find A when $r = 5$.

Example continued on next page.

SOLUTION:

$$A = \pi r^2$$
$$= (3.14)(5)^2$$
$$= (3.14)(25)$$
$$= 78.50$$

Final answer: $A = 78.5$ square units

Do the following practice set. Check your answers with the answers in the right-hand margin.

PRACTICE SET 3-2B

Evaluate each formula (see examples 1, 2, or 3).

1.   Given $I = PRT$. Find I when $P = \$8000$, $R = 0.15$, and $T = 4$ years.
2.   Given $I = E/R$. Find I when $E = 100$ volts and $R = 0.08$ ohms.
3.   Given $A = \pi r^2$. Find A when $\pi = 3.14$ and $r = 3$.

1. $I = \$4800$
2. $I = 1250$ amperes
3. $A = 28.26$

EXERCISE 3-2

Evaluate each expression for $x = 3$, $y = 5$, $a = 6$, $b = 4$.

1. $5x$

2. ax

3. $5x - 14$

4. $97 + 3xy$

5. $2y - 16$

6. $5y - b$

7. $ab - xy$

8. $x - y - a$

9. $\dfrac{xy}{a}$

10. $\dfrac{3x}{a} + \dfrac{2y}{a}$

11. $\dfrac{by}{2y} + \dfrac{ab}{4}$

12. $\dfrac{xy}{b} - \dfrac{a}{b}$

Evaluate each expression for $x = -3$, $y = 5$, $a = -6$, $b = -4$.

13. $4(a + 6)$

14. $3x + (4 + y)$

15. $6(3x - a)$

16. $5x - (x - a)$

17. $2a - (5 + x)$

18. $a(x - y)$

19. $-b + [4a - (2x)]$

20. $(a + b)(x + y)$

21. $xy[3 - (2 + x) + ab]$

22. $2x + (x - (a + b))$

23. $\dfrac{x}{y}(4 + (xy - ab))$

24. $\dfrac{(3a - 3b)}{xy}$

Evaluate each expression for $a = 2$, $b = -4$, $x = 10$, $y = -8$, $c = 4$.

25. x^2

26. y^2

27. $x^2 + y^2$

28. $-y^2$

29. $x^2 - y^2$

30. $a + (b^2 - a)$

31. $ax^2 + by^2 + c$

32. $-ax^2 - by^2 - c$

33. $b^2 - 4ac$

34. $-b + (b^2 - 4ac)$

35. $\dfrac{x^2}{a^2} + \dfrac{y^2}{b^2}$

36. $\dfrac{-b - b^2 - 4ac}{2a}$

Evaluate each of the following formulas by using the values of the letters given.

37. $C = \dfrac{5}{9}(F - 32)$ $(F = 68°)$

38. $A = \dfrac{1}{2}bh$ $(b = 20, h = 16)$

39. $A = \dfrac{1}{2}bh$ $(b = 6, h = 8)$

40. $F = \dfrac{9}{5}C + 32$ $(C = 20°)$

41. $C = \dfrac{5}{9}(F - 32)$ $(F = 81°)$—to the nearest tenth.

42. $C = \dfrac{en}{R + rn}$ $(e = 2.06, n = 12, r = 0.3,$ $R = 53.4$—to the nearest hundredth$)$

43. $S = \dfrac{N}{2}(a + l)$ $(N = 16, a = 5, l = 8)$

44. $PD = 0.7854D^2LN$ $(D = 3, L = 3.5, N = 6)$

45. $S = \dfrac{1}{2}gt^2$ $(g = 32.16$ feet per second squared; $t = 3.4$ seconds—to the nearest tenth$)$

46. $A = p(1 + rt)$ $(p = 5, r = 10, t = 3)$

47. $A = p(1 + r)^N$ $(p = 5, r = 10, N = 3)$

48. $S = \dfrac{a(1 - r^N)}{1 - r}$ $(a = 2, r = 3, N = 2)$

49. $l = ar^{N-1}$ $(a = 2.5, r = 0.5, N = 4)$

50. $l = \dfrac{N}{2}[2a + (N-1)d]$ $\left(N = 6, a = 3, d = -\dfrac{5}{2}\right)$

Solve the following problems.

51. To find voltage, amperage, or resistance in electrical circuits, we use the formula (known as Ohm's law): $I = E/R$, where $E =$ voltage, $I =$ amperes and $R =$ resistance (ohms).
  a. How many amperes are needed for a circuit that contains 12 volts with a resistance of 30 ohms?
  b. How much current (amperes) will flow through a heat gun that has a resistance of 8.6 ohms and is connected across a 120-volt circuit?
  c. How much current will flow during the starting process of an automobile that contains a 12-volt battery and the starting motor has a resistance of 0.05 ohms?

52. Using the "power formula" $P = EI$, where $P =$ watts, $E =$ volts, and $I =$ amperes,
  a. How many watts of power will an electric blow dryer use if it is connected to 110 volts and takes 9.1 amperes?
  b. How many watts of power will an electric motor use if it is connected to a 220-volt circuit and is drawing 50 amperes?

53. To convert from degrees Celsius to degrees Fahrenheit, we use the formula $F = \frac{9}{5}C + 32$.
  a. If the melting point of cast iron is 1260°C,

54. To convert from degrees Fahrenheit to degrees Celsius, we use the formula, $C = \frac{5}{9}(F - 32)$.

Exercise 3-2 continued on next page.

what is its equivalent temperature in degrees Fahrenheit?

b. If the melting point of copper is 1060°C, what is its equivalent temperature in degrees Fahrenheit?

55. The formula $H = DV/375$ is used to determine the horsepower required for flight, where D = weight of object and V = speed. How much horsepower is needed for a 2700-lb object to be put into flight if its speed is 250 miles per hour?

57. Piston displacement of an automobile engine can be found by the formula

$$PD = 0.7854 \, D^2 LN$$

where D = diameter of a cylinder, L = length of stroke (inches), N = number of cylinders, and PD = piston displacement (cubic inches).

a. What is the displacement of a 4-cylinder engine with a bore of 2.99 inches and a stroke of 3.23 inches?

b. What is the displacement (in cubic centimeters) of a 4-cylinder engine with a bore of 7.6 centimeters and a stroke of 7.7 centimeters?

59. The following equation is used to determine the intensity of light from its source to a particular object.

$$I = \frac{6400}{d^2}$$

where I = intensity of a certain light (in candle power) and d = distance an object is from the light source (in feet).

a. What is the intensity if $d = 4$ feet?

b. An energy-saving program at Monroe Community College mandated that all offices must have a maximum intensity of 72 candlepower. In W. Setek's office, the distance from the light source to his desk

a. If alcohol boils at 179°F, what is the equivalent Celsius reading?

b. If a reading of -4°F was taken, what is the equivalent Celsius reading?

56. The Society of Automotive Engineers (SAE) uses the formula $HP = D^2 N/2.5$, where D = diameter of a cylinder and N = number of cylinders, to approximate the horsepower of a gasoline engine. (*Note:* This formula does not accurately determine the horsepower but is used simply as a rating.)

a. How much horsepower will a 4-cylinder engine produce if the bore is 2.95 inches? (*Note:* Bore means the diameter of an engine cylinder.)

b. How much horsepower will an 8-cylinder engine produce if the bore is 3.30 inches?

58. Tony is a firefighter for the city of Orange, California. While studying for a promotion exam, he came across the following equation.

$$h = \frac{1}{2}p + 26$$

where h = horizontal range of water (in feet) coming from a $\frac{3}{4}$-inch nozzle and p = nozzle pressure (in pounds per square foot—psi).

a. How far will a stream of water travel if the nozzle being used is $\frac{3}{4}$-inch and the nozzle pressure is 50 psi?

b. If the nozzle pressure for a $\frac{3}{4}$-inch nozzle was only 18 psi, what is the range of water?

60. If a ball is thrown upward with an initial velocity of 120 feet per second, its height, h (in feet), after t seconds is given by the formula,

$$h = 120 \, t - 16t^2 \qquad \text{(h is equal to maximum height in feet)}.$$

If a baseball player hits a fly ball with an initial velocity of 120 feet per second, how high will the ball travel if it takes the outfielder exactly 6 seconds to catch it from the time it is hit?

is 9.5 feet. Does the corresponding intensity level comply with the program?

61. In industrial management, managers are constantly seeking ways in which to improve the productivity of their workers. The following formula allows managers to measure the productivity of their workers.

$$\text{Productivity} = \frac{\text{hours spent producing}}{\text{total hours worked}} \times \frac{\text{earned hours}}{\text{hours spent producing}}$$

If a factory worker was able to produce a product in 6 hours of an 8-hour day (during the remaining 2 hours he was not being utilized due to a malfunction in his machine), and he "earned" 5 hours, what was his percent of productivity? (Adapted from *Purchasing World*, January 1981.)

62. The front end of a VW Rabbit is "tapered" to allow for more visibility to the driver. A formula that gives the taper, T, in inches is

$$T = \frac{12(w - x)}{L}$$

where $w =$ longer height, $x =$ shorter height, and $L =$ length.

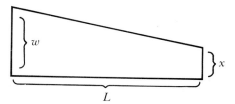

Determine the taper of a VW Rabbit if $w = 3$ feet, $L = 3\frac{3}{4}$ feet, and $x = 2\frac{1}{3}$ feet (round to the nearest hundredth).

63. The following formula is used to determine the amount of money accumulated, A, over a period of time.

$$A = P(1 + rt)$$

where $P =$ principal (amount of money invested), $r =$ rate of interest, $t =$ time.
If Barbara invests $20,000 at 14 percent for $2\frac{1}{2}$ years, how much money will she accumulate at the end of the $2\frac{1}{2}$ years?

64. The following formula is used in psychology to measure the intelligence quotient (IQ) of a person.

$$IQ = \frac{MA}{CA} \times 100$$

where $MA =$ mental age (derived from tests) and $CA =$ chronological age (your actual age). What is the IQ of a 30-year-old female if her mental age is 27?

65. The following formula may be used to approximate the pulse rate of a person after he or she has stopped running.

$$P = \frac{1580 + 58x^2}{10 + x^2}$$

where $P =$ pulse rate (beats per second) and $x =$ time in minutes after running.
a. What is the pulse rate of a person immediately after he or she has stopped running?
b. What is the pulse rate of a person 5 minutes after he or she has stopped running?
c. If a normal pulse rate is between 60 and 80 beats per second, how long after a run will it take a person to attain a normal pulse rate?

3-3 SIMPLIFYING ALGEBRAIC EXPRESSIONS

A. Combining Like Terms of an Expression

In the expression $5x + 2x$ notice the following.

There are two terms ($5x$ and $2x$).	$5x + 2x$
The variable is the same in both terms (x).	$5\,x + 2\,x$

Whenever the terms of an expression contain the same variable, and also the same exponent for each corresponding variable, the terms are regarded as *like terms*. Thus, $5x$ and $2x$ in the preceding expression are like terms. Other examples follow.

$$\text{Examples of pairs} \atop \text{of like terms} \quad \begin{cases} 5y & \text{and} & y \\ 4xy & \text{and} & 10xy \\ 6(a+b) & \text{and} & 3(a+b) \\ 2x^2y & \text{and} & 5x^2y \end{cases}$$

Whenever the terms of an expression contain variables that are *not* the same, the terms are regarded as *unlike terms*. Some examples follow.

$$\text{Examples of pairs} \atop \text{of unlike terms} \quad \begin{cases} 2a & \text{and} & 3x \\ 5\,a & \text{and} & 2\,ab \\ 3x^2y & \text{and} & 5xy^2 \end{cases}$$

NOTE:

> Although both terms, $5a$ and $2ab$, contain the variable, a, the terms are *not* like terms; the terms must be *exactly alike*.

From our work in arithmetic we learned that only like quantities can be added or subtracted. The same holds true for algebra. When dealing with the terms of an expression, we can only add or subtract like terms. To do this, we simply add or subtract the coefficients of the like terms. For example:

$5x + 2x$ can be written as $(5 + 2)x$ and is equal to $7x$.	$\begin{aligned} 5x + 2x &= (5 + 2)x \\ &= 7x \end{aligned}$
$5x - 2x$ can be written as $(5 - 2)x$ and is equal to $3x$.	$\begin{aligned} 5x - 2x &= (5 - 2)x \\ &= 3x \end{aligned}$
$-5x + 2x$ can be written as $(-5 + 2)x$ and is equal to $-3x$.	$\begin{aligned} -5x + 2x &= (-5 + 2)x \\ &= -3x \end{aligned}$
$-5x - 2x$ can be written as $(-5 - 2)x$ and is equal to $-7x$.	$\begin{aligned} -5x - 2x &= (-5 - 2)x \\ &= -7x \end{aligned}$

This method of adding or subtracting like terms is called combining like terms. In algebra, we combine like terms by adding or subtracting their coefficients according to their sign. That is, we express the sum or difference of like terms as a single term.

When dealing with an expression that contains a mix of both like and unlike terms, we often find it helpful first to *collect* like terms before we combine them. By collecting like terms, we mean writing like terms next to each other to facilitate combining them. For example, in the expression, $2x + 5y - 3x + a - y$:

First, collect like terms.

$$2x + 5y - 3x + a - y$$
$$= 2x - 3x + 5y - y + a$$

NOTE:

Always include the sign that comes before the term as part of the term.

Then, combine like terms according to their sign.

$$2x - 3x + 5y - y + a$$
$$= \quad -x \quad + \quad 4y \quad + a$$

Thus, the expression, $2x + 5y - 3x + a - y$, is simplified to $-x + 4y + a$ by combining like terms.

PROCEDURE:

To combine like terms in an algebraic expression:

1. Collect like terms by writing the like terms next to each other. (Be sure to include the sign that comes immediately before each term when rearranging the terms.)

2. Combine all like terms according to their sign.

EXAMPLE 1: Simplify the expression $4x + 3x$ by combining terms.

SOLUTION:

$$4x + 3x$$
$$= \quad 7x$$

Final answer: $7x$

EXAMPLE 2: Simplify the expression $14x + 3y - 23x$ by combining like terms.

SOLUTION: First, collect like terms, then combine.

$$14x + 3y - 23x$$
$$= 14x - 23x + 3y$$
$$= \quad -9x \quad + 3y$$

Final answer: $-9x + 3y$

EXAMPLE 3: Simplify the expression $5a + 8b + 2a - 3b$ by combining like terms.

SOLUTION:

$$5a + 8b + 2a - 3b$$
$$= \underbrace{5a + 2a} + \underbrace{8b - 3b}$$
$$= \qquad 7a \quad + \quad 5b$$

Final answer: $7a + 5b$

EXAMPLE 4: Simplify the expression $5ab - 8x^2y + 5 + 3ab - 6 + 6x^2y$ by combining like terms.

SOLUTION:

$$5ab - 8x^2y + 5 + 3ab - 6 + 6x^2y$$
$$= \underbrace{5ab + 3ab} - \underbrace{8x^2y + 6x^2y} + \underbrace{5 - 6}$$
$$= \qquad 8ab \quad - \quad 2x^2y \quad - \quad 1$$

Final answer: $8ab - 2x^2y - 1$

Do the following practice set. Check your answers with the answers in the right-hand margin.

PRACTICE SET 3-3A

Combine like terms (see example 1).

1. $3x + 6x$
2. $2xy - 9xy$
3. $5x^2y + 2x^2y - 6x^2y$

Combine like terms (see example 2).

4. $3a + 5a + m$
5. $5a - 6b - 10a$
6. $3x - 2xy + 6xy$

Combine like terms (see example 3).

7. $3x + 8 + 5x - 6$
8. $6m + 3mn - 9m - 8mn$
9. $3y^2 + 5y + 8y - 4y^2$

1. $9x$
2. $-7xy$
3. x^2y

4. $8a + m$
5. $-5a - 6b$
6. $3x + 4xy$

7. $8x + 2$
8. $-3m - 5mn$
9. $-y^2 + 13y$

Combine like terms (see example 4).

10. $3x - 8y - 5x + 6y$	10. $-2x - 2y$
11. $5x^2 - 8x + 3x - 9x^2$	11. $-4x^2 - 5x$
12. $9a - 5b - 12c + 5a - 6c + 4b$	12. $14a - b - 18c$

B. Removing Grouping Symbols

Frequently in algebra, it will be necessary to simplify an algebraic expression such as $5x + 3(2x + 4)$. To simplify such an expression, we must first remove all grouping symbols and then combine like terms. For example, to simplify $5x + 3(2x + 4)$:

First, remove parentheses by applying the distributive property.

$$5x + 3(2x + 4)$$
$$5x + 3(2x) + 3(4)$$

Simplify $3(2x)$, by using the associative property of multiplication.

$$5x + 3(2x) + 3(4)$$
$$5x + (3 \cdot 2)x + 12$$
$$5x + 6x + 12$$

NOTE:

$3(2x)$ also means $2x + 2x + 2x$ which is equal to $6x$. Thus, $3(2x) = 6x$.

Now, combine like terms.

$$5x + 6x + 12$$
$$(5x + 6x) + 12$$
$$11x + 12$$

Thus,

$5x + 3(2x + 4)$ is simplified to $11x + 12$.

The following procedure can be used to simplify algebraic expressions completely.

PROCEDURE:

To simplify an algebraic expression:

1. Remove all symbols of grouping.

2. Combine like terms.

EXAMPLE 1: Simplify the expression $x - 5(x + y)$.

Example continued on next page.

SOLUTION: First, remove grouping symbols.

$$x - 5(x + y)$$

$$x - 5(x + y)$$

$$x - 5x - 5y$$

Next, combine like terms.

$$x - 5x - 5y$$

$$\underbrace{x - 5x} - 5y$$

$$-4x - 5y$$

Final answer: $-4x - 5y$

EXAMPLE 2: Simplify the expression $5(4 - 3a) + 9a$.

SOLUTION: Remove grouping symbols.

$$5(4 - 3a) + 9a$$

$$5(4 - 3a) + 9a$$

$$20 - 15a + 9a$$

Combine like terms.

$$20 - 15a + 9a$$

$$20 - \underbrace{15a + 9a}$$

$$20 - \quad 6a$$

Final answer: $20 - 6a$

EXAMPLE 3: Simplify $4(3x + 8) - 2(4 - 6x)$

SOLUTION:

$$4(3x + 8) - 2(4 - 6x)$$

$$= 12x + 32 - 8 + 12x$$

$$= \underbrace{12x + 12x} + \underbrace{32 - 8}$$

$$= \quad 24x \quad + \quad 24$$

Final answer: $24x + 24$

EXAMPLE 4: Simplify $4[3a + 2(a - 4) + 5] - 6a$.

SOLUTION:

$$4[3a + 2(a - 4) + 5] - 6a$$
$$= 4[3a + 2a - 8 + 5] - 6a$$
$$= 4[5a - 3] - 6a$$
$$= 20a - 12 - 6a$$
$$= 20a - 6a - 12$$
$$= 14a - 12$$

Final answer: $14a - 12$

Do the following practice set. Check your answers with the answers in the right-hand margin.

PRACTICE SET 3-3B

Simplify each expression (see examples 1 or 2).

1. $a + 5(a + b)$ 2. $3(4 - x) + 5x$ 3. $9x - 4(2x - 6)$

1. $6a + 5b$
2. $12 + 2x$
3. $x + 24$

Simplify each expression (see example 3).

4. $(x + 2y) - (x - 2y)$

5. $3(5x - 4) + 2(3 - 2x)$

6. $2(a - 6b) - 4(3a - b)$

4. $4y$
5. $11x - 6$
6. $-10a - 8b$

Simplify each expression (see example 4).

7. $5[2x + 3(x - 4) + 4]$

8. $4[3a - 5(2a - 2) + 4] - 2a$

9. $-3[2x + 6 - 4(x - 1)]$

7. $25x - 40$
8. $-30a + 56$
9. $6x - 30$

EXERCISE 3-3

Combine like terms.

1. $9x + 2x$ 2. $3y - 2y$

3. $6t - 100t$ 4. $5a - 16a$

5. $6m + 3m + 5m$ 6. $25x + 4x + 31x$

7. $9x + 9x + 2x - 3x$

8. $3a - 7a + 4a$

9. $5x + 3y - x + 9y$

10. $6y + 3ab + 14y - 6ab$

11. $13c - 6d + 14d - 4c$

12. $m + 2a - 6m - 2a$

13. $3w - 5w + 6x - 9x$

14. $-9r - 6s - 6r - 12s$

15. $2(x + y) + 3(x + y)$

16. $9(2a) - 14(2a)$

17. $2xy - 3x + 4xy$

18. $5mn + 6n - 9mn$

19. $5x^2 + 3x - 2x^2 + 9x$

20. $14x^2 + 11x - 3xy + 2y^2$

Combine like terms.

21. $12x^2y - 15x^2y$

22. $11x^2y^2 - 16x^2y^2$

23. $3x^2y + 5xy^2 - 6x^2y + 3xy^2$

24. $2x^2 + 3x^3 - 6x^2 - 4x^3$

25. $5a^2b + 3ab^2 - 6a^2b$

26. $3x - 4x^2 + 2x^3 - 6x + 5x^2 + 6x^3$

27. $13 + 2ab - b + 5a - 18 + 3ab$

28. $4xy - 3x + 2yx - 10x + 3y$

29. $6xy^2 - 3x^2y^2 + 2x^2y - 6x$

30. $x - 6x^2 + 3y^2 + 2x - 8y^2$

Simplify each of the following expressions by removing the grouping symbols and combining like terms.

31. $x + (2x - y)$

32. $x + 5(x + y)$

33. $3x - 4(x - 6)$

34. $3(5 + x) - 2x$

35. $(x - 3) + (x - 4)$

36. $(2x + 6y) - (3x - 3y)$

37. $9(2x + 6) + 3(5x - 2)$

38. $4(6 - 3x) - 5(5x - 3)$

39. $(4a - b)3 - (3a + 4b)2$

40. $5(2x - 6) - 4(2x + 8) - 9$

41. $5(x^2 - 3x + 16) + 2x + 4$

42. $9(x^2 + 4x - 6) - (3x^2 - 2x + 4)$

43. $2(x^2 + 3x - 6) + 5(x^2 - 6x + 4)$

44. $4(x^2 - x) + 5 - 3(2x^2 + x) + 4$

45. $(3x^2 - 6x - 9)3 + 2(5x^2 + 6x - 4)$

46. $4[5x + 3(2x + 5)] - 6$

47. $5[4x - 6 + 5(x - 4) + 6]$

48. $3a - [4 + (2a - 6)] - 2(a - 8)$

49. $14 - 3[5x - 6(3x - 8)]$

50. $27 - 4[5a - 9(3a - 10)]$

Simplify by removing the grouping symbols and combining like terms.

51. $-5(x - 3y) + 9(2x + 8y)$

52. $-4(3a - 6b) - 4(3a - 6b)$

53. $-7(4x - 5y) - 3(6x - 8y)$

54. $24 - 6[3x - 4(x + y) - 8y]$

55. $5x - [3x + 2(4x - 6a) + 3a]$

56. $5[3(x + y) - 4(3x - 6y) - 8] + 2x - 3y$

57. $2[-(2x - 3y) + (x - y)] - 6(3x + 4y)$

58. $-2[-(2x - 5y) - (2x + 3y)] - 4(2x - 3y)$

59. $-5[-6\{3(2x - 4)\} - 5(2x + 3)]$

60. $-3[-4\{-2(3x - 8) - 6(2x - 4)\} - 6x]$

REVIEW EXERCISES

In problems $1 - 12$, identify each of the following as a factor, variable, constant, coefficient, or term.

1. The x in $14x + 12$

2. The y in $(3 - y)$

3. $12x$

4. $\dfrac{3x}{2}$

5. The 5 in $x + 5$

6. The -6 in $-6 + 2y$

7. The 5 in $5x + 8$

8. The -4 in $b - 4ac$

9. The $(x + y)$ in $3(x + y)$

10. The -3 in $-3(x - y)$

11. The 8 in $3x + 8y$

12. The $(x - 8)$ in $(x + 4)(x - 8)$

13. What is the coefficient of $-y^2$?

14. In the expression $8x^4 + 5x^3 - x^2 + 7x - 6$:
    a. State the number of terms.
    b. State the terms.
    c. State the coefficient for each term.

15. Write the product $(5)(x)(x)(x)(x)(y)(y)(y)$ by using exponents.

16. Write the term $8x^3y^4$ without using exponents.

Evaluate each of the following expressions.

17. $4a - 3b$ when $a = 6$, $b = 3$.

18. $3x - 2(y + z)$ when $x = 3$, $y = 4$, $z = 8$.

19. $\dfrac{3y - 10x}{4}$ when $x = 3$, $y = 2$.

20. $9z + 5(2x - (3y + 4))$ when $x = 5$, $y = 2$, $z = 1$.

21. $4x^2y^3 - 3xy$ when $x = 3$, $y = -2$.

Evaluate each of the following formulas.

22. $I = \dfrac{E}{R}$ when $E = 2000$ volts, $R = 0.08$ ohms.

23. $l = ar^{n-1}$ when $a = 3$, $r = \dfrac{1}{2}$, $n = 5$.

24. $A = \pi r^2$ when $\pi \approx 3.14$, $r = 4$.

25. $F = \dfrac{9}{5}C + 32$ when $C = 15°$.

Simplify each of the following expressions.

26. $3x + (2x - y)$

27. $(3a - b)4 - (3a + 4b)2$

28. $2(3x^2 - 6x - 8) + 3(5x^2 - 6x + 4)$

29. $4[-(2x - 3y) + (x - y)] - 3(2x - 5y)$

CHAPTER 4

Solving Algebraic Equations and Inequalities in One Variable

4-1 INTRODUCTION TO ALGEBRAIC EQUATIONS

Whenever we have a mathematical statement that shows an equality, we refer to the statement as an *equation*.

In arithmetic, the equation, $8 + 6 = 14$, means:

> The *numerical expression* on the left of the equal sign $(8 + 6)$ is equal in value to the expression on the right of the equal sign (14).

$$8 + 6 = 14$$

$8 + 6 = 14$ is called an *arithmetic equation*.

In algebra, the equation, $x + 6 = 14$, means:

> The *algebraic expression* on the left of the equal sign $(x + 6)$ is equal in value to the expression on the right of the equal sign (14).

$$x + 6 = 14$$

$x + 6 = 14$ is called an *algebraic equation*.

NOTE:

> An equation consists of three parts.
>
> 1. The equal sign $(=)$.
> 2. The expression on the left of the equal sign (called the *left side* or *left member*).
> 3. The expression on the right of the equal sign (called the *right side* or *right member*).
>
> $$x + 6 \quad = \quad 14$$
> left   equal   right
> side   sign   side

The numerical equation, $8 + 6 = 14$, is referred to as a true sentence. (The sum of $8 + 6$ is indeed equal to 14.) An equation may also be a false sentence (e.g., $8 + 6 = 10$) or an *open sentence* (e.g., $x + 6 = 14$).

In the case of the open sentence, $x + 6 = 14$, we do not know if the equation is true or false until we replace the variable x with a value. For example:

The equation, $x + 6 = 14$ can only be true when the variable, x, is given the value of 8 (i.e., $x = 8$).

$$x + 6 = 14$$
$$8 + 6 = 14$$
$$14 = 14$$

The number, $x = 8$, is said to *satisfy* the equation and is called the *solution of the equation*.

In our work with solving equations, we will be dealing with equations that are open sentences. Our goal is to find the value that makes the open sentence a true sentence. The process of taking an equation that is an open sentence and finding its solution is called *solving an equation*.

4-2 SOLVING EQUATIONS WITH ONE OPERATION

A. Introduction

If we *add* the same number to both sides of an equation, the sums remain equal. For example:

Given the equation $15 = 15$. If we add 3 to both sides of the equation,

$$15 = 15$$
$$\frac{15}{+3} = \frac{15}{+3}$$

Then, the sums are equal.

$$\begin{array}{cc} 15 & 15 \\ +3 = +3 \\ \hline 18 & 18 \end{array}$$

If we *subtract* the same number from both sides of an equation, the differences remain equal. For example:

Given the equation $15 = 15$. If we subtract 3 from both sides of the equation,

$$15 = 15$$
$$\frac{15}{-3} = \frac{15}{-3}$$

Then, the differences are equal.

$$\begin{array}{cc} 15 & 15 \\ -3 = -3 \\ \hline 12 & 12 \end{array}$$

If we *multiply* both sides of an equation by the same nonzero number, the products remain equal. For example:

Given the equation $15 = 15$. If we multiply both sides of the equation by 3,

$$15 = 15$$
$$(15)(3) = (15)(3)$$

Then, the products are equal.

$$(15)(3) = (15)(3)$$
$$45 \qquad 45$$

If we *divide* both sides of an equation by the same nonzero number, the quotients remain equal. For example:

Given the equation $15 = 15$. If we divide both sides of the equation by 3,

$$15 = 15$$

$$\frac{15}{3} = \frac{15}{3}$$

Then, the quotients are equal.

$$\frac{15}{3} = \frac{15}{3}$$

$$5 \qquad 5$$

Thus, if we add, subtract, multiply, or divide (except by zero) the same number to both sides of an equation, the equality is maintained. In other words, whatever we do to one side of an equation, we must do to the other side in order to preserve the equality.

Our goal in solving simple algebraic equations is to get the variable alone on one side of the equal sign with a coefficient of 1. That is, we must get the equation in the form

$$x = \text{something} \qquad \text{or} \qquad \text{something} = x$$

To do this, we use *opposite operations*.

In arithmetic, we learned that addition and subtraction are opposite or inverse operations. Thus, to "undo" subtraction, we add. And, to undo addition, we subtract. For example, in the equation, $x - 6 = 18$:

Since we are subtracting 6 from x,

$$x - 6 = 18$$

We can *add* 6 to both sides of the equation to get x by itself.

$$\begin{array}{r} x - 6 = 18 \\ +6 \quad +6 \\ \hline x + 0 = 24 \end{array}$$

In the equation $x + 6 = 18$:

Since we are adding 6 to x,

$$x + 6 = 18$$

We can subtract 6 from both sides of the equation to get x by itself.

$$\begin{array}{r} x + 6 = 18 \\ -6 \quad -6 \\ \hline x + 0 = 12 \end{array}$$

We also learned in arithmetic that multiplication and division are opposite or inverse operations. Thus, to undo multiplication, we divide. And to undo division, we multiply. For example, in the equation $6x = 18$:

Since we are multiplying x by 6,

$$6x = 18$$

We can divide both sides of the equation by 6 to get x by itself.

$$6x = 18$$

$$\frac{6x}{6} = \frac{18}{6}$$

$$\frac{\overset{1}{6}x}{\underset{1}{6}} = \frac{\overset{3}{18}}{\underset{1}{6}}$$

$$x = 3$$

In the equation, $x/6 = 18$:

Since we are dividing x by 6,

$$\frac{x}{6} = 18$$

We can multiply both sides of the equation by 6 to get x by itself.

$$\frac{x}{6} = 18$$

$$6 \cdot \frac{x}{6} = 6 \cdot 18$$

$$\frac{^1 6}{1} \cdot \frac{x}{6}_1 = 108$$

$$x = 108$$

This is the first step in solving algebraic equations. We must identify what operation connects the variable with the number and then use opposite operations to get the variable by itself on one side of the equal sign. Remember, whatever we do to one side of the equation, we must always do to the other side.

Do the following practice set. Check your answers with the answers in the right-hand margin.

PRACTICE SET 4-2A

For each problem
(a) State the operation that connects the variable to the number.
(b) State what must be done to get the variable by itself.

1. $x + 4 = 5$

2. $x - 9 = 8$

3. $3x = 12$

4. $\frac{x}{5} = 21$

5. $12 + x = 8$

6. $9 = 6x$

1. (a) Addition
    (b) Subtract 4 from both sides.
2. (a) Subtraction
    (b) Add 9 to both sides.
3. (a) Multiplication
    (b) Divide both sides by 3.
4. (a) Division
    (b) Multiply both sides by 5.
5. (a) Addition
    (b) Subtract 12 from both sides.
6. (a) Multiplication
    (b) Divide both sides by 6.

B. Solving Equations with One Operation

In part A, we learned the process behind solving simple algebraic equations. For example, in the equation, $x - 6 = 18$:

Since subtraction connects the variable and the number,

$$x - 6 = 18$$

We must add 6 to both sides of the equation to get x all by itself on one side of the equal sign.

$$\begin{array}{r} x - 6 = 18 \\ +6 \quad +6 \\ \hline x + 0 = 24 \end{array}$$

Notice that we are left with the equation, $x = 24$. This resulting equality is our *solution*. However, we do not know if $x = 24$ is the correct solution. To check this, we simply replace the value of the variable for the variable in the given equation and perform the indicated operation. If the left side of the equation equals the right side of the equation, then the solution is correct. For example, to check the solution $x = 24$

For the given equation, $x - 6 = 18$, substitute 24 for x and perform the subtraction.

$$x - 6 = 18$$

Since the left side of the equation is equal to the right side of the equation ($18 = 18$), the solution $x = 24$ is correct.

Left side	Right side
$x - 6$	18
$(24) - 6$	
18	

When solving algebraic equations, we *must* check our solutions.

We now offer the following procedure for solving simple algebraic equations.

PROCEDURE:

To solve an algebraic equation with one operation:

1. Determine what operation connects the variable and the number.

2. If the operation connecting the variable and the number is
   Addition → subtract the number from *both sides* of the equation.
   Subtraction → add the number to *both sides* of the equation.
   Multiplication → divide *both sides* of the equation by the number.
   Division → multiply *both sides* of the equation by the number.

3. Check the solution by substituting it back into the given equation.

EXAMPLE 1: Solve $x + 18 = 23$ for x.

SOLUTION: Since x and 18 are connected by addition, subtract 18 from both sides of the equation.

$$x + 18 = 23$$

$$\begin{array}{r} x + 18 = 23 \\ -18 -18 \\ \hline x = 5 \end{array}$$

Check:

$x + 18$	23
$5 + 18$	23
23	

Final answer: $x = 5$

EXAMPLE 2: Solve $3.16 = x - 4$ for x.

SOLUTION: Since x and 4 are connected by subtraction, add 4 to both sides of the equation.

$$3.16 = x - 4$$

$$\begin{array}{r} 3.16 = x - 4 \\ +4 \qquad +4 \\ \hline 7.16 = x \end{array}$$

Check:

3.16	$x - 4$
3.16	$7.16 - 4$
	3.16

Final answer: $x = 7.16$

EXAMPLE 3: Solve $-12 = \dfrac{y}{4}$ for y.

SOLUTION: Since y and 4 are connected by division, multiply both sides of the equation by 4.

$$-12 = \frac{y}{4}$$

$$-12(4) = \frac{y}{4}(4)$$

$$-48 = y$$

Check:

-12	$\dfrac{y}{4}$
-12	$\dfrac{-48}{4}$
	-12

Final answer: $y = -48$

EXAMPLE 4: Solve $-5x = 30$.

SOLUTION: Since x and -5 are connected by multiplication, divide both sides of the equation by -5.

$$-5x = 30$$

$$\frac{-5x}{-5} = \frac{30}{-5}$$

$$x = -6$$

Example continued on next page.

Check:

$$\begin{array}{c|c} -5x & 30 \\ \hline -5(-6) & 30 \\ 30 & \end{array}$$

Final answer: $x = -6$

Do the following practice set. Check your answers with the answers in the right-hand margin.

PRACTICE SET 4-2B

Solve and check (see example 1 or 2).

1. $x + 8 = 9$

2. $x - \dfrac{1}{2} = 3\dfrac{1}{2}$

3. $1.16 = x - 1.16$

1. $x = 1$
2. $x = 4$
3. $x = 2.32$

Solve and check (see example 3).

4. $\dfrac{x}{2} = 8$

5. $-7 = \dfrac{x}{4}$

6. $\dfrac{x}{10} = 5.16$

4. $x = 16$
5. $x = -28$
6. $x = 51.6$

Solve and check (see example 4).

7. $5x = 10$

8. $9x = -63$

9. $36 = -12x$

7. $x = 2$
8. $x = -7$
9. $x = -3$

EXERCISE 4-2

For each equation
(a) Determine what operation connects the variable and number.
(b) State what must be done to get the variable by itself on one side of the equal sign.

1. $x + 6 = 5$

2. $2x = 8$

3. $x - 5 = 12$

4. $\dfrac{x}{10} = 10$

5. $13 = x - 5$

6. $40 = 5x$

7. $23 + x = 19$

8. $20 = \dfrac{x}{4}$

9. $2x + 5 = 7$

Solve and check.

10. $x + 3 = 8$ 11. $x + 9 = 12$ 12. $9 + x = 13$

13. $x - 8 = 9$ 14. $x - 3 = 1$ 15. $x - 12 = 13$

16. $x + 4 = -8$ 17. $x + 16 = -9$ 18. $5 + x = -5$

19. $x - 8 = -15$ 20. $x - 3 = -12$ 21. $x - 27 = -8$

22. $23 = x + 4$ 23. $18 = x + 3$ 24. $15 = x + 16$

25. $14 = x - 8$ 26. $20 = x - 9$ 27. $-3 = x - 5$

28. $-14 = -18 + x$ 29. $-31 = -3 + x$ 30. $-10 = -29 + x$

Solve and check.

31. $2x = 16$ 32. $5x = 15$ 33. $4x = 20$

34. $\dfrac{x}{4} = 2$ 35. $\dfrac{x}{9} = 3$ 36. $\dfrac{x}{10} = 1$

37. $2x = -18$ 38. $3x = -21$ 39. $9x = -27$

40. $\dfrac{x}{4} = -6$ 41. $\dfrac{x}{3} = -8$ 42. $\dfrac{x}{6} = -9$

43. $-3x = 24$ 44. $-6x = -36$ 45. $-2x = 4$

46. $\dfrac{x}{-3} = -18$ 47. $-30 = \dfrac{x}{-2}$ 48. $3 = \dfrac{x}{-8}$

49. $\dfrac{x}{2} = \dfrac{1}{2}$ 50. $\dfrac{x}{-4} = -\dfrac{3}{8}$ 51. $-\dfrac{3}{9} = \dfrac{x}{27}$

4-3 SOLVING EQUATIONS WITH MORE THAN ONE OPERATION

Many algebraic equations will require more than one operation to remove all numbers that are connected to the variable. For example, in the equation, $3x - 4 = 8$:

3x is connected to the 4 by subtraction. $3x - 4 = 8$

x is connected to the 3 by multiplication. $3x - 4 = 8$

To solve such an equation, we must first remove all numbers that are connected to the variable by addition and subtraction. Then, we must remove all numbers that are connected to the variable by multiplication and division. For example, to solve the equation, $3x - 4 = 8$:

First, add 4 to both sides of the equation.

$$\begin{array}{rcl} 3x - 4 &=& 8 \\ +4 && +4 \\ \hline 3x + 0 &=& 12 \end{array}$$

Now, divide both sides of the equation by 3.

$$3x = 12$$

$$\frac{3x}{3} = \frac{12}{3}$$

$$x = 4$$

Check.

$3x - 4$	8
$3(4) - 4$	8
$12 - 4$	
8	

Thus,

4 is the correct solution.

PROCEDURE:

To solve an equation that involves more than one operation:

1. Perform the order of operations in reverse. That is:
    a. Undo all addition and subtraction by using opposite (inverse) operations.
    b. Undo all multiplication and division by using opposite (inverse) operations.

2. Check the solution.

EXAMPLE 1: Solve $6x + 1 = 25$ for x.

SOLUTION: First subtract 1 from both sides.

$$6x + 1 = 25$$

$$
\begin{array}{r}
6x + 1 = 25 \\
-1 \quad -1 \\
\hline
6x = 24
\end{array}
$$

Next, divide both sides by 6.

$$6x = 24$$

$$\frac{6x}{6} = \frac{24}{6}$$

$$x = 4$$

Check:

$6x + 1$	25
$6(4) + 1$	25
$24 + 1$	
25	

Final answer: $x = 4$

EXAMPLE 2:   Solve $-16 = 5x - 6$ for x.

SOLUTION: Add 6 to both sides.

$$-16 = 5x - 6$$

$$\begin{array}{r} -16 = 5x - 6 \\ \underline{+6 \qquad +6} \\ -10 = 5x \end{array}$$

Divide both sides by 5.

$$-10 = 5x$$

$$\frac{-10}{5} = \frac{5x}{5}$$

$$-2 = x$$

Check:

$$\begin{array}{c|c} -16 & 5x - 6 \\ \hline -16 & 5(-2) - 6 \\ & -10 - 6 \\ & -16 \end{array}$$

Final answer: $x = -2$

EXAMPLE 3:   Solve $\dfrac{3x}{8} + 6 = 9$ for x.

SOLUTION: Subtract 6 from both sides.

$$\frac{3x}{8} + 6 = 9$$

$$\begin{array}{r} \dfrac{3x + 6 =}{8} \qquad 9 \\ \underline{-6 \qquad -6} \\ \dfrac{3x}{8} = 3 \end{array}$$

Multiply both sides by 8.

$$\frac{3x}{8} = 3$$

$$(8)\frac{3x}{8} = 3(8)$$

$$3x = 24$$

Example continued on next page.

Divide both sides by 3.

$$3x = 24$$

$$\frac{3x}{3} = \frac{24}{3}$$

$$x = 8$$

Check:

$\frac{3x}{8} + 6$	9
$\frac{3(8)}{8} + 6$	9
$\frac{24}{8} + 6$	
$3 + 6$	
9	

Final answer: $x = 8$

EXAMPLE 4:   Solve $-x + 3 = 12$ for x.

SOLUTION: Subtract 3 from both sides.

$$-x + 3 = 12$$

$$\begin{array}{r} -x + 3 = 12 \\ -3 \quad -3 \\ \hline -x = 9 \end{array}$$

Multiply or divide both sides by -1.

$$-x = 9 \qquad\qquad -x = 9$$

$$-x(-1) = 9(-1) \qquad \frac{-x}{-1} = \frac{9}{-1}$$

$$x = -9 \qquad\qquad x = -9$$

Check:

$-x + 3$	12
$-(-9) + 3$	12
$9 + 3$	
12	

Final answer: $x = -9$

NOTE:

> The coefficient of x in the equation $-x + 3 = 12$ is -1. Since we must solve for x (i.e., positive $1x$), we have to multiply or divide both sides of the equation by -1.

Do the following practice set. Check your answers with the answers in the right-hand margin.

PRACTICE SET 4-3

Solve and check (see example 1 or 2).

1. $3x - 4 = 11$ 2. $7x + 6 = -15$ 3. $-11 = -5x + 4$

    1. $x = 5$
    2. $x = -3$
    3. $x = 3$

Solve and check (see example 3).

4. $\dfrac{x}{3} - 4 = 5$ 5. $\dfrac{3x}{4} + 9 = 15$ 6. $-11 = 4 + \dfrac{5x}{6}$

    4. $x = 27$
    5. $x = 8$
    6. $x = -18$

Solve and check (see example 4).

7. $-x + 4 = 9$ 8. $-x - 5 = 6$ 9. $6 = 5 - x$

    7. $x = -5$
    8. $x = -11$
    9. $x = -1$

EXERCISE 4-3

Solve and check.

1. $2x + 4 = 10$ 2. $3x + 9 = 15$ 3. $4x + 1 = 29$
4. $3x - 8 = 16$ 5. $9x - 4 = 5$ 6. $2x - 23 = 2$
7. $5x + 16 = -9$ 8. $4x + 5 = -27$ 9. $3x + 3 = -31$
10. $9x - 5 = -23$ 11. $7x - 8 = -29$ 12. $12x - 10 = -70$
13. $34 = 7x + 48$ 14. $23 = 9x + 32$ 15. $-10 = 2x + 40$
16. $-19 = 5x - 4$ 17. $-23 - 2x = 21$ 18. $-14 - 5x = -14$

Solve and check.

19. $\dfrac{x}{2} + 15 = 20$ 20. $\dfrac{x}{3} + 4 = 9$ 21. $\dfrac{x}{5} + 8 = 11$

22. $\dfrac{x}{3} - 4 = 12$

23. $\dfrac{x}{6} - 5 = -10$

24. $\dfrac{x}{7} - 9 = -3$

25. $\dfrac{2x}{3} + 5 = 15$

26. $\dfrac{3x}{4} + 8 = 23$

27. $\dfrac{5x}{7} + 7 = -13$

28. $\dfrac{4x}{5} - 9 = 11$

29. $\dfrac{7x}{9} - 16 = -58$

30. $\dfrac{2x}{5} - 18 = -12$

31. $23 = \dfrac{3x}{8} + 2$

32. $-12 = \dfrac{2x}{7} - 10$

33. $-18 + \dfrac{5x}{8} = 32$

34. $\dfrac{8x}{7} + 21 = -3$

35. $\dfrac{4x}{3} - 14 = 14$

36. $-15 + \dfrac{7x}{3} = -57$

Solve and check.

37. $-x + 8 = 17$

38. $-x + 9 = -18$

39. $-x + 4 = -1$

40. $-x - 3 = 12$

41. $-x - 6 = -18$

42. $-x - 15 = -1$

43. $12 - x = 17$

44. $-15 - x = 16$

45. $-21 - x = -8$

46. $\dfrac{-x}{4} + 3 = 6$

47. $\dfrac{-x}{3} - 4 = 6$

48. $\dfrac{-x}{7} - 3 = -5$

49. $14 = \dfrac{-3x}{4} + 5$

50. $21 - \dfrac{2x}{3} = 29$

51. $-18 - \dfrac{5x}{4} = 7$

4-4 SOLVING EQUATIONS BY COMBINING LIKE TERMS

In algebra, we are often presented with equations in which like terms appear on one side of the equation. To solve such equations, we simply combine like terms first and then solve for the variable by using opposite operations. For example, to solve the equation, $4x + 2x = 12$:

First, combine like terms.

$$4x + 2x = 12$$
$$6x = 12$$

Then, divide both sides of the equation by 6.

$$6x = 12$$
$$\dfrac{6x}{6} = \dfrac{12}{6}$$
$$x = 2$$

Check:

$4x + 2x$	12
$4(2) + 2(2)$	12
$8 + 4$	
12	

Thus,

$x = 2$ is the correct solution.

PROCEDURE:

> To solve an equation that contains like terms on one side of the equal sign:
>
> 1.   Combine all like terms.
>
> 2.   Solve the equation by using opposite operations.
>
> 3.   Check the solution.

EXAMPLE 1:   Solve $2x + 3x = 15$ for x.

SOLUTION: Combine like terms.

$$2x + 3x = 15$$
$$\underbrace{2x + 3x} = 15$$
$$5x \quad\;\; = 15$$

Solve the equation.

$$5x = 15$$
$$\frac{5x}{5} = \frac{15}{5}$$
$$x = 3$$

Check:

$2x + 3x$	15
$2(3) + 3(3)$	15
$6 \;+\; 9$	
15	

Final answer: $x = 3$

EXAMPLE 2:   Solve $7x + 5 - 2x = -10$

SOLUTION: Combine like terms.

$$7x + 5 - 2x = -10$$
$$7x + 5 - 2x = -10$$
$$\underbrace{7x - 2x} + 5 = -10$$
$$5x \quad\;\; + 5 = -10$$

Example continued on next page.

Solve the equation.

$$5x + 5 = -10$$
$$5x = -10 - 5$$
$$5x = -15$$
$$x = -\frac{15}{5}$$
$$x = -3$$

Check:

$7x + 5 - 2x$	-10
$7(-3) + 5 - 2(-3)$	-10
$-21 + 5 + 6$	
$-21 + 11$	
-10	

Final answer: $x = -3$

Do the following practice set. Check your answers with the answers in the right-hand margin.

PRACTICE SET 4-4

Solve and check (see example 1 or 2).

1. $5x - 2x = 12$
2. $-24 = 3x + 9x$
3. $7x + 9 - 8x = -5$

1. $x = 4$
2. $x = -2$
3. $x = 14$

EXERCISE 4-4

Solve and check.

1. $3x + x = 12$
2. $5x + 3x = 16$
3. $27 = 12x - 3x$
4. $18x - 3x = 30$
5. $20x - 22x = 144$
6. $100 = 22y - 47y$
7. $3x + 4x - 2x = 155$
8. $4x + 3x + 4x = 121$
9. $120 = 6x - 5x - 3x$
10. $-142 = 4x - 16x + 2$
11. $-x + 6x + 12 = 22$
12. $12x - 16 - 2x = -84$
13. $-12x + 3x + 17 = 89$
14. $-123 = -2x + 7x - 18$
15. $-8 + 4x - 5x = -10$
16. $27 = 5x - 13 - 13x$

17. $\dfrac{3x}{4} - \dfrac{2x}{4} = 25$

18. $\dfrac{2x}{3} + \dfrac{7x}{3} = 15$

19. $\dfrac{3x}{5} - \dfrac{7x}{5} - 9 = -119$

20. $-120 + \dfrac{3x}{7} - \dfrac{5x}{7} = -10$

4-5   SOLVING EQUATIONS IN WHICH THE VARIABLE APPEARS ON BOTH SIDES OF THE EQUATION

Thus far, all the equations we have discussed have had the variable on only one side of the equal sign. Since a variable represents a number, and since any number may be added or subtracted to or from both sides of an equation without changing the answer, we may then add or subtract terms containing the variable to or from both sides of the equation. Following this, we then simply combine like terms and solve the equation. For example, to solve $3x = 8 + 2x$:

First, get all the terms containing x, the variable, on one side of the equation.

$$3x = 8 + 2x$$

Since $2x$ is being added to 8,

$$3x = 8 + 2x$$

Subtract $2x$ from both sides of the equation.

$$\begin{array}{r} 3x = 8 + 2x \\ -2x \qquad\;\; -2x \end{array}$$

Combine like terms
$(3x - 2x = x$ and $2x - 2x = 0)$.

$$\begin{array}{r} 3x = 8 + 2x \\ -2x \qquad\;\; -2x \\ \hline x = 8 + 0 \\ x = 8 \end{array}$$

Check the solution.

$3x$	$8 + 2x$
$3(8)$	$8 + 2(8)$
24	$8 + 16$
	24

Thus,

$$x = 8 \text{ is the correct solution}$$

PROCEDURE:

To solve an equation in which the variable appears on both sides of the equal sign:

1.   Isolate the variable; that is, get the variable on one side of the equal sign by using opposite operations.

2.   Combine like terms.

3.   Solve the equation.

4.   Check the solution.

EXAMPLE 1: Solve $5x = 16 - 3x$ for x.

SOLUTION: Get the variable on one side of the equal sign and combine like terms.

$$5x = 16 - 3x$$

$$
\begin{array}{rcrr}
5x & = & 16 & - 3x \\
+3x & & & +3x \\
\hline
8x & = & 16 &
\end{array}
$$

Solve the equation.

$$8x = 16$$

$$\frac{8x}{8} = \frac{16}{8}$$

$$x = 2$$

Check:

$$
\begin{array}{c|c}
5x & 16 - 3x \\
\hline
5(2) & 16 - 3(2) \\
10 & 16 - 6 \\
& 10
\end{array}
$$

Final answer: $x = 2$

EXAMPLE 2: Solve $6x - 20 = -8 + 2x$.

SOLUTION: Get the variable on one side of the equal sign and combine like terms.

$$6x - 20 = -8 + 2x$$

$$
\begin{array}{rclr}
6x - 20 & = & -8 & + 2x \\
-2x & & & -2x \\
\hline
4x - 20 & = & -8 &
\end{array}
$$

Solve the equation.

$$4x - 20 = -8$$

$$4x = -8 + 20$$

$$4x = 12$$

$$x = \frac{12}{4}$$

$$x = 3$$

Check:

$6x - 20$	$-8 + 2x$
$6(3) - 20$	$-8 + 2(3)$
$18 - 20$	$-8 + 6$
-2	-2

Final answer: $x = 3$

EXAMPLE 3:   Solve $8x + 4x + 10 = 5x - 20 + 2$ for x.

SOLUTION: Before solving this equation for x, we will find it very helpful first to combine like terms on each separate side of the equal sign.

$$\underline{8x + 4x} + 10 = 5x \underbrace{- 20 + 2}$$
$$12x \quad + 10 = 5x - \quad 18$$

Now let us solve the equation.

$$8x + 4x + 10 = 5x - 20 + 2$$
$$12x + 10 = 5x - 18$$
$$12x - 5x = -18 - 10$$
$$7x = -28$$
$$x = \frac{-28}{7}$$
$$x = -4$$

Check:

$8x + 4x + 10$	$5x - 20 + 2$
$8(-4) + 4(-4) + 10$	$5(-4) - 20 + 2$
$-32 - 16 + 10$	$-20 - 20 + 2$
$-48 + 10$	$-40 + 2$
-38	-38

Final answer: $x = -4$

Do the following practice set. Check your answers with the answers in the right-hand margin.

PRACTICE SET 4-5

Solve and check (see example 1 or 2).

1. $4x = 9 + x$

2. $10 - x = 4x$

3. $12 + 2x = 4x + 16$

1. $x = 3$

2. $x = 2$

3. $x = -2$

Solve and check (see example 3).

4. $4x + 2x + 7 = 8x + 3$	4. $x = 2$
5. $5x - 3x + 11 = 27 - 6x$	5. $x = 2$
6. $9x - 4 = 7x - 2x - 16$	6. $x = -3$

EXERCISE 4-5

Solve and check.

1. $5x = 10 + 3x$ 　　　　　　　　　2. $10x = 22 - x$

3. $12x = 5x + 49$ 　　　　　　　　4. $6x = 5x + 16$

5. $8x - 50 = 18x$ 　　　　　　　　6. $3x + 21 = -4x$

7. $9x - 28 = 5x$ 　　　　　　　　　8. $2x = 7x - 20$

9. $10x + 8 = x - 10$ 　　　　　　10. $4x + 28 = 7x + 13$

11. $5x + 12 = 4x - 6$ 　　　　　　12. $17x - 1 = x + 15$

13. $2 + 7x = 11 - 2x$ 　　　　　　14. $8x - 16 = 3x - 96$

15. $-4x - 21 = 7x - 27$ 　　　　16. $18 - 16x = 6 - 4x$

17. $14x - 6 = 7x - 2x + 21$ 　　18. $8x - 17 - 3x = 73 - 4x$

19. $15 - 5x + 8 = 3x - 4 + x$ 　20. $4x - 6x - 15 = 8x - 89 - 6$

4-6　SOLVING EQUATIONS CONTAINING SYMBOLS OF GROUPING

Algebraic equations frequently contain symbols of grouping. To solve such equations, we must first remove these grouping symbols *before* we begin to solve the equation. For example, to solve $6(3x - 4) - 4x = 4$:

First, remove the parentheses (using the distributive property).

$$6(3x - 4) - 4x = 4$$
$$6(3x) + 6(-4) - 4x = 4$$
$$18x - 24 - 4x = 4$$

Next, combine like terms.

$$18x - 24 - 4x = 4$$
$$(18x - 4x) - 24 = 4$$
$$14x - 24 = 4$$

Now, add 24 to both sides of the equation.

$$14x - 24 = \quad 4$$
$$\underline{\quad + 24 \quad +24}$$
$$14x + 0 = \quad 28$$

Finally, divide both sides by 14.

$$14x = 28$$
$$\frac{14x}{14} = \frac{28}{14}$$
$$x = 2$$

Check the solution.

$$\begin{array}{c|c} 6(3x-4)-4x & 4 \\ \hline 6(3(2)-4)-4(2) & 4 \\ 6(6-4)-8 & \\ 6(2)-8 & \\ 12-8 & \\ 4 & \end{array}$$

Thus,

$$x = 2 \text{ is the correct solution.}$$

PROCEDURE:

To solve an equation that contains grouping symbols:

1. Remove all grouping symbols.

2. Combine all like terms.

3. Isolate the variable.

4. Solve the equation.

5. Check the solution.

EXAMPLE 1: Solve $12x - 4(2x + 5) = 8$ for x.

SOLUTION: Remove the parentheses.

$$12x - 4(2x + 5) = 8$$

$$12x - 4(2x + 5) = 8$$
$$12x - 8x - 20 = 8$$

Combine like terms.

$$12x - 8x - 20 = 8$$
$$\underbrace{12x - 8x} - 20 = 8$$
$$4x \quad - 20 = 8$$

Solve the equation.

$$4x - 20 = 8$$
$$4x = 8 + 20$$
$$4x = 28$$
$$x = \frac{28}{4}$$
$$x = 7$$

Example continued on next page.

Check:

$$\begin{array}{c|c} 12x - 4(2x + 5) & 8 \\ \hline 12(7) - 4(2(7) + 5) & 8 \\ 84 - 4(14 + 5) & \\ 84 - 4(19) & \\ 84 - 76 & \\ 8 & \end{array}$$

Final answer: $x = 7$

EXAMPLE 2: Solve $-12 - 12(x + 3) = 6(x - 5) - 8(x + 6)$ for x.

SOLUTION: Remove the parentheses.

$$-12 - 12(x + 3) = 6(x - 5) - 8(x + 6)$$

$$-12 - 12(x + 3) = 6(x - 5) - 8(x + 6)$$

$$-12 - 12x - 36 = 6x - 30 - 8x - 48$$

Combine like terms.

$$-12 - 12x - 36 = 6x - 30 - 8x - 48$$

$$\underbrace{-12 - 36} - 12x = \underbrace{6x - 8x} \, \underbrace{-30 - 48}$$

$$-48 \quad - 12x = \quad -2x \qquad -78$$

Solve the equation.

$$-48 - 12x = -2x - 78$$
$$-12x + 2x = -78 + 48$$
$$-10x = -30$$
$$x = \frac{-30}{-10}$$
$$x = 3$$

Check:

$$\begin{array}{c|c} -12 - 12(x + 3) & 6(x - 5) - 8(x + 6) \\ \hline -12 - 12((3) + 3) & 6((3) - 5) - 8((3) + 6) \\ -12 - 12(6) & 6(-2) - 8(9) \\ -12 - 72 & -12 - 72 \\ -84 & -84 \end{array}$$

Final answer: $x = 3$

EXAMPLE 3: Solve $2x - 2[-2(3x - 4)] - 9 = 5x - (3x - 6) + 17$ for x.

SOLUTION:

$$2x - 2[-2(3x - 4)] - 9 = 5x - (3x - 6) + 17$$
$$2x - 2[-6x + 8] - 9 = 5x - 3x + 6 + 17$$
$$2x + 12x - 16 - 9 = 5x - 3x + 6 + 17$$
$$14x - 25 = 2x + 23$$
$$14x - 2x = 23 + 25$$
$$12x = 48$$
$$x = \frac{48}{12}$$
$$x = 4$$

Check:

$2x - 2[-2(3x - 4)] - 9$	$5x - (3x - 6) + 17$
$2(4) - 2[-2(3(4) - 4)] - 9$	$5(4) - (3(4) - 6) + 17$
$8 - 2[-2(12 - 4)] - 9$	$20 - (12 - 6) + 17$
$8 - 2[-2(8)] - 9$	$20 - (6) + 17$
$8 - 2[-16] - 9$	$20 - 6 + 17$
$8 + 32 - 9$	$14 + 17$
$40 - 9$	31
31	

Final answer: $x = 4$

Do the following practice set. Check your answers with the answers in the right-hand margin.

PRACTICE SET 4-6

Solve and check (see example 1).

1. $5x + 3(4 + 2x) = -32$
2. $3 - 5x + 3(9 - 5x) = 50$
3. $15 + 5(x - 4) + 7(3x - 6) = -177$

Solve and check (see example 2 or 3).

4. $3 - 5x = 3(9 - 5x) - 4$
5. $15 + 5(x - 4) = 7(3x - 6) - (x + 8)$
6. $2(3x + 6) - 6 = 2x + [2(3x - 6) - 4]$

1. $x = -4$
2. $x = -1$
3. $x = -5$

4. $x = 2$
5. $x = 3$
6. $x = 11$

EXERCISE 4-6

Solve and check.

1. $x + (x + 4) = 10$
2. $x + (x - 6) = 10$
3. $2x - (x + 3) = 5$
4. $2x - (x - 4) = 5$
5. $(5x + 13) + 14 = 32$
6. $(-9 - 8x) - 8x = 55$
7. $2(x + 4) = 12$
8. $3(x - 8) = 0$
9. $4(3x - 4) = 92$
10. $7(4x - 16) = -252$
11. $-3(14 + 2x) = 0$
12. $-5(-8 - 4x) = 20$

Solve and check.

13. $4(x + 5) = 2x$
14. $8(x - 6) = 0$
15. $4(x - 2) = 3(x + 5)$
16. $7(x + 4) = 2(x - 6)$
17. $4(x + 2) = 2(x - 7)$
18. $5(x - 2) = 8(2 + x)$
19. $3(2x - 4) = 4(6 - 3x)$
20. $2(4x - 6) = 3(x + 6)$
21. $14x - 8 = 3(2x - 16)$
22. $-10x - 18 = -8(2x + 6)$

Solve and check.

23. $4(2x - 6) + 5 = -3$
24. $5(3x - 4) + 16 = 26$
25. $3(2x - 8) + 4x = 16$
26. $2(5x + 10) - 12x = 0$
27. $3x - 2(4x + 8) + 4 = 48$
28. $8x - 5(4 - 6x) + 15 = -75$
29. $4 - 3x + 2(8x - 4) = 35$
30. $5 - 2x + 3(4 + 6x) = -81$
31. $2 + 4(x - 6) + 5(3x + 4) = -42$
32. $4x + (3x - 5) - 2(x + 2) = 36$
33. $6(x - 5) - 4x + 3(3 + 5x) = -140$
34. $7x - 4(3 - 2x) - (2x - 5) = 21$

Solve and check.

35. $2(x + 4) - 16 = 11 - 3(x - 3)$
36. $5(x - 2) + 18 = 6 + 4(3x - 6)$
37. $x + 3(5x - 2) = 8 - 2(x - 4)$
38. $3x - 4(2x + 3) = 2(3x - 6) - 6x$
39. $3x + (2x - 5) = 13 - 2(x + 2)$
40. $4x + (3x + 6) = 12 - 4(x - 3)$
41. $2x + 3(8 - 2x) = 2x - (x - 2) - 2$
42. $4x + 4(2 - x) = 2 - 3x + 4(2 + x)$
43. $2(5 - 4x) + 6x - 8 = 3(8 - 3x) + 8$
44. $-4(2 - x) - 3 = 2(x - 6) + 4(2 - 3x) + 7$

Solve and check.

45. $3x + 5[4(2x - 6)] + 16 = 25$
46. $15 - 6x + 5[-2(x - 6) + 3x - 8] + 10 = 0$
47. $4(2x - 6) + 3[(3x - 4) - 4(2x - 8) + 6x] - 4 = 1$

48. $8 - 2(3x - 7) - 4x + 3[2x - (4 - x) - 6] = -8$

49. $2[2(3x - 4) + 6x - 4(2x + 8)] - 4x + 16 = -16$

50. $-5\{3[2(4x - 6) - 8x] - 10 + 12x\} - 4 = -74$

51. $3(2x - 5) - 2(4x + 4) - (3x + 10) = -(2x + 5) - 9$

52. $3(4x - 3) + 2[(3x - 5) - x] = 2(10 - 6x) + 4x + 9$

53. $2(3x - 6) - 4[3(2x + 8)] = 7[-3(x - 6)] + 6$

54. $-3\{2[3(4x - 6)] - 16x\} = -2[-3(2x - 4)] - 12$

4-7   SOLVING FORMULAS

In Section 3-2, we learned how to evaluate formulas. For example, to calculate the distance a person travels, we would use the formula

$$D = rt$$

where D = distance, r = rate of speed, and t = time traveled.
    Thus:

If we travel at a rate of 60 miles per hour	$D = r \cdot t$
$(r = 60)$ for 3 hours $(t = 3)$, we would travel	$D = (60)(3)$
180 miles.	$D = 180$

Suppose, however, we know the distance (D) and the time (t) and we need to know the rate of speed (r). To find the rate, we can *solve* the formula for r. To do this, we use our knowledge of solving equations.

Solve the formula, $D = rt$ for r.	
Since t is connected to r by multiplication,	$D = rt$
Divide both sides of the formula by t, to get rid of t.	$\dfrac{D}{t} = \dfrac{r^1 \not{t}}{\not{t}_1}$
	$\dfrac{D}{t} = r \cdot$

Thus,

$$r = \frac{D}{t}$$

As can be seen from this example, solving formulas for a certain "part" is the same as solving equations for a certain variable.

EXAMPLE 1:   Solve $x - y = 8$ for x.

Example continued on next page.

SOLUTION:

$$x - y = 8$$

$$\begin{array}{r} x - y = 8 \\ \underline{+y \qquad +y} \\ x = 8 + y \end{array}$$

Final answer: $x = 8 + y$

EXAMPLE 2: Solve $C = 2\pi r$ for r.

SOLUTION:

$$C = 2\pi r$$

$$\frac{C}{2\pi} = \frac{2\pi r}{2\pi}$$

$$\frac{C}{2\pi} = r$$

Final answer: $r = \dfrac{C}{2\pi}$

EXAMPLE 3: Solve $A = \dfrac{1}{2}bh$ for h.

NOTE:

$$\frac{1}{2}bh = \frac{1}{2} \cdot \frac{b}{1} \cdot \frac{h}{1}$$

$$= \frac{bh}{2}$$

SOLUTION:

$$A = \frac{1}{2}bh$$

$$A = \frac{bh}{2}$$

$$2A = bh$$

$$\frac{2A}{b} = h$$

Final answer: $h = \dfrac{2A}{b}$

EXAMPLE 4:   Solve $A = p(1 + rt)$ for t.

SOLUTION:

$$A = p(1 + rt)$$
$$A = p(1 + rt)$$
$$A = p + prt$$
$$A - p = prt$$
$$\frac{A - p}{pr} = t$$

Final answer: $t = \dfrac{A - p}{pr}$

Do the following practice set. Check your answers with the answers in the right-hand margin.

PRACTICE SET 4-7

Solve each of the following formulas for the indicated letter (see Example 1, 2, 3, or 4).

1. $A + B = 12$,   solve for B

2. $I = PRT$,   solve for R

3. $V = \dfrac{1}{3} Bh$,   solve for h

4. $T = m(g - b)$,   solve for g

1. $B = 12 - A$
2. $R = I/PT$
3. $h = 3V/B$
4. $g = (T + mb)/m$

EXERCISE 4-7

Solve each of the following formulas for the indicated letter.

1. $H + L = 19$,   for H

2. $B - A = -3$,   for B

3. $F = MA$,   for M

4. $D = 60t$,   for t

5. $C = \pi D$,   for π

6. $A = BH$,   for H

7. $V = LWH$,   for W

8. $C = 2\pi r$,   for r

9. $y = mx + b$,   for b

10. $Ax + By + C = 0$,   for y

11. $A = \dfrac{1}{2} bh$,   for b

12. $P = \dfrac{1}{3} mgh$,   for g

13. $F = \dfrac{9}{5} C + 32$,   for C

14. $F = \dfrac{1}{4} N + 40$,   for N

15. $S = \dfrac{N}{2}(a + l)$,   for a

16. $A = \dfrac{h}{2}(b + c)$,   for c

17. $C = \dfrac{5}{9}(F - 32)$,   for F

18. $P = 2(l + w)$,   for w

19. $A = p(1 + rt)$,   for t

20. $l = \dfrac{N}{2}[2a + (N - 1)d]$,   for d

4-8   LINEAR INEQUALITIES IN ONE VARIABLE

A.   Introduction: The Symbols of Inequality

1.   LESS THAN SYMBOL ($<$)

In algebra, the less than symbol is used to indicate all numbers that are smaller in value than a given number. For example, $x < 4$ means all numbers that are less than 4. To show this graphically, we do the following.

Construct a number line and circle point 4.

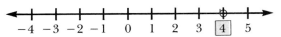

Since $x < 4$, shade all points to the left of 4. (We do not include the point 4, however, since we only want numbers that are *less* than 4.)

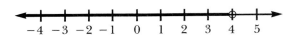

The shaded part represents all numbers that are smaller in value than 4 (e.g., 3, 2, 1, 0, −1, etc.). The open circle at point 4 shows that we do not include 4.

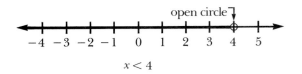

$x < 4$

2.   LESS THAN OR EQUAL TO SYMBOL ($\leq$)

In algebra, the less than or equal to symbol ($\leq$) is used to indicate all numbers that are smaller than or equal to the value of a given number. For example, $x \leq 4$ means all numbers that are less than or equal to 4. To show this graphically, we do the following.

Construct a number line and circle point 4.

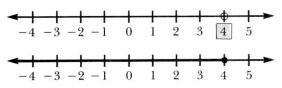

Since $x \leq 4$, shade all points to the left of 4 (including point 4).

The shaded part represents all numbers that are less than or equal to 4 (e.g., 4, $3\frac{1}{2}$, 3, −5, etc.). We show that 4 is included by a closed circle at point 4.

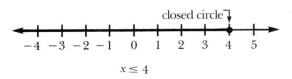

$x \leq 4$

3.   GREATER THAN SYMBOL (>)

In algebra, the greater than symbol is used to indicate all numbers that are larger in value than a given number. For example, $x > 2$, means all numbers that are greater than 2. To show this graphically, we do the following.

Construct a number line and circle point 2.

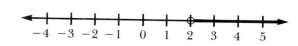

Since $x > 2$, shade all points to the right of 2. (We do not include point 2, however, since we only want numbers that are *greater than* 2.)

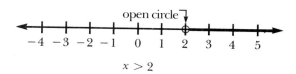

The shaded part represents all numbers that are greater in value than 2 (e.g., $2\frac{1}{2}$, $2\frac{3}{4}$, 3, 5, 8, etc.). The open circle at point 2 shows that we do not include 2.

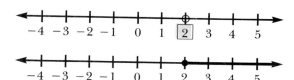

4.   GREATER THAN OR EQUAL TO SYMBOL (≥)

In algebra, the greater than or equal to symbol ($\geq$) is used to indicate all numbers that are greater than or equal to the value of a given number. For example, $x \geq 2$ means all numbers that are greater than or equal to 2. To show this graphically, we do the following.

Construct a number line and circle point 2.

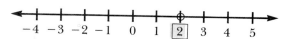

Since $x \geq 2$, shade all points to the right of 2 (including point 2).

The shaded part represents all numbers that are greater than or equal to 2 (e.g., 2, $2\frac{1}{4}$, $2\frac{3}{8}$, 5, $6\frac{1}{2}$, etc.). We show that 2 is included by a closed circle at point 2.

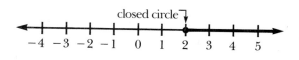

THE SENSE OF AN INEQUALITY SYMBOL

The direction to which an inequality symbol "points" is sometimes called the sense (or order). For example:

$x > y$ and $a > b$ have the same sense since both inequality symbols point in the same direction.

$x > y$
$a > b$
same sense

$x > y$ and $a < b$ have the opposite sense since each inequality symbol points in the opposite direction.

$x > y$
$a < b$
opposite sense

STATEMENTS OF INEQUALITIES

Whenever we use inequalities in a mathematical statement, the statement is simply called a statement of inequality (or just an inequality). Inequalities are very similar to equations in their makeup; they consist of algebraic expressions that are separated by an inequality symbol. For example:

■ The expression $2x + 3 < 6$ is an inequality.

■ The less than symbol separates the inequality into two sides (the left side and the right side).

$$2x + 3 \boxed{<} 6$$
left side   right side

■ The left side contains the algebraic expressions $2x + 3$.

$$\boxed{2x + 3} < 6$$

■ The right side contains the number 6.

$$2x + 3 < \boxed{6}$$

■ This inequality means that the quantity on the left, $2x + 3$, is smaller in value than the quantity on the right, 6.

$$2x + 3 < 6$$
"$(2x + 3)$ is less than 6"

B.   Solving Inequalities in One Variable

The procedure for solving inequalities is very similar to that for solving equations. The only difference is that we must be careful to maintain the correct sense (or order) of the inequality. For example, in the inequality, $6 > 2$:

If we *add the same positive number* (3) to both sides of the inequality, we obtain a true statement.

$$\begin{array}{rr} 6 > & 2 \\ +3 & +3 \\ \hline 9 > & 5 \quad (\text{True}) \end{array}$$

Notice that the sense is *not* changed.

$$\begin{array}{rr} 6 \boxed{>} & 2 \\ +3 & +3 \\ \hline 9 \boxed{>} & 5 \end{array}$$

Also:

If we *add the same negative number* (-3) to both sides of the inequality, we obtain a true statement.

$$\begin{array}{rr} 6 > & 2 \\ +(-3) & +(-3) \\ \hline 3 > & -1 \quad (\text{True}) \end{array}$$

Notice that the sense is *not* changed.

$$\begin{array}{rr} 6 \boxed{>} & 2 \\ +(-3) & +(-3) \\ \hline 3 \boxed{>} & -1 \end{array}$$

In the inequality, $9 < 16$:

If we *subtract the same positive number* (7) from both sides, we obtain a true statement.

$$\begin{array}{rr} 9 < & 16 \\ -7 & -7 \\ \hline 2 < & 9 \quad (\text{True}) \end{array}$$

Notice that the sense is *not* changed.

$$\begin{array}{rr} 9 \boxed{<} & 16 \\ -7 & -7 \\ \hline 2 \boxed{<} & 9 \end{array}$$

Also:

If we *subtract the same negative number* (-7) from both sides, we obtain a true statement.

$$\begin{array}{rcr} 9 & < & 16 \\ -(-7) & & -(-7) \\ \hline 16 & < & 23 \end{array} \quad \text{(True)}$$

Notice that the sense is *not* changed.

$$\begin{array}{rcr} 9 & < & 16 \\ -(-7) & & -(-7) \\ \hline 16 & < & 23 \end{array}$$

Thus, if the same signed number is added or subtracted to or from both sides of an inequality, the sense (or order) of the inequality is not changed. In general,

If

$$a < b$$

then

$$a + c < b + c$$

Let us now consider what happens when we multiply or divide both sides of an inequality by the same quantity. In the inequality, $2 < 5$:

If we multiply both sides by the same positive number (4), we obtain a true statement of inequality.

$$4(2) < 4(5)$$
$$8 < 20 \quad \text{(True)}$$

Notice that the sense is *not* changed.

$$4(2) < 4(5)$$
$$8 < 20$$

HOWEVER:

If we multiply both sides by the same negative number (-4), we do *not* obtain a true statement of inequality.

$$-4(2) < -4(5)$$
$$-8 < -20 \quad \text{(False)}$$

To make this statement true, we must *reverse the sense.*

$$-8 < -20 \quad \text{(False)}$$
$$-8 > -20 \quad \text{(True)}$$

In the inequality, $15 > 5$:

If we divide both sides by the same positive number (5), we obtain a true statement of inequality.

$$\frac{15}{5} > \frac{5}{5}$$
$$3 > 1 \quad \text{(True)}$$

Notice that the sense is not changed.

$$\frac{15}{5} > \frac{5}{5}$$
$$3 > 1$$

HOWEVER:

If we divide both sides by the same negative number (-5), we do *not* obtain a true statement of inequality.

$$\frac{15}{-5} > \frac{5}{-5}$$

$-3 > -1$ (False)

To make this statement true, we must *reverse the sense*.

$-3 > -1$ (False)

$-3 < -1$ (True)

Thus, when multiplying or dividing both sides of an inequality by the same *positive number*, the sense (or order) of the inequality is not changed.

However, when multiplying or dividing both sides of an inequality by the same *negative number*, the *sense* (or order) of the inequality *must be reversed*.

In general,

If

$$a < b$$

and c is positive, then

$$(a)(c) < (b)(c).$$

However, if

$$a < b$$

and c is negative, then

$$(a)(c) > (b)(c).$$

We summarize these expressions in the following chart.

When solving inequalities:

1. The same number may be added to both sides.

2. The same number may be subtracted from both sides.

3. a. Both sides may be multiplied by the same *positive* number.
   b. If both sides are multiplied by the same *negative* number, the sense of the inequality must be reversed.

4. a. Both sides may be divided by the same *positive* number.
   b. If both sides are divided by the same *negative* number, the sense of the inequality must be reversed.

PROCEDURE:

To solve an inequality in one variable:

1. Solve the inequality for the variable in the same manner as you would solve an equation.

2. Reverse the sense of the inequality symbol, if you multiply or divide both sides of an inequality by a negative number.

EXAMPLE 1:   Solve the inequality $x - 6 > 2$.

SOLUTION: To solve this inequality, we simply add 6 to both sides of the inequality.

$$x - 6 > 2$$
$$x > 2 + 6$$
$$x > 8$$

Final answer: $x > 8$

The solution, $x > 8$, means that the original inequality, $x - 6 > 2$, is only true when we replace x with numbers that are greater in value than 8. For example,

When $x = 9$: $x - 6 > 2$ When $x = 10.3$: $x - 6 > 2$
$\qquad\qquad(9) - 6 > 2$ $\qquad\qquad\qquad\quad(10.3) - 6 > 2$
$\qquad\qquad\ 9 - 6 > 2$ $\qquad\qquad\qquad\quad\ 10.3 - 6 > 2$
$\qquad\qquad\quad\ 3 > 2$ (True) $\qquad\qquad\qquad\qquad\ 4.3 > 2$ (True)

This solution is easily shown by its graph. Note the open circle at point 8.

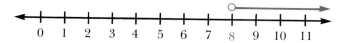

EXAMPLE 2:   Solve and graph $x + 4 \leq 6$.

Example continued on next page.

SOLUTION: Solve for x by subtracting 4 from both sides.

$$x + 4 \leq 6$$
$$x \leq 6 - 4$$
$$x \leq 2$$

Graph the solution. (Note the closed circle at 2.)

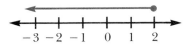

Final answer: $x \leq 2$

EXAMPLE 3: Solve and graph $6x - 6 \geq 12$.

SOLUTION: Solve for x.

$$6x - 6 \geq 12$$
$$6x \geq 12 + 6$$
$$6x \geq 18$$
$$x \geq \frac{18}{6}$$
$$x \geq 3$$

Graph the solution.

Final answer: $x \geq 3$

EXAMPLE 4: Solve and graph $3x + 9 < 5 + x$.

SOLUTION: Solve for x.

$$3x + 9 < 5 + x$$
$$3x - x < 5 - 9$$
$$2x < -4$$
$$x < \frac{-4}{2}$$
$$x < -2$$

Graph.

Final answer: $x < -2$

EXAMPLE 5:   Solve and graph $\dfrac{x}{8} \geq -3$.

SOLUTION: Solve for x.

$$\frac{x}{8} \geq -3$$

$$(8)\left(\frac{x}{8}\right) \geq (-3)(8)$$

$$x \geq -24$$

Graph.

$$-25\ -24\ -23\ -22\ -21$$

Final answer: $x \geq -24$

EXAMPLE 6:   Solve and graph $3(2x - 4) - 9x > 0$.

SOLUTION: Solve for x.

$$3(2x - 4) - 9x > 0$$
$$6x - 12 - 9x > 0$$
$$6x - 9x - 12 > 0$$
$$-3x - 12 > 0$$
$$-3x > 12$$
$$x < \frac{12}{-3}$$
$$x < -4$$

Notice when we divided by -3 we had to reverse the sense of the inequality.

Graph.

$$-7\ -6\ -5\ -4$$

Final answer: $x < -4$

EXAMPLE 7:   Solve and graph $4(3x - 6) \leq 6 + 2x$.

Example continued on next page.

SOLUTION: Solve for x.

$$4(3x - 6) \leq 6 + 2x$$
$$12x - 24 \leq 6 + 2x$$
$$12x - 2x \leq 6 + 24$$
$$10x \leq 30$$
$$x \leq \frac{30}{10}$$
$$x \leq 3$$

Graph.

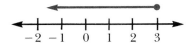

Final answer: $x \leq 3$

Do the following practice set. Check your answers with the answers in the right-hand margin.

PRACTICE SET 4-8B

Solve and graph the following inequalities (see example 1, 2, 3, or 4).

1. $x - 8 \leq 3$
2. $5x - 6 > 4$
3. $2x - 5 \geq -9 - 2x$

Solve and graph the following inequalities (see example 5).

4. $\dfrac{x}{3} > 6$

5. $-4 < -\dfrac{x}{5}$

1. $x \leq 11$

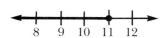

2. $x > 2$

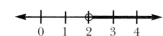

3. $x \geq -1$

4. $x > 18$

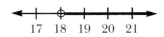

5. $20 > x$ or $x < 20$

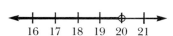

6. $3 - \dfrac{2x}{3} \le -5$

Solve and graph the following inequalities (see example 6 or 7).

7. $2(5x - 3) > 14$
8. $4x - 6 \le 3x$
9. $3(4x - 2) \ge 2(3x + 3)$

6. $x \ge 12$

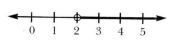

7. $x > 2$

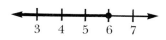

8. $x \le 6$

9. $x \ge 2$

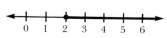

EXERCISE 4-8

Solve and graph the answers to each of the following inequalities.

1. $x - 4 < 3$

2. $x + 8 \ge 7$

3. $3x + 6 \le 12$

4. $4x - 9 < -25$

5. $15 > 5x$

6. $13 \le x + 11$

7. $9x - 12 > -30$

8. $\dfrac{x}{2} > 5$

9. $4 < \dfrac{2x}{3}$

10. $-10 \ge \dfrac{3x}{5} + 5$

Solve and graph the answers to each of the following inequalities.

11. $5x < 10 + 3x$

12. $5x + 16 < 6x$

13. $5x \ge 9x - 28$

14. $2x > 7x - 20$

15. $7x + 2 < 11 - 2x$

16. $18 - 16x \ge 6 - 4x$

17. $7x - 27 > -4x - 21$

18. $73 - 4x \le 8x - 17 - 3x$

19. $8x - 89 - 6 \ge 4x - 6x - 15$

20. $3x - 4 + x < 15 - 5x + 8$

Solve and graph the answers to each of the following inequalities.

21. $3x + 4(x - 5) \ge -48$

22. $4x + 3(2x - 8) \le 16$

23. $2x - 5(x + 4) > 1$

24. $6(2 - 4x) + 14 < -2$

25. $-12x - (7x - 10) > 28$

26. $17 \le 10 - 3(4x - 5)$

27. $\dfrac{x}{5} + \dfrac{1}{4} \ge \dfrac{27}{4}$

28. $\dfrac{x}{3} - x < 8$

29. $3[2x + 5(3 - 6x)] - 4 > -43$

30. $-2[4 - 2(x - 8) + 3x] - 2x \le 0$

Solve and graph the answers to each of the following inequalities.

31. $4(x - 2) \ge 3(x + 5)$

32. $3(2x - 4) < 4(6 - 3x)$

33. $7(x + 4) > 2(x - 6)$

34. $-8(2x + 6) \le -10x - 18$

35. $14x - 8 > 3(2x - 16)$

36. $x + 3(5x - 2) < 8 - 2(x - 4)$

37. $4x + (3x + 6) \ge 12 - 4(x - 3)$

38. $2x + 3(8 - 2x) \le 2x - (x - 2) - 2$

39. $2(x - 6) + 4(2 - 3x) + 7 > -4(2 - x) - 3$

40. $3(8 - 3x) + 8 \ge 2(5 - 4x) + 6x - 8$

REVIEW EXERCISES

Solve and check.

1. $x + 3 = 6$

2. $x - 4 = 5$

3. $8 + x = 9$

4. $17 = x + 12$

5. $-18 = x - 4$

6. $21 = 3 - x$

7. $9x = 45$

8. $3x = 16$

9. $7x = 42$

10. $38 = 2x$

11. $-14 = 5x$

12. $-6x = -36$

13. $\dfrac{1}{8}x = 10$

14. $\dfrac{1}{3}x = 12$

15. $\dfrac{1}{4}x = 8$

16. $\dfrac{5}{7}x = 10$

17. $\dfrac{4}{9}x = -2$

18. $-\dfrac{5}{6}x = 11$

19. $2x + 7 = 5$

20. $4x + 8 = -2$

21. $5x - 6 = -11$

22. $4 + 3x = -10$

23. $6 - 7x = -15$

24. $-9 - 4x = -21$

25. $8x - 15 - 5x = -20$

26. $4x + 8 + 2x = 14$

27. $4 + 2x - 6 = 10$

28. $5x = 2x + 14$

29. $5x + 34 = 9x$

30. $6x = 3x - 21$

31. $3x + 4 = 5x - 6$

32. $17 - 6x = 8x - 4$

33. $2x - 8 = 5 - 3x$

34. $\dfrac{2}{3}x - 6 = 9$

35. $\dfrac{x}{5} + 8 = 10$

36. $2 + \dfrac{1}{9}x = 6$

37. $-\dfrac{x}{5} + 7 = 8$

38. $-12 = \dfrac{3x}{5} + 4$

39. $7 = 5 - \dfrac{2x}{7}$

40. $\dfrac{2x}{7} + \dfrac{3x}{7} - 4 = 16$

41. $\dfrac{3x}{5} - \dfrac{2x}{5} + 8 = 9$

42. $5 + \dfrac{8x}{15} = \dfrac{2x}{15} + 4$

43. $6 - \dfrac{3x}{8} = \dfrac{2x}{8} + 3$

44. $2x - 6 = 21x + 5 - 4x$

45. $3x + 2x - 8 = 15x - 6$

46. $8 + 5x + 4 = 11x - 6 + 3x$

47. $7 - 2x - 6 = 5x - 6 + x$

48. $9 - 3x + 2x = 12 + 4x - 8$

49. $6x - 3 - 8x = 2x + 5 - 3x$

50. $x - 5x - 6 + 3 = 2x - 4 + 5x$

51. $13 - 2x + 1 + 6x = 8x - 5 + 3x$

52. $2(x + 4) = 16$

53. $3(2x + 6) = 9$

54. $-5(2x + 4) = 10$

55. $-6(3 + 2x) = 6$

56. $3(6x - 1) = -39$

57. $8(4 - 3x) = -8$

58. $7x - (x - 8) = 14$

59. $2x + (2x - 6) = 18$

60. $18 - 2(3x - 6) = 10$

61. $15 + (3x - 6) = 21$

62. $5 + 2(3x + 9) = 2x + 4$

63. $6 - (2x + 4) = 8 - 3x$

64. $4x - 3(5x + 6) = x - 6$

65. $7x + 4(x - 6) = 3x + 8$

66. $2(5 - 6x) + 3x = 4x + 5$

67. $3(2 - 2x) + 6x = 5 - 3x$

68. $6(2x + 5) - 3x = 4(2 - 3x) + 7$

69. $9x - (2x + 5) = 6 + (2x + 1) - 3$

70. $4 - 14(x - 2) = 6(x - 4) - 10(x + 6)$

71. $4 + 6(x + 4) = 2(3x - 6) + 5(3 - x)$

Solve and graph the answer on the number line.

72. $x - 6 > 3$

73. $14 < x + 8$

74. $4x \geq -8$

75. $-5x < 20$

76. $\dfrac{x}{4} \leq -1$

77. $2x + 7 > 15$

78. $8x - 5 < -37$

79. $3x + 4x - 6 \geq 15$

80. $8x - 10x + 5 < -1$

81. $5x - 8 \leq 17 - 10x$

82. $\dfrac{2x}{3} - 4 \leq x + 5$

83. $3x - 4(2x + 3) \geq 0$

84. $-2(3x + 4) < 2(5 - 2x)$

85. $6 + 3(-x - 4) > -3(5x - 6)$

Solve the following formulas for the indicated variable.

86. $H = \dfrac{DV}{375}$, for V

87. $S = \dfrac{1}{2}gt^2$, for g

88. $S = \dfrac{N}{2}(a + l)$, for l

89. $HP = \dfrac{D^2N}{2.5}$, for N

CHAPTER 5

Using Linear Equations in One Variable to Solve Problems

5-1 TRANSLATING VERBAL EXPRESSIONS AND SENTENCES TO THE LANGUAGE OF ALGEBRA

A. Translating English Phrases into Algebraic Expressions

One of our goals in studying algebra is to have the capability of solving problems. Thus, it will be important that we be able to translate English phrases into algebraic expressions. Stating a problem first in words is usually easier. However, it then becomes necessary to represent these "word expressions" by using the language of algebra.

For example, in arithmetic:

The English phrase *five times six* can be translated into the numerical expression 5×6.

$$\underbrace{\text{Five}}\ \underbrace{\text{times}}\ \underbrace{\text{six}}$$
$$5 \quad \times \quad 6$$

In algebra:

The English phrase, *five times a certain number*, can be translated into the algebraic expression $5x$.

$$\underbrace{\text{Five}}\ \underbrace{\text{times}}\ \underbrace{\text{a certain number}}$$
$$5 \quad \cdot \qquad (x)$$

NOTE:

> We let the variable, x, represent the unknown quantity, "a certain number."

Translating English phrases into algebraic expressions is not difficult if we know the meaning of certain key English terms. Thus, we provide the following chart as a review.

Translation Chart

The English Word or Phrase	Means to	Example Arithmetic	Algebra
Plus The sum of More than Increased by Added to	Add (+)	$4 + 2$	$x + y$
Minus Decreased by Reduced by Less Take away (2) subtracted from (4) (2) less than (4)	Subtract (−)	$4 - 2$	$x - y$
Multiply Times Product of Multiplied by Of	Multiply or ( )	$4 \cdot 2$ $4(2)$ $(4)2$ $(4)\,(2)$	$x \cdot y$ $x(y)$ $(x)y$ $(x)\,(y)$ xy
Quotient (4) Divided by (2) (2) Divides (4)	Divide ÷ or —	$4 \div 2$ $\frac{4}{2}$	$x \div y$ $\frac{x}{y}$
Twice Double Twice as much	Multiply by 2 $2 \cdot$ or $2(\ )$	$2 \cdot 4$ $2(4)$	$2 \cdot x$ $2(x)$ or $2x$
Squared—the square of	Raise to the second power	2^2	x^2
Cubed—the cube of	Raise to the third power	2^3	x^3
A certain number A variable An unknown	Use a letter to represent the quantity		x, y, z a, b, etc.

EXAMPLE 1: Translate the phrase, "six more than twice a number" into an algebraic expression.

SOLUTION:

$$\underset{6}{\underbrace{\text{Six}}}\ \underset{+}{\underbrace{\text{more than}}}\ \underset{2x}{\underbrace{\text{twice a number}}}$$

Final answer: $6 + 2x$

EXAMPLE 2:   Translate the phrase, "a number decreased by 6" into an algebraic expression.

SOLUTION:

$$\underbrace{\text{A number}}_{x} \underbrace{\text{decreased by}}_{-} \underbrace{6}_{6}$$

Final answer: $x - 6$

EXAMPLE 3:   Translate the phrase, "six more than the square of twice a number" into an algebraic expression.

SOLUTION:

$$\underbrace{\text{Six}}_{6} \underbrace{\text{more than}}_{+} \underbrace{\text{the square of twice a number}}_{(2x)^2}$$

Final answer: $6 + (2x)^2$

EXAMPLE 4:   Translate the phrase, "the square of x decreased by the quotient of y and 6" into an algebraic expression.

SOLUTION:

$$\underbrace{\text{The square of } x}_{x^2} \underbrace{\text{decreased by}}_{-} \underbrace{\text{the quotient of } y \text{ and } 6}_{\frac{y}{6}}$$

Final answer: $x^2 - \dfrac{y}{6}$

Do the following practice set. Check your answers with the answers in the right-hand margin.

PRACTICE SET 5-1A

Translate into an algebraic expression (see example 1, 2, 3, or 4).

1.	A number less 2.	1.	$x - 2$
2.	The product of an unknown and 7.	2.	$7x$
3.	The sum of a number and 5.	3.	$x + 5$

4. The product of a number squared, and 3.

5. The quantity of a number plus 6, squared, less 4.

6. The quotient of 2 and 3, quantity cubed.

4. $(x^2)(3)$ or $3x^2$

5. $(x + 6)^2 - 4$

6. $\left(\frac{2}{3}\right)^3$

B.   Translating English Sentences into Algebraic Equations

In part A, we learned how to translate English phrases into algebraic expressions. We will use the same ideas that were considered in that section for our discussion on translating English sentences into algebraic equations. For example, the English sentence, "5 times a number is equal to 30" is translated to $5x = 30$.

$$\underbrace{5 \text{ times a number}}_{5 \cdot x} \underbrace{\text{is equal to}}_{=} 30$$

$$5x = 30$$

NOTE:

> The following words (or phrases) all mean equal ($=$):
>
> | | |
> |---|---|
> | is equal to | the same as |
> | equals | the result is |
> | is | leaves |
> | equal to | makes |
> | are | was |
> | were | the product is |
> | the sum is | the quotient is |
> | the difference is | yields |

Thus, we translate English sentences into algebraic equations in the exact same manner as we translate English phrases into algebraic expressions. The only difference is that we must include the equal sign as part of our translation.

EXAMPLE 1: Translate "6 more than twice a number is 8" into an algebraic equation.

SOLUTION:

$$\underbrace{6 \text{ more than}}_{6 \quad +} \underbrace{\text{twice a number}}_{2x} \underbrace{\text{is}}_{=} 8$$

Final answer: $6 + 2x = 8$

EXAMPLE 2: Translate "5 times a certain number, divided by 3, is equal to 10" into an algebraic equation.

SOLUTION:

$$\underbrace{\text{5 times a certain number,}}_{5x} \quad \underbrace{\text{divided by 3,}}_{\div\ 3} \quad \underbrace{\text{is equal to}}_{=} \quad \underbrace{10}_{10}$$

Final answer: $5x \div 3 = 10$ or $\cdot \dfrac{5x}{3} = 10$

EXAMPLE 3: Translate "5 times the sum of a number and 9 is equal to 100" into an algebraic equation.

SOLUTION:

$$\underbrace{\text{5 times}}_{5\ \cdot} \quad \underbrace{\text{the sum of a number and 9}}_{(x+9)} \quad \underbrace{\text{is equal to}}_{=} \quad \underbrace{100}_{100}$$

Final answer: $5(x + 9) = 100$

NOTE:

> The expression "the sum of a number and 9" is treated as one quantity and is placed within parentheses.

EXAMPLE 4: Translate "the difference of 6 and 3 times a number, all divided by 12, yields a quotient of 5" into an algebraic equation.

SOLUTION:

$$\underbrace{\text{The difference of 6 and 3 times a number,}}_{(6-3x)} \quad \underbrace{\text{all divided by 12,}}_{\div\ 12} \quad \underbrace{\text{yields a quotient of 5.}}_{=\ 5}$$

Final answer: $(6 - 3x) \div 12 = 5$ or $\dfrac{(6 - 3x)}{12} = 5$

Do the following practice set. Check your answers with the answers in the right-hand margin.

PRACTICE SET 5-1B

Translate each of the following English sentences into an algebraic equation (see example 1 or 2).

1. Five more than a number is equal to 8.

2. Eleven times a number equals 21.

3. Twice a number, reduced by 7, is equal to 29.

1. $x + 5 = 8$

2. $11x = 21$

3. $2x - 7 = 29$

Translate each of the following English sentences into an algebraic equation (see example 3 or 4).

4. Twice the sum of a number and 6 is equal to 40.

5. The difference of 12 and a number, divided by 4, is equal to 10.

6. When the sum of 3 and twice a number is multiplied by 5, the result is 25.

4. $2(x + 6) = 40$

5. $(12 - x)/4 = 10$

6. $(3 + 2x)5 = 25$

EXERCISE 5-1

Translate into an algebraic expression.

1. Three plus a number.

2. A number minus 6.

3. A number plus another number.

4. The sum of a number and itself.

5. Three times x, minus 2.

6. The quantity of 12 divided by 2, times a number.

7. Twelve, divided by 2 times a number.

8. x minus 4 times y.

9. The product of 9 and x.

10. The product of 9 and x *subtracted from* 12.

11. Two multiplied by the quantity $(3 + x)$.

12. a less b.

13. x plus y, minus the *quantity* of 2 times x plus 4.

14. x plus y, minus 2 times x, plus 4.

15. The quotient of two *different* numbers.

16. The product of a number squared, and 5.

17. The difference of two numbers, divided by 2.

18. x times the difference of 5 and y.

19. Five times the quantity K less 3, divided by x, less y.

20. The quotient of x and y, quantity squared.

21. The quotient of x squared and y squared.

22. x, divided by y squared.

Translate each of the following English sentences into an algebraic equation.

23. Three more than a number equals 23.

24. Twenty-seven more than twice a number is 48.

25. Six times a number, reduced by 5, is equal to 41.

26. A number divided by 6 yields a quotient of 3.

27. Nineteen subtracted from 3 times a number yields a difference of 8.

28. Three times a number decreased by 1 equals 17.

29. Nine more than 6 times a number is equal to 63 plus twice a number.

30. Five minus twice a number is equal to 15 minus the same unknown number.

31. A number minus a second (different) number is equal to 5 times the first number.

32. One half a number equals 9.

33. Eighteen more than twice a number is equal to 1/3 the same number.

34. Three times a number, reduced by 3, is equal to 5 times the same number increased by 8.

35. A number divided by 9 is equal to 3 times the same number decreased by 5.

36. The sum of a number and 6 is equal to 38.

37. The difference of twice a number and 13 equals 29.

Translate each of the following English sentences into an algebraic equation.

38. The sum of a number and 10 is equal to 12.

39. The difference of a number and 5 is equal to 23.

40. Twenty-seven is the product of a number and 9.

41. The difference of twice a number and 9 equals 31.

42. The quotient of a number and 3 is equal to 20.

43. Twice the sum of a number and 5 is equal to 80.

44. The difference of 16 and a number, divided by 4, is equal to 5.

45. A number divided by 6, plus 23, is equal to 7.

46. Six, divided by a number plus 23, is equal to 5.

47. The sum of 4 and 3 times a number, divided by 8, is equal to 15.

48. When the sum of 8 and twice a number is multiplied by 4, the product is 20.

49. When the sum of 8 and 3 times a number is divided by 4, the quotient is 20.

50. When the difference of 8 and twice a number is multiplied by 2, the product is 14.

51. When 12 is added to 3 times a number and this sum is then multiplied by 4, the product is equal to 96.

52. When 12 is subtracted from 3 times a number and this difference is then divided by 3, the quotient is 10.

5-2   SOLVING NUMBER PROBLEMS

Equipped with the skill of translating verbal sentences into algebraic equations, we now transfer this knowledge to solving specific types of word problems. We begin by solving number problems.

EXAMPLE 1: Solve the following problem. A number decreased by 8 is 21. Find the number.

SOLUTION: We first translate this sentence into an equation.

$$\underbrace{\text{A number}}_{x} \ \underbrace{\text{decreased by 8}}_{-8} \ \underbrace{\text{is 21.}}_{=21}$$

Now we solve and check.

Solve:
$$x - 8 = 21$$
$$x = 21 + 8$$
$$x = 29$$

Check:

$x - 8$	21
$29 - 8$	21
21	

Final answer: $x = 29$

EXAMPLE 2: Solve the following problem. Twelve plus twice an unknown number is equal to 10. Find the number.

SOLUTION: First, we translate.

$$\underbrace{\text{Twelve plus}}_{12+} \ \underbrace{\text{twice an unknown number}}_{2x} \ \underbrace{\text{is equal to 10.}}_{=10}$$

Now, we solve and check.

Solve:
$$12 + 2x = 10$$
$$2x = 10 - 12$$
$$2x = -2$$
$$x = \frac{-2}{2}$$
$$x = -1$$

Check:

$12 + 2x$	10
$12 + 2(-1)$	10
$12 - 2$	
10	

Final answer: $x = -1$

EXAMPLE 3: Solve the following problem. If $\frac{2}{5}$ of a number is increased by 6, the result is 10. Find the number.

SOLUTION: Translate.

$$\underbrace{\text{If } \frac{2}{5} \text{ of a number}}_{\frac{2}{5}x} \ \underbrace{\text{is increased by 6,}}_{+6} \ \underbrace{\text{the result is 10.}}_{=10}$$

Example continued on next page.

Solve and check.

Solve: $\dfrac{2}{5}x + 6 = 10$

$\dfrac{2x}{5} = 10 - 6$

$\dfrac{2x}{5} = 4$

$2x = 4(5)$

$2x = 20$

$x = \dfrac{20}{2}$

$x = 10$

Check: | $\dfrac{2x}{5} + 6$ | 10 |
|---|---|
| $\dfrac{2(10)}{5} + 6$ | 10 |
| $\dfrac{20}{5} + 6$ | |
| $4 + 6$ | |
| 10 | |

Final answer: $x = 10$

EXAMPLE 4:   Solve the following problem. Five times a number is equal to 14 more than 3 times the same number. Find the number.

SOLUTION: Translate.

$$\underbrace{\text{Five times a number}}_{5x} \; \underbrace{\text{is equal to}}_{=} \; \underbrace{14 \text{ more than 3 times the same number.}}_{14 + 3x}$$

Solve and check.

Solve: $5x = 14 + 3x$

$5x - 3x = 14$

$2x = 14$

$x = \dfrac{14}{2}$

$x = 7$

Check: | $5x$ | $14 + 3x$ |
|---|---|
| $5(7)$ | $14 + 3(7)$ |
| 35 | $14 + 21$ |
| | 35 |

Final answer: $x = 7$

EXAMPLE 5:   Solve the following problem. The larger of two numbers is 3 times the smaller number. The sum of these two numbers is 12. Find the numbers.

SOLUTION: Before we can translate this sentence into an equation, we must represent the two unknown numbers by a variable.

"The larger of two numbers is 3 times the smaller number."

Let x = the smaller number

Now, $3x$ = the larger number

Translate.

"The sum of these two numbers is 12."

$$(x + 3x) \qquad = 12$$

Solve and check.

Solve: $(x + 3x) = 12$

$4x = 12$

$x = \dfrac{12}{4}$

$x = 3$

$3x = 9$

Check:

$x + 3x$	12
$3 + 3(3)$	12
$3 + 9$	
12	

Final answer: Smaller number is 3; larger number is 9

EXAMPLE 6: Solve the following problem. If a pair of pants cost $12 more than a shirt, what is the cost of each if two pairs of pants and four shirts total $96?

SOLUTION: We proceed as in example 5.

". . . pants cost $12 more than a shirt . . ."

Let x = the cost of the shirt

Now, $x + 12$ = the cost of a pair of pants

Translate.

". . . two pairs of pants and four shirts total $96?"

$$2(x + 12) \qquad + \qquad 4x \qquad = 96$$

Solve and check.

Solve: $2(x + 12) + 4x = 96$

$2x + 24 + 4x = 96$

$2x + 4x = 96 - 24$

$6x = 72$

$x = \dfrac{72}{6}$

$x = 12$

and $x + 12 = 24$

Check:

$2(x + 12) + 4x$	96
$2(12 + 12) + 4(12)$	96
$2(24) + 48$	
$48 + 48$	
96	

Final answer: Shirt is $12; pair of pants is $24

Do the following practice set. Check your answers with the answers in the right-hand margin.

PRACTICE SET 5-2

Solve and check (see example 1, 2, or 3).

1. Three times a number plus 4 is equal to 19. Find the number.

2. If a number is multiplied by 6, and this product is then decreased by 4, the result is −10. Find the number.

3. Two thirds of a number reduced by 4 is 4. Find the number.

Solve and check (see example 4).

4. Twice a number is equal to 15 decreased by 3 times the same number. Find the number.

5. If 20 is added to 5 times a number, the result is the same as 10 times the same number. What is the number?

6. Four times a number is increased by 1. This sum is equal to 7 times the same number increased by 25. Find the number.

Solve and check (see example 5).

7. One number is 4 times a second number. If their sum is 45, find the numbers.

8. One number is twice a second number, what are the numbers if their difference is 12?

9. One number is $\frac{3}{5}$ that of another number. If their difference is 4, what are the numbers?

Solve and check (see example 6).

10. A sweater costs $9 more than a shirt. Three shirts and two sweaters cost $108. Find the cost of each.

1. $x = 5$

2. $x = -1$

3. $x = 12$

4. $x = 3$

5. $x = 4$

6. $x = -8$

7.   Let x = first number
     Then, $4x$ = second number
     Thus, $x = 9$
     And $4x = 36$

8.   Let x = first number
     Then, $2x$ = second number
     Thus, $x = 12$
     And $2x = 24$

9.   Let x = first number
     Then, $3x/5$ = second number
     Thus, $x = 10$
     And $3x/5 = 6$

10. Let x = cost of shirt
     Then, $x + 9$ = cost of sweater
     Thus, one shirt is $18 and one sweater is $27.

11. Given two numbers. The second number is 10 more than the first number. If twice the first number is added to twice the second number, the result is equal to 40. Find the numbers.

12. Chuck is 15 years *older* than Mike. Three times Chuck's age, decreased by 15, is equal to 4 times Mike's age. Find Chuck's and Mike's age.

11.
$$\text{Let } x = \text{first number}$$
$$\text{Then, } x + 10 = \text{second number}$$
$$\text{Thus, } x = 5$$
$$\text{And } x + 10 = 15$$

12. Let x = Mike's age
Then, $x + 15 =$ Chuck's age
Thus, Mike is 30 and Chuck is 45.

EXERCISE 5-2

Translate each of the following sentences into an algebraic equation and solve the equation.

1. Ten more than a number is equal to 12. Find the number.

2. Six times a number is equal to -12. Find the number.

3. Twenty-one subtracted from a number has a difference of 4. What is the number?

4. If a number is tripled, the product is -18. What is the number?

5. If a number is reduced by 14, the difference is 16. What is the number?

6. Negative twenty-nine is the sum of -7 and a number. Find the number.

7. Twice a number minus 7 is equal to 19. Find the number.

8. The sum of twice a number and 2 is equal to -32. Find the number.

9. Twenty-eight is the difference of 3 times a number and 16. Find the number.

10. A number divided by 2, increased by 8, has a sum of 12. Find the number.

11. The quotient of a number and 3 is added to 11. This sum is equal to 9. Find the number.

12. When -8 is added to 3 times a number divided by 4, the result is 16. Find the number.

13. One number is twice a second number. If their sum is 36, find the numbers.

14. Twice a number increased by 3 times the same number has a sum of -15. What is the number?

15. One number is 4 times a second number. If their difference is 18, find the numbers.

16. Five times a number, reduced by twice the same number, has a difference of -27. Find the number.

17. The sum of twice a number, 4 times the number and 5, is equal to 23. Find the number.

18. Seven times a number is equal to 12 more than 3 times the number. What is the number?

19. Twice a number, increased by 15, is equal to 3 times the number. Find the number.

20. If a number is multiplied by 6, the product is the same as when twice the number is decreased by 12. Find the number.

21. When 4 times a number is subtracted from 120, the result is equal to 5 times the number. Find the number.

22. When 27 is subtracted from 8 times a number, the result is 5 times the number. Find the number.

23. Five times a number, increased by 12, is equal to 8 times the number less 3. Find the number.

24. Given two numbers, the second one is 8 more than the first. If 3 times the first number is increased by 5 times the second number, the sum is 80. Find the numbers.

25. Given two numbers, the second one is 12 more than the first. Twice the second number minus 4 times the first number is equal to 16. Find the numbers.

26. Given three numbers, the second one is 5 smaller than the first number, while the third number is 3 larger than the second. Twice the third number minus the sum of 3 times the second number, and twice the first number, has a difference of −4. Find the numbers.

27. Milt is 12 years older than Mike. Five times Milt's age is equal to 3 times Mike's age, increased by 116. Find the age of each person.

28. A pair of pants costs $22 more than a shirt. If two pairs of pants and three shirts cost $134, find the cost of each.

5-3 STRATEGIES FOR SOLVING WORD PROBLEMS

In Section 5-2, we learned how to solve a special type of word problem—number problems. For the most part, number problems are relatively easy to solve since they virtually consist of nothing more than a direct translation from English to algebra. Once this translation has been completed, we then solve the problem, using our knowledge of solving algebraic equations.

In the next four sections of this chapter, we will present other types of word problems (coin, mixture, investment, and motion), which may prove a bit more challenging. However, we are confident that by learning the techniques and strategies of good problem solving, you will be able to master these problems.

STRATEGIES FOR SOLVING WORD PROBLEMS

1. Read the problem, slowly and carefully, at least twice.
2. Determine the unknown quantity in the problem, and let a variable, say, x, represent this quantity.
3. Represent *any other* unknown quantity in the problem, *in terms of x*.
4. Set up a chart to organize your work.
5. Determine the relationship that is stated in the problem. Write the equation that *represents* this relationship.
6. Solve the equation.
7. Check the answer in the original problem.

5-4 SOLVING COIN PROBLEMS

In this section, we discuss a type of word problem that is commonly referred to as a *coin problem*. In order to solve coin problems successfully, we must be aware of three facts.

1. The *number* of coins we have.
2. The *value* of each coin (i.e., How much is each coin worth?).
3. The *total value* of *all* coins.

For example, if we have five quarters, then:

1. The *number* of quarters we have is 5.

five quarters
5

2. The *value* of each quarter is 25¢.

five quarters
5   25¢

3. The *total value* of all five quarters is 125¢ (or $1.25).

five quarters
5 · 25¢
125¢

NOTE:

> When solving coin problems, you will find it helpful to write all monetary values in the same unit. For example, use either all cents or all dollars.

As another example, consider twelve nickels.

1. The *number* of nickels is 12.

twelve nickels
12

2. The *value* of each nickel is 5¢.

twelve nickels
12   5¢

3. The *total value* of all 12 nickels is 60¢.

twelve nickels
12 · 5¢
60¢

Using these two examples, we can place the given information into a chart.

Type of Coin	Number of Coins	Value of Each Coin	Total Value of Each Coin
Quarter	5	25¢	$(5)(25) = 125$¢
Nickel	12	5¢	$(12)(5) = 60$¢

From the table, note that we have the following relationship.

(The number of coins) · (The value of each coin) = (The total value of the coins)

$$N \qquad \cdot \qquad V \qquad = \qquad T$$

We now use this relationship to solve coin problems.

EXAMPLE 1: Kelly received a play purse from her grandfather that contained 25 coins worth $2.00. If the purse contains only nickels and dimes, how many of each coin are in the purse?

STEP 1: READ THE PROBLEM CAREFULLY.

STEP 2: DETERMINE THE UNKNOWN QUANTITY AND SET x EQUAL TO IT.

There are *two* unknown quantities.

■ The number of nickels in the purse
■ The number of dimes in the purse

Let x = the number of nickels in the purse

STEP 3: REPRESENT ANY OTHER UNKNOWN QUANTITY IN TERMS OF x.

The other unknown quantity is the number of dimes in the purse. Since there are a total of 25 coins in the purse, we have

$25 - x$ = the number of dimes in the purse

STEP 4: SET UP A CHART.

We now organize our work by setting up a chart that shows the $N \cdot V = T$ relationship.

Type of Coin	Number of Each Coin	Value of Each Coin	Total Value
Nickels	x	5¢	$5x$
Dimes	$25 - x$	10¢	$10(25 - x)$

STEP 5: DETERMINE THE RELATIONSHIP OF THE PROBLEM.

The relationship used in this problem is:

$$\text{The total value of the nickels} + \text{The total value of the dimes} = \text{The total value of all 25 coins}$$

$$5x \quad + \quad 10(25 - x) \quad = \quad 200¢$$

NOTE:

1. The total value of the nickels and dimes is readily identified from the chart.
2. The total value of all 25 coins is $2.00, which comes from the problem but is expressed in cents.

STEP 6: SOLVE THE EQUATION.

The equation to be solved is $5x + 10(25 - x) = 200$

$$5x + 10(25 - x) = 200$$
$$5x + 250 - 10x = 200$$
$$5x - 10x = 200 - 250$$
$$-5x = -50$$
$$x = \frac{(-50)}{(-5)}$$
$$x = 10$$

Thus,

$$x = 10 \quad \text{and} \quad 25 - x = 15$$

STEP 7: CHECK.

- Ten nickels are worth 50¢.
- Fifteen dimes are worth 150¢.
- The total is 25 coins that are worth 200 cents.

$(5¢)(10) = 50¢$

$(10¢)(15) = 150¢$

$$
\begin{array}{rl}
10 \text{ nickels} = & 15¢ \\
15 \text{ dimes} = & 150¢ \\
\hline
25 \text{ coins} = & 200¢
\end{array}
$$

Final answer: 10 nickels; 15 dimes

EXAMPLE 2: LuAnn has $7.70 consisting of dimes and quarters. The number of quarters is two more than twice the number of dimes. How many of each kind does she have?

STEP 1: READ THE PROBLEM CAREFULLY.

STEP 2: DETERMINE THE UNKNOWN QUANTITY, AND SET x EQUAL TO IT.

There are two unknown quantities.

- The number of dimes
- The number of quarters

Let x = The number of dimes

STEP 3: REPRESENT ANY OTHER UNKNOWN IN TERMS OF x.

The other unknown quantity is the number of quarters. The number of quarters is

"... *two more than* *twice the number of dimes*"
$2 +$ $2x$

Therefore,

$2 + 2x$ = The number of quarters

Example continued on next page.

STEP 4: SET UP A CHART.

		$N \cdot V = T$	
Type of Coin	The Number of Each Coin	The Value of Each Coin	Total Value
Dimes	x	10¢	$10x$
Quarters	$2 + 2x$	25¢	$25(2 + 2x)$

STEP 5: DETERMINE THE RELATIONSHIP OF THE PROBLEM.

The relationship used in this problem is:

The total value of dimes + The total value of quarters = $7.70

$$10x \qquad + \qquad 25(2 + 2x) \qquad = 770$$

STEP 6: SOLVE THE EQUATION.

$$10x + 25(2 + 2x) = 770$$
$$10x + 50 + 50x = 770$$
$$10x + 50x = 770 - 50$$
$$60x = 720$$
$$x = \frac{720}{60}$$
$$x = 12$$

Thus,

$$x = 12 \quad \text{and} \quad 2 + 2x = 26$$

STEP 7: CHECK.

- Twelve dimes are worth 120¢. $(12)(10¢) = 120¢$
- Twenty-six quarters are worth 650¢. $(26)(25¢) = 650¢$
- The total is $7.70. $120¢ + 650¢ = 770¢$
                                                 or 7.70

Final answer: 12 dimes; 26 quarters

EXAMPLE 3: Garfield Senior High School's music department sold tickets for a concert recital. Tickets cost $1 if purchased in advance and $2.50 if purchased at the door. If the total number of tickets sold was 488, and the amount of money received was $800, how many tickets of each kind were sold?

STEP 1: READ THE PROBLEM CAREFULLY.

STEP 2: DETERMINE THE UNKNOWN QUANTITY, AND SET x EQUAL TO IT.

There are two unknown quantities.

- The number of tickets sold in advance
- The number of tickets sold at the door

Let x = the number of tickets sold in advance

STEP 3: REPRESENT ANY OTHER UNKNOWN IN TERMS OF x.

Since 488 tickets were sold, and x represents the number of tickets sold in advance,

$488 - x$ = the number of tickets sold at the door

STEP 4: SET UP A CHART.

		$N \cdot V = T$	
Type of Ticket	Number Sold	Price of Each Ticket	Total Value
Advance	x	$1.00	$1x$
Door	$488 - x$	$2.50	$2.5(488 - x)$

STEP 5: DETERMINE THE RELATIONSHIP OF THE PROBLEM.

The relationship used in this problem is:

The total value of tickets sold in advance	+	The total value of tickets sold at the door	=	The total amount of money received
x	+	$2.5(488 - x)$	=	$800

STEP 6: SOLVE THE EQUATION.

$$x + 2.5(488 - x) = 800$$
$$x + 1220 - 2.5x = 800$$
$$x - 2.5x = 800 - 1220$$
$$-1.5x = -420$$
$$x = \frac{(-420)}{(-1.5)}$$
$$x = 280$$

Thus,

$$x = 280 \quad \text{and} \quad 488 - x = 208$$

STEP 7: CHECK.

- 280 tickets at $1 each are $280. $(280)(\$1) = \$280$
- 208 tickets at $2.50 each are $520. $(208)(\$2.50) = \$520$

Example continued on next page.

■ The total receipts are $800. $280 + $520 = $800

Final answer: 280 advance tickets; 208 door tickets

EXERCISE 5-4

Solve the following problems.

1. Mou-ta has an equal number of dimes and nickels in his pocket that total 90¢. How many of each coin are there?

2. Rosita has 17 coins in her change purse that have a total value of $1.15. If these coins consist of only nickels and dimes, how many of each coin are there?

3. Dale has 3 times as many quarters as dimes, totaling $6.80. How many of each coin does he have?

4. Pamela has 5 times as many nickels as quarters, totaling $4.00. How many of each coin does she have?

5. In her cash drawer, Nancy has $10.40 in nickels, dimes, and quarters. The number of nickels is twice the number of quarters and the number of dimes is 3 times the number of quarters. How many of each coin does she have?

6. Ron, the meter bandit, broke open a parking meter. When he organized the coins, he found that the number of nickels was 5 times the number of dimes and that there were 15 more quarters than dimes. If the total amount of money taken from the meter was $19.95, how many of each coin were there?

7. A vending machine employee collected $24.75 from one vending machine. The number of quarters was 23 more than the number of dimes and there were 12 more dimes than nickels. How many of each coin were there?

8. A coin vender must fill a dollar change machine with $50.00 worth of coins. The number of quarters he puts in is 50 more than the number of dimes, and the number of dimes he puts in is twice the number of nickels. How many of each coin does he put into the machine?

9. At a recent concert, 10,000 tickets were sold. The tickets had sold for $8.50, $6.50, and $4.50. There were an equal number of $8.50 and $6.50 tickets sold and twice as many $4.50 tickets sold as $8.50 tickets. If the total receipts were $60,000, how many of each kind were sold?

10. A movie theatre sells matinee tickets for $1.50, children under 12 tickets for $3.00, and adult tickets for $4.50. At the end of one day, the theatre sold 2384 tickets. There were 5 times as many adult tickets sold than matinee tickets and 200 more children tickets than matinee tickets. If the total receipts were $9024, how many of each kind were sold?

11. Gladys bought 60 postage stamps and paid $12.70 for them. Some were 15 cent stamps and some were 25 cent stamps. How many of each kind did she buy?

12. Dick, the postal clerk, sold 80 stamps for $19.10. Some were 20 cent stamps and some were 30 cent air grams. How many of each kind did he sell?

13. Carmen spent $33.74 for 100 stamps. She bought only 28 cent, 35 cent, and 40 cent airmail stamps. If there were twice as many 28 cent stamps as 40 cent stamps, how many of each kind did she buy?

14. Julio spent $7.50 for 45 stamps. He bought only 10 cent, 20 cent, and 50 cent stamps. If there were 15 more 20 cent stamps than 50 cent stamps, how many of each kind did he buy?

5-5 SOLVING MIXTURE PROBLEMS

Mixture problems are similar to coin problems, in that they both deal with combining two or more items (each having a certain unit value) to form a mixture. Mixture problems are solved basically the same as coin problems.

NOTE:

> Coin problems involve the mixing of a *specific* item—money. Mixture problems involve the mixing of *general* ingredients (or items), for example, chemicals, candy, "mixed" nuts, and so on.

EXAMPLE 1: Rochester Bakery is having a sale on two of its biggest selling cookies—peanut butter and chocolate chip. Peanut butter cookies are selling for 95¢ per pound and chocolate chip cookies are selling for 70¢ per pound. If a customer requests a mixture of the two cookies, the price is 80¢ per pound. How many pounds of each kind must be used if a customer orders a mixture of 20 pounds?

STEP 1: READ THE PROBLEM CAREFULLY.

STEP 2: DETERMINE THE UNKNOWN QUANTITY AND SET x EQUAL TO IT.

There are two unknowns.

- The number of pounds of peanut butter cookies that are needed for the mixture
- The number of pounds of chocolate chip cookies that are needed for the mixture

Let x = the number of pounds of peanut butter cookies needed

STEP 3: REPRESENT ANY OTHER UNKNOWN IN TERMS OF x.

Since the mixture requires 20 pounds and x equals the number of pounds of peanut butter cookies,

$20 - x$ = the number of pounds of chocolate chip cookies

STEP 4: SET UP A CHART.

Type of Cookies	Number of Pounds Needed	Price per Pound	Total Value
Peanut butter	x	95¢	$95x$
Chocolate chip	$20 - x$	70¢	$70(20 - x)$
Mixture	20	80¢	$80(20)$

Example continued on next page.

NOTE:

> The last row of our chart indicates the mixture we need.

STEP 5: DETERMINE THE RELATIONSHIP IN THE PROBLEM.

The relationship in this problem is:

$$\underset{95x}{\text{The total value of peanut butter cookies}} + \underset{70(20-x)}{\text{The total value of chocolate chip cookies}} = \underset{80(20)}{\text{The total value of the mixture}}$$

STEP 6: SOLVE THE EQUATION.

$$95x + 70(20 - x) = 80(20)$$
$$95x + 1400 - 70x = 1600$$
$$95x - 70x = 1600 - 1400$$
$$25x = 200$$
$$x = \frac{200}{25}$$
$$x = 8$$

Thus,

$$x = 8 \quad \text{and} \quad 20 - x = 12$$

STEP 7: CHECK.

■ Eight pounds of peanut butter cookies at 95¢ per pound is $7.60. $8(95¢) = \$7.60$

■ Twelve pounds of chocolate chip cookies at 70¢ per pound is $8.40. $12(70¢) = \$8.40$

Compare ⎡→■ Twenty pounds of the mixture at 80¢ per pound is $16.00. $20(80¢) = \$16.00$

⎣→■ The total value is $16.00. $\$7.60 + \$8.40 = \$16.00$

Final answer: 8 pounds peanut butter cookies; 12 pounds chocolate chip cookies

---

EXAMPLE 2:   Hi A contains 15 percent pure apple juice. Alaskan Treat contains 40 percent pure apple juice. How much of each kind of drink should be used to make a 100-quart mixture that is 25 percent pure apple juice?

STEP 1:   READ THE PROBLEM CAREFULLY.

STEP 2:   DETERMINE THE UNKNOWN QUANTITY AND SET x EQUAL TO IT.

There are two unknowns.

- The number of quarts of Hi A needed
- The number of quarts of Alaskan Treat needed

Let x = number of quarts of Hi A needed

STEP 3:   REPRESENT ANY OTHER UNKNOWN IN TERMS OF x.

Since we want a 100-quart mixture, and x equals the number of quarts of Hi A needed.

$100 - x$ = the number of quarts of Alaskan Treat needed

STEP 4:   SET UP A CHART.

Type of Drink	Number of Quarts	Part Pure Apple Juice	Amount of Pure Apple Juice
Hi A	x	15%	15% of x
Alaskan Treat	$100 - x$	40%	40% of $(100 - x)$
Mixture	100	25%	25% of 100

STEP 5:   DETERMINE THE RELATIONSHIP IN THE PROBLEM.

The relationship in this problem is:

The total number of quarts of pure apple juice contained in Hi A.	+	The total number of quarts of Alaskan Treat needed with 40% pure apple juice.	=	The total number of quarts of the mixture with 25% pure apple juice.
15% of x	+	40% of $(100 - x)$	=	25% of 100
or $0.15x$	+	$0.40(100 - x)$	=	$0.25(100)$

STEP 6:   SOLVE THE EQUATION.

$$0.15x + .40(100 - x) = 0.25(100)$$
$$0.15x + 40 - 0.40x = 25$$
$$0.15x - 0.40x = 25 - 40$$
$$-.25x = -15$$
$$x = \frac{(-15)}{(-0.25)}$$
$$x = 60$$

Thus,

$$x = 60 \quad \text{and} \quad 100 - x = 40$$

Example continued on next page.

STEP 7: CHECK.

■ Fifteen percent of 60 quarts is 9. $(0.15)(60) = 9$

■ Forty percent of 40 quarts is 16. $(0.40)(40) = 16$

Compare

■ Twenty-five percent of 100 quarts is 25. $(0.25)(100) = 25$

■ The total part of apple juice is 25. $9 + 16 = 25$

Final answer: 60 quarts of Hi A; 40 quarts of Alaskan Treat

EXAMPLE 3: A brand of gasoline contains a 20 percent solution of pure alcohol. Jim, the brilliant chemist, claims his car can run on gasoline that contains about a 35 percent solution of pure alcohol. Approximately, how much pure alcohol must Jim add to 15 gallons of this brand of gasoline to produce a mixture that is 35 percent pure alcohol?

STEP 1: READ THE PROBLEM CAREFULLY.

STEP 2: DETERMINE THE UNKNOWN QUANTITY AND SET x EQUAL TO IT.

There is only one unknown quantity—the number of gallons of pure alcohol that must be added to produce a mixture that is 35 percent pure alcohol.

Let x = the number of gallons of pure alcohol needed

NOTE:

Jim originally has 15 gallons of gasoline. When he adds x number of gallons of pure alcohol to get his desired mixture, the total number of gallons of the mixture is $15 + x$.

STEP 4: SET UP A CHART.

Type of Solution	Number of Gallons	Part Pure Alcohol	Amount of Pure Alcohol
Original gasoline	15	20%	20% of 15
Pure alcohol	x	100%	100% of x
Mixture	$15 + x$	35%	35% of $(15 + x)$

STEP 5: DETERMINE THE RELATIONSHIP IN THE PROBLEM.

The relationship in this problem is:

The amount of pure alcohol in original gasoline	+	The amount of pure alcohol that is being added	=	The amount of alcohol that is required for the mixture
(20% of 15)	+	(100% of x)	=	(35% of $(15 + x)$)

or $(0.20)(15)$ + $(1.00)(x)$ = $(0.35)(15 + x)$

STEP 6: SOLVE THE EQUATION.

$$(0.20)(15) + (1.00)(x) = (0.35)(15 + x)$$
$$3 + x = 5.25 + 0.35x$$
$$x - 0.35x = 5.25 - 3$$
$$.65x = 2.25$$
$$x = \frac{2.25}{0.65}$$
$$x = 3.46$$

STEP 7: CHECK.

To check this, we must satisfy the following fraction.

$$\frac{\text{The total number of gallons of pure alcohol}}{\text{The total number of gallons of entire mixture}} = 35\%$$

■ The total number of gallons of pure alcohol comes from our equation in step 5.

$$0.35(15 + x)$$
$$= 0.35(15 + 3.46)$$
$$= 0.35(18.46)$$
$$= 6.461$$

■ The total number of gallons of the entire mixture can be found by adding 15 gallons of gasoline (which is the initial amount) to 3.46 (which is the amount of alcohol that will be added to the gasoline).

$$15 + 3.46 = 18.46$$

Now,

$$\frac{6.461}{18.46} = 0.3499 \approx 0.35 \text{ or } 35\%$$

Final answer: 3.46 gallons

EXERCISE 5-5

Solve the following problems.

1. Fenby's Groceries sells a homemade natural food snack consisting of sunflower seeds and wheat germ nuggets. The sunflower seeds cost Fenby 75¢ per pound and the wheat germ nuggets cost him $1.25 per pound. How much of each kind must Fenby mix if he wants to make a mixture of 50 pounds to sell at $1 per pound?

2. How many pounds of macaroni that sells for 60¢ per pound and how many pounds of beans that sell for 90¢ per pound must be mixed to produce 30 pounds of a mixture, which would sell for 70¢ per pound?

3. Prinello Oil Company wants to mix regular unleaded gasoline, which sells for $1.50 per gallon and super unleaded gasoline, which sells for $1.68 per gallon, to produce 400 gallons of a "super regular" gasoline that would sell for $1.56 $\frac{3}{10}$ per gallon. How many gallons of each kind must the company mix?

4. A & M Garage mixes 10W30 oil worth 60 cents a quart and 10W40 oil worth $1.05 per quart to produce a mixture that sells for 75 cents per quart. If they sell 300 gallons, how many quarts of each kind do they use? (*Hint*: Be certain to use the same units of measure.)

5. A generic brand of mixed nuts contains cashews, filberts, almonds, and peanuts. The cashews cost $3.00 per pound, the filberts cost $1.80 per pound, the almonds cost $1.90 per pound, and the peanuts cost $1.50 per pound.
   a. How many pounds of cashews must be added to a mixture consisting of 20 pounds of filberts, 10 pounds of almonds, and 30 pounds of peanuts to produce a mixture that would sell for $2.20 per pound?
   b. Is there less than 50 percent peanuts in this mixture?

6. Union Meat Market makes up a special meat loaf that consists of ground beef, veal, and pork. The ground beef costs the store $2.20 per pound, the veal costs $4.10 per pound, and the pork costs $1.80 per pound. How much veal should the store add to 60 pounds of ground beef and 40 pounds of pork to be able to sell the meat loaf for $2.10 per pound?

7. Hazel's special grass seed mix contains 35 percent Kentucky blue grass, whereas Crossman's grass seed mix contains 80 percent Kentucky blue grass. How many ounces of each brand must be mixed to produce a 3-pound mixture that contains 65 percent Kentucky blue grass?

8. Apple Drink contains 19 percent pure apple juice and Apple Tang contains 35 percent pure apple juice. How much of each kind is needed to produce 128 fluid ounces of a drink that will contain 25 percent pure apple juice?

9. A 1-ounce serving of Wheatmate cereal contains 50 percent of the U.S. required daily allowance (U.S. RDA) of Vitamin A. An equal amount of Cornhusk cereal contains only 15 percent of the U.S. RDA of Vitamin A. How many ounces of Wheatmate must be added to 8 ounces of Cornhusk in order to produce a mixture that will yield 25 percent of the U.S. RDA of Vitamin A?

10. One liter of 80 proof Wild Turkey contains 40 percent pure alcohol, whereas 1 liter of 101 proof Wild Turkey contains 50.5 percent pure alcohol. How many ounces of the 80 proof Turkey must be mixed with 5 ounces of the 101 proof Turkey to produce a mixture that contains 45 percent pure alcohol?

11. A certain brand of Scotch contains 45 percent pure alcohol. If Cynthia wishes to dilute 16 ounces of this Scotch with water so that her drink contains 30 percent pure alcohol, how much water must she add?

12. A certain antifreeze solution contains 60 percent pure alcohol. Donny needs to dilute this antifreeze with water so that the antifreeze will contain 45 percent pure alcohol. How much water will he need to dilute 6 quarts of antifreeze?

13. Jim must "cut" his wood finish with lacquer thinner before he applies the first coat of finish to his hardwood floors. If the wood

14. Gallo's Florist sells a "Friday bunch" for $17.10. The bunch consists of roses at $22 per dozen and carnations at $15 per dozen. What

finish he uses contains 75 percent volatile (the main ingredients of the finish), how many quarts of thinner will Jim need to "cut" 2 *gallons* of wood finish to 50 percent volatile?

is the total number of dozens of each flower used if the shop makes up 100 bunches?

5-6 SOLVING INVESTMENT PROBLEMS

In solving investment problems, we use the interest formula

$$I = P \cdot R \cdot T$$

where $I =$ interest (the amount of money earned from an investment), $P =$ principal (the amount of money invested, given in dollars), $R =$ rate of interest (the percent paid on the principal; always expressed as a percent and always for 1 year), and $T =$ time (the length of time the principal has been invested).

For example:

If Mr. Pascale invests $5000 at 15% for $2\frac{1}{2}$ years, his interest earned is $1875.

$$I = P \cdot R \cdot T$$
$$I = (5000)(.15)(2.5)$$
$$I = \$1875$$

In this section, we discuss only *annual* investments; that is, time will be for one year. Thus:

The formula is simplified to $I = P \cdot R$

$$I = P \cdot R \cdot T$$
$$I = P \cdot R \cdot (1)$$
$$I = P \cdot R$$

EXAMPLE 1: Laurie invested a sum of money at 12 percent and a second sum, twice as much as the first, at 15 percent. The total annual interest she earned was $1050. How much did she invest at each rate?

STEP 1: READ THE PROBLEM CAREFULLY.

STEP 2: DETERMINE THE UNKNOWN QUANTITY AND SET *x* EQUAL TO IT.

There are two unknown quantities.

■ The amount of money (or principal) invested at 12 percent

■ The amount of money (or principal) invested at 15 percent

Let $x =$ amount of money invested at 12%

STEP 3: REPRESENT ANY OTHER UNKNOWNS IN TERMS OF *x*.

The amount of money invested at 15 percent was

"*twice as much as the first*"

$2x$

Example continued on next page.

Thus,

$$2x = \text{amount of money invested at } 15\%$$

STEP 4: SET UP A CHART THAT SHOWS THE $I = P \cdot R$ RELATIONSHIP.

	P $\cdot$	R =	I
Investment	*Principal*	*Rate*	*Total Interest*
12% investment	x	12%	$0.12x$
15% investment	$2x$	15%	$0.15(2x)$

STEP 5: DETERMINE THE RELATIONSHIP IN THE PROBLEM.

The relationship in this problem is:

The amount of interest earned + from the 12% investment	The amount of interest earned = $1050 from the 15% investment
$0.12x$ +	$0.15(2x)$ = 1050

STEP 6: SOLVE THE EQUATION.

$$0.12x + 0.15(2x) = 1050$$
$$0.12x + 0.3x = 1050$$
$$0.42x = 1050$$
$$x = \frac{1050}{0.42}$$
$$x = 2500$$

Thus,

$$x = 2500 \quad \text{and} \quad 2x = 5000$$

STEP 7: CHECK.

■ The interest earned on $2500 at 12% is $300.
$$I = P \cdot R$$
$$I = (2500)(.12)$$
$$I = 300$$

■ The interest earned on $5000 at 15% is $750.
$$I = P \cdot R$$
$$I = (5000)(0.15)$$
$$I = 750$$

■ The total interest earned from both investments is $1050.
$$300 + 750 = 1050$$

Final answer: $2500 was invested at 12%; $5000 was invested at 15%

EXAMPLE 2: Mr. Moran invested $5000, part at 6 percent and the other part at 11 percent. The total annual interest he received was $475. How much did he invest at each rate?

STEP 1: READ THE PROBLEM CAREFULLY.

STEP 2: DETERMINE THE UNKNOWN QUANTITY AND SET x EQUAL TO IT.

There are two unknowns.

- The amount of money (principal) invested at 6 percent
- The amount of money (principal) invested at 11 percent

Let x = the amount of money invested at 6 percent

STEP 3: REPRESENT ANY OTHER UNKNOWN IN TERMS OF x.

Since Mr. Moran invested $5000, and x represents the amount invested at 6%,

$5000 - x$ = the amount invested at 11 percent

STEP 4: SET UP A CHART.

	P $\cdot$	R =	I
Investment	*Principal*	*Rate*	*Interest*
6% investment	x	6%	$0.06x$
11% investment	$5000 - x$	11%	$0.11(5000 - x)$

STEP 5: DETERMINE THE RELATIONSHIP IN THE PROBLEM.

The relationship in this problem is:

The total interest received from 6% investment	+	The total interest received from 11% investment	=	The total interest he received ($475)
$0.06x$	+	$0.11(5000 - x)$	=	475

STEP 6: SOLVE THE EQUATION.

$$0.06x + 0.11(5000 - x) = 475$$
$$0.06x + 550 - 0.11x = 475$$
$$.06x - 0.11x = 475 - 550$$
$$-.05x = -75$$
$$x = \frac{(-75)}{(-0.05)}$$
$$x = 1500$$

Example continued on next page.

Thus,

$$x = 1500 \quad \text{and} \quad 5000 - x = 3500$$

STEP 7: CHECK.

■ The total interest received from 6 percent investment is $90.

$$I = P \cdot R$$
$$I = (1500)(0.06)$$
$$I = 90$$

■ The total interest received from 11 percent investment is $385.

$$I = P \cdot R$$
$$I = (3500)(0.11)$$
$$I = 385$$

■ The total interest received from both investments
is $475.

$$90 + 385 = 475$$

Final answer: $1500 was invested at 6%; $3500 was invested at 11%

EXAMPLE 3:   Mrs. Mayer has $10,000 to invest. She has decided to invest part of it in treasury cer-
tificates and the remaining part in a family stock. The treasury certificates yield an
annual rate of interest of 13 percent, and the family stock yields 5 percent. If she wants
a total yield of 9 percent, how much should she invest in each?

STEP 1:   READ THE PROBLEM CAREFULLY.

STEP 2:   DETERMINE THE UNKNOWN QUANTITY AND SET x EQUAL TO IT.

There are two unknown quantities.

The amount of money to be invested at 13 percent

■ The amount of money to be invested at 5 percent

Let x = amount of money to be invested at 13 percent

STEP 3:   REPRESENT ANY OTHER UNKNOWN IN TERMS OF x.

Since Mrs. Mayer has $10,000 to invest, and x represents the amount to be invested
at 13 percent,

$10,000 - x$ = the amount to be invested at 5 percent

STEP 4:   SET UP A CHART.

| Investment | P · | R = | I |
	Principal	Rate	Interest
Treasury certificates	x	13%	$0.13x$
Family stock	$10,000 - x$	5%	$0.05(10,000 - x)$
Combination	10,000	9%	$0.09(10,000)$

NOTE:

> This is very similar to a mixture problem. We are combining the two investments to yield a total rate of 9 percent.

STEP 5: DETERMINE THE RELATIONSHIP IN THE PROBLEM.

The relationship in this problem is:

The interest received at 13%	+	The interest received at 5%	=	The total interest received at 9%
$0.13x$	+	$0.05(10,000 - x)$	=	$0.09(10,000)$

STEP 6: SOLVE THE EQUATION.

$$0.13x + 0.05(10000 - x) = 0.09(10000)$$
$$0.13x + 500 - 0.05x = 900$$
$$0.13x - 0.05x = 900 - 500$$
$$0.08x = 400$$
$$x = \frac{400}{0.08}$$
$$x = 5000$$

Thus,

$$x = 5000 \quad \text{and} \quad 10000 - x = 5000$$

STEP 7: CHECK.

■ The interest received from $5000 at 13 percent is $650.

$I = P \cdot R$
$I = (5000)(0.13)$
$I = 650$

■ The interest received from $5000 at 5 percent is $250.

$I = P \cdot R$
$I = (5000)(.05)$
$I = 250$

Compare

■ The total interest received from $10,000 at 9 percent is $900.

$I = P \cdot R$
$I = (10,000)(.09)$
$I = 900$

■ The total interest received from both investments is $900.

$650 + 250 = 900$

Final answer: $5000 in treasury certificates; $5000 in family stock

---

EXERCISE 5-6

Solve the following problems.

1. Bill invested part of $1000 at 6 percent and the remaining part at 10 percent. If the total

2. A sum of $25,000 is invested, part at 5 percent and the remainder at 12 percent. The total

annual income from these investments is $86, how much was invested at each rate?

3. Mrs. Merz invested a sum of money at 5 percent. She also invested $500 more than the first investment at 7 percent. At the end of the year, her total statement of interest income was $73.40. How much did she invest at each rate?

NOTE:

> At the end of each year, anyone who has money invested in a savings account receives a *statement of interest income* from the bank. This statement reflects the amount of money earned from those investments and must be reported to the Internal Revenue Service (IRS) for tax purposes.

annual income from both investments was $2510. Find the amount invested at each rate.

4. Barbara has two separate savings accounts at one bank. One account earns $5\frac{3}{4}$ percent and the other earns 14 percent. The amount of money invested in the 14 percent account is 3 times the amount invested in the $5\frac{3}{4}$ percent account. Using the statement of interest income below, how much money does Barbara have invested at each rate? (*Note:* Use the total earnings reported to IRS as the total interest earned from both accounts.)

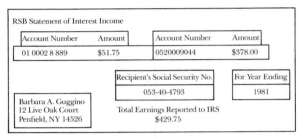

5. Dr. Faux, an investor, inherited a large sum of money. A co-worker gave him a tip to invest this money in a stock that *could* yield as high as 22 percent. Being the conservative type, and also not knowing if the tip was good, Faux decided to play it safe and invested part of the money into a savings account at 6 percent and $2500 *less* into the stock. If the total amount of interest he received at the end of one year was $270, and the stock yielded 18 percent, how much did Dr. Faux invest at each rate?

6. Ms. Hustler earned a considerable amount of money "under the table." She invested part of it into treasury certificates from which she received a return of 12 percent. She then invested $500 less than her first investment in a growth stock that earned $8\frac{1}{2}$ percent. If the total amount of interest earned was $613.50, how much did she invest at each rate?

7. Mr. Johnson invested a sum of money at 8 percent, a second sum, $1000 more than the first, at 15 percent, and yet a third sum, $300 less than the first, at 12 percent. His total annual interest was $569.
   a. How much did he invest at each rate?
   b. What was the total amount of money that Mr. Johnson invested?

8. Mrs. Petrie invested a sum of money that returned 14 percent. She invested a second sum, which was $500 more than twice the first sum at 10 percent. If her total annual income was $322, how much did she invest at each rate?

9. Mr. Danko invested $6000, part at 6 percent and the rest at 11 percent, so that the total return amounted to 9 percent of the principal. How much did he invest at each rate?

10. Mrs. Herbert invested $2500, part at $6\frac{1}{2}$ percent and the remaining part at 9 percent, so that the total return amounted to $8\frac{1}{2}$ percent of the principal. How much did she invest at each rate?

11. Margo is confronted with a problem. She has $3000 to invest and would like very much to invest it in a $2\frac{1}{2}$-year-term account with an annual yield of 14 percent. However, she does not want the money tied up for such a long period of time. Thus, she decides to invest part of it into a savings account with an annual yield of only 6 percent. How should she divide this investment so that her overall annual yield is 12 percent?

12. Rose has $1500 to invest. She wants to invest part in a savings bond with a yield of 9 percent and the remaining part in a credit union account with a yield of $6\frac{1}{2}$ percent. How much should she invest in each so her total annual yield is 8 percent?

13. Ron, the investor, invested $800 in two parts. The first part was invested at 8 percent and the second part at 6 percent. The amount of interest he received from the 8 percent investment was only $1 more than the amount of interest he received from the 6 percent investment. Find the amount invested at each rate.

14. Ronolog, Inc., invested $5000 in two promising businesses. The first business made a 7 percent profit, whereas the second business had a 2 percent loss. Ronolog's net profit for the year was a measly $35. Find the amount of money invested in each business.

5-7 SOLVING MOTION PROBLEMS

In solving motion problems, we must be aware of three quantities.

- The distance traveled D
- The rate of speed R
- The time it takes to travel T

The relationship between these three quantities is given as

$$\text{Distance} = \text{rate} \cdot \text{time}$$
$$D = R \cdot T$$

For example:

If you travel at a constant rate of speed of 60 miles per hour, for 4 hours, the distance traveled is 240 miles.

$D = R \cdot T$
$D = (60)(4)$
$D = 240$

It is most helpful, when solving motion problems, first to draw a diagram. This assists us in "visualizing the motion." Then, set up a chart showing the $D = R \cdot T$ relationship.

EXAMPLE 1: Two trucks leave the same truck stop at the same time and travel in opposite directions. One truck travels at 50 miles per hour and the second truck travels at 65 miles per hour. In how many hours will the trucks be 805 miles apart?

Example continued on next page.

STEP 1: READ THE PROBLEM CAREFULLY. DRAW AND LABEL A DIAGRAM.

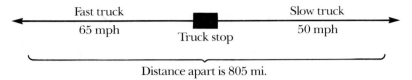

STEP 2: DETERMINE THE UNKNOWN QUANTITY AND SET IT EQUAL TO *x*.

The question being asked is

"In how many hours will the trucks be 805 miles apart?"

Thus,

we are looking for *time*.

Let *x* = the number of hours in which the trucks will be 805 miles apart

STEP 3: THERE IS NO OTHER UNKNOWN QUANTITY.

STEP 4: SET UP A CHART.

We now set up a chart that shows the $D = R \cdot T$ relationship for each truck.

	R $\cdot$	**T** =	**D**
Vehicle	*Rate*	*Time*	*Distance*
Truck 1	50 mph	*x*	50*x*
Truck 2	65 mph	*x*	65*x*

STEP 5: DETERMINE THE RELATIONSHIP IN THE PROBLEM.

The relationship in this problem is:

The distance of the first truck	+	The distance of the second truck	=	The total distance
50*x*	+	65*x*	=	805.

STEP 6: SOLVE THE EQUATION.

$$50x + 65x = 805$$
$$115x = 805$$
$$x = \frac{805}{115}$$
$$x = 7$$

STEP 7: CHECK.

■ Truck 1 travels at 50 miles per hour for 7 hours; thus, it travels 350 miles.

$D = R \cdot T$
$D = 50 \cdot 7$
$D = 350$

■ Truck 2 travels at 65 miles per hour for 7 hours; thus, it travels 455 miles.

$$D = R \cdot T$$
$$D = 65 \cdot 7$$
$$D = 455$$

■ The total distance must, therefore, equal 805 miles.

$$350 + 455 = 805$$

Final answer: 7 hours

EXAMPLE 2:   A helicopter and an airplane leave air bases that are 300 miles apart, and travel toward each other. The rate of speed of the airplane is 30 miles per hour more than the rate of speed of the helicopter. In 2 hours, the aircrafts will pass each other. Find the rate of each aircraft.

STEP 1: READ THE PROBLEM CAREFULLY. (DRAW A DIAGRAM.)

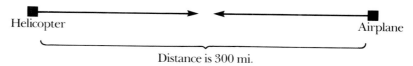

Helicopter   Airplane

Distance is 300 mi.

STEP 2: DETERMINE THE UNKNOWN QUANTITY AND SET x EQUAL TO IT.

The question being asked is

"Find the rate of each aircraft."

Thus, there are two unknowns:

■ The rate of speed of the helicopter

■ The rate of speed of the airplane

Let x = the rate of the helicopter

STEP 3: REPRESENT ANY OTHER UNKNOWN IN TERMS OF x.

"The rate of speed of the airplane is 30 miles per hour more than the rate of speed of the helicopter"

$$30 + x$$

Thus,

$x + 30$ = the rate of the airplane

STEP 4: SET UP A CHART.

Aircraft	R Rate	T Time	D Distance
Helicopter	x	2	$2x$
Airplane	$x + 30$	2	$2(x + 30)$

Example continued on next page.

STEP 5: DETERMINE THE RELATIONSHIP IN THE PROBLEM.

The relationship in the problem is:

$$\underbrace{\text{The distance traveled by the helicopter}}_{2x} + \underbrace{\text{The distance traveled by the airplane}}_{2(x + 30)} = \underbrace{300 \text{ miles}}_{300}$$

STEP 6: SOLVE THE EQUATION.

$$2x + 2(x + 30) = 300$$
$$2x + 2x + 60 = 300$$
$$2x + 2x = 300 - 60$$
$$4x = 240$$
$$x = \frac{240}{4}$$
$$x = 60$$

Thus,

$$x = 60 \quad \text{and} \quad x + 30 = 90$$

STEP 7: CHECK.

■ The distance traveled by the helicopter is 120 miles.

$$D = R \cdot T$$
$$D = (60)(2)$$
$$D = 120$$

■ The distance traveled by the airplane is 180 miles.

$$D = R \cdot T$$
$$D = (90)(2)$$
$$D = 180$$

■ The total distance traveled by both aircraft is 300 miles.

$$120 + 180 = 300$$

Final answer: helicopter: 60 mph; airplane: 90 mph

EXAMPLE 3: Mike and Chuck decide to have a race. Mike, driving a VW Rabbit, travels at a rate of speed of 50 miles per hour. Chuck, driving a Porsche, travels at a rate of speed of 80 miles per hour. Mike has a 3-hour head start. How long will it take Chuck to catch up with Mike?

STEP 1: READ THE PROBLEM CAREFULLY. (DRAW A DIAGRAM.)

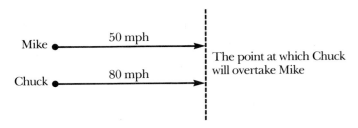

STEP 2:   DETERMINE THE UNKNOWN QUANTITY AND SET x EQUAL TO IT.

The question being asked is

"How long will it take Chuck to catch up with Mike?"

Thus, we are looking for *time*.

Let x = the time it takes Chuck to catch up with Mike

STEP 3:   REPRESENT ANY OTHER UNKNOWN IN TERMS OF x.

There is one more unknown; the length of time Mike travels when Chuck overtakes him. Since Mike has a 3-hour head start over Chuck, and x is the time Chuck travels,

$x + 3$ = the time Mike will have traveled when Chuck overtakes him

STEP 4:   SET UP A CHART.

Vehicle	**R** ·	**T** =	**D**
	Rate	Time	Distance
Rabbit (Mike)	50 mph	$x + 3$	$50(x + 3)$
Porsche (Chuck)	80 mph	x	$80x$

STEP 5:   DETERMINE THE RELATIONSHIP IN THE PROBLEM.

Since both Mike and Chuck are traveling the same distance, the relationship in this problem is:

The distance Mike travels = The distance Chuck travels

$$50(x + 3) \qquad = \qquad 80x$$

STEP 6:   SOLVE THE EQUATION.

$$50(x + 3) = 80x$$
$$50x + 150 = 80x$$
$$150 = 80x - 50x$$
$$150 = 30x$$
$$\frac{150}{30} = x$$
$$5 = x$$

Thus,

$$x = 5 \quad \text{and} \quad x + 3 = 8$$

STEP 7:   CHECK.

■   Mike travels 400 miles.

$$D = R \cdot T$$
$$D = 50 \cdot 8$$
$$D = 400$$

Example continued on next page.

■ Chuck travels 400 miles.

$D = R \cdot T$
$D = 80 \cdot 5$
$D = 400$

■ The two distances traveled are equal.

$400 = 400$

Final answer: 5 hours

EXERCISE 5-7

Solve the following problems.

1. Two trains leave the same station at the same time and travel in opposite directions. The first train travels 35 miles per hour and the second train travels 50 miles per hour. In how many hours will the trains be 340 miles apart?

2. Mr. and Mrs. Squire leave their house for work at the same time. If Mr. Squire travels at an average rate of speed of 45 miles per hour and Mrs. Squire averages 28 miles per hour, in how many hours will they be 36 miles apart?

3. Two trucks are at a terminal 725 miles apart. They start traveling at the same time toward each other. The second truck travels 25 miles per hour more than the first truck. If the trucks pass each other in 5 hours, what is the rate of speed of each truck?

4. A helicopter and an airplane leave their respective air bases at the same time; the air bases are 212 miles apart. The helicopter travels 38 miles per hour slower than the airplane. If the two aircrafts pass each other in 1 hour, what is their respective rate of speed?

5. Mike and Mary have an argument. Mike gets in his car and decides to drive to Howard's house in Virginia (a distance of 475 miles). An hour later, Mary decides she must stop Mike. Knowing Mike will travel at an average rate of speed of 60 miles per hour, Mary decides she must travel at an average rate of 70 miles per hour in order to catch up with Mike before he reaches Howard's house.
   a. How long will it take Mary to catch up with Mike?
   b. How far from Howard's house will Mary catch up with Mike?

6. A Pilot Freight Carrier truck leaves the Rochester terminal traveling 48 miles per hour. Two hours later, Noreen, the dispatcher, remembers she forgot to give the drivers their manifest. Noreen immediately gets in her car and chases them traveling at 80 miles per hour.
   a. How long will it take Noreen to catch up with the truck?
   b. If the first weigh station is 250 miles from Rochester, will Noreen catch up with the truck before it reaches the weigh station?

7. A battleship spots an enemy cruiser 6 miles straight ahead, traveling at approximately 75 miles per hour. A torpedo is launched from the battleship and travels toward the cruiser at a rate of speed of 45 miles per hour.
   a. How long will it take the torpedo to hit the cruiser?
   b. How far will the torpedo travel before it hits the cruiser?

8. Two avid motorists, John and Rose, decide to have a race to the store, a distance of 3 miles. Due partly to John's inexperience, Rose gives him a $\frac{3}{4}$ mile head start. Rose knows John will be able to average only 35 miles per hour and thus quickly calculates that she will have to average 50 miles per hour if she hopes to overtake him before he gets to the store.
   a. How long will it take Rose to overtake John?
   b. How far from the store will she do this?

9. Susan, the traveling saleswoman, made a trip of 430 miles in 8 hours. During the first 6 hours of the trip, she had good weather. However, she then ran into a snowstorm that forced her to decrease her average rate of speed by 45 miles per hour for the remainder of the trip. What was her rate of speed for each part of the trip?

10. Nancy made a trip of 116 miles by train and taxi. She traveled 4 hours by train and 1 hour by taxi. If the train averaged 26 miles per hour more than the taxi, find the rate of speed of each.

11. Melanie made a trip of 561 miles in 9 hours. Before 11 A.M., she was able to average 65 miles per hour. At 11 A.M., she stopped 1 hour for lunch. After lunch, she was only able to average 53 miles per hour. At what time did she begin her trip and at what time did she end it?

12. A "road team" traveled 838 miles in 16 hours. They were able to average 58 miles per hour before 4 P.M. and only 43 miles per hour after 5 P.M. (They made a 1 hour stop for supper.) At what times did they begin and end their travels?

13. A bicyclist spent 5 hours on a trip through the country and back. He rode out at the rate of speed of 22 miles per hour and rode back at the rate of speed of 33 miles per hour. How far into the country did he travel?

14. Mr. Cody drove away from his repair garage at a rate of speed of 14 miles per hour. Along the way, his truck got a flat tire. He did not have a spare so he decided to walk back to his garage. If he walked at a rate of speed of 6 miles per hour, and the round trip took 1 hour, how far from his garage did Cody's truck get a flat tire?

15. A skydiver jumps out of an airplane and free falls for 20 seconds at a rate of speed of 176 feet per second (120 miles per hour). Upon opening his parachute, his rate of speed decreases to 22 feet per second (15 miles per hour). The entire jump lasts 87 seconds. What was the altitude of the plane when the skydiver jumped? (*Note:* Be careful with *units.*)

REVIEW EXERCISES

Solve the following number problems.

1. The difference of a number and 8 is equal to 5. Find the number.

2. Negative twenty-one is the product of 7 times a number. Find the number.

3. Thirty-two is the quotient of a number and −8. Find the number.

4. Eight less a number is equal to 5. Find the number.

5. When a number is subtracted from 15, the result is −14. Find the number.

6. The difference of 3 and twice a number divided by 3 is equal to 7. Find the number.

7. Negative sixteen is the difference of 6 times a number and 4 times the same number. Find the number.

8. When 8 is subtracted from the sum of 3 times a number and twice the number, the difference is 17. What is the number?

9. One number is $\frac{1}{4}$ of another. If the sum of the two numbers is 125, what are the numbers?

10. The difference of $\frac{3}{5}$ of one number and $\frac{7}{5}$ of the same number is equal to -40. Find the number.

11. If 6 times a number is decreased by 8, the result is equal to 5 times the number reduced by 9. Find the number.

12. Two more than 5 times a number is equal to 12 less than 3 times the number. Find the number.

13. Eight increased by the product of 3 and a number is equal to 24 decreased by 5 times the number. Find the number.

Solve the following word problems.

14. Mrs. O'Brien is about to have her property sold. Together, her house and land have been appraised at $138,000. The house is worth 3 times as much as the value of the land. What is the appraised value for each?

15. Mr. and Mrs. Guggino had a garage sale. At the end of the sale, they collected $1620 in five-dollar, ten-dollar, and twenty-dollar bills. If there were 5 times as many twenty-dollar bills as five-dollar bills and 3 times as many ten-dollar bills as five-dollar bills, how many bills of each denomination were there?

16. A health food store sells dried apricots for $1.75 per pound and blanched nuts for $3.00 per pound. How many pounds of each must be used to produce a mixture of 50 pounds that would sell for $2.30 per pound?

17. Mr. Pascale is in charge of finances at a casino in Las Vegas. He decided to invest $100,000 in two separate accounts. One account is a short-term account that will earn 16 percent annually and the other account is a long-term account that will yield 12 percent annually. How much did he invest in each account if the total amount of interest earned for 1 year was $14,450?

18. At 2 P.M., a Pilot Freight Carrier truck leaves the Rochester, N.Y., terminal headed for Winston-Salem, N.C., traveling at an average rate of speed of 40 miles per hour. Three hours later, another Pilot truck leaves the same terminal and travels the same route but travels at an average rate of speed of 52 miles per hour. How long will it take the second truck to overtake the first truck? At what time will this occur?

19. One Sunday afternoon, in late October, Mr. Gilligan decided to winterize his truck. Checking the radiator, he noticed that the 5-gallon solution was only 15 percent antifreeze. How much pure antifreeze must Mr. Gilligan add so that his truck's radiator contains a 50 percent solution?

20. On Sunday, March 22, 1981, the U.S. Postal Service increased the cost of first class postage (for mailing letters) from 15¢ to 18¢. On the Friday prior to that date, Dick, a postal clerk, sold 100 times more 3¢ stamps than 15¢ stamps and 40 times more new 18¢ stamps than 15¢ stamps. The total amount from the sale of these stamps was $351.90. What was the number of stamps sold for each type?

21. Mr. Fratangelo decided to have the side walls of his house insulated and then have the entire house sided with vinyl siding. The total job cost $8050. The cost of the siding was $2\frac{1}{2}$ times more than the cost of the insulation. Find the cost of each.

22. Mrs. Hileman invested a sum of money in one stock with a yield of 3 percent and a $1200 more than the first sum in a second stock with a yield of 8 percent. If the total annual interest that Mrs. Hileman received was $145.50, how much did she invest at each rate? What was the total amount of money invested?

23. Ken and Jacki each have their own favorite route to travel from Jacki's parents' house to his own home (both routes are the exact same distance). One night, they decided to see whose route was quicker. Both left Jacki's parents' house at the exact same time. Jacki made it home in 21 minutes and traveled at an average rate of speed of 15 miles per hour faster than Ken. If Ken arrived 7 minutes *after* Jacki, how fast was each traveling?

CHAPTER 6

Polynomials

6-1 INTRODUCTION

The fundamental operations in arithmetic are addition, subtraction, multiplication, and division. In this chapter, we develop these four fundamental operations further as they apply to algebraic expressions. Before doing this, though, we must first develop some new vocabulary.

In Chapter 3, we learned that algebraic expressions are made up of one or more terms that are connected by the operation of addition or subtraction. Each variable of the expression must have a positive integer exponent in order for the expression to be called a polynomial. Moreover, special names are given to polynomials that consist of one term, two terms, or three terms; consider the following examples.

- A polynomial of one term is called a **monomial.**

$$\left.\begin{array}{c} 4 \\ -3x \\ 2x^2 \end{array}\right\} \text{monomials}$$

- A polynomial of two terms is called a **binomial.**

$$\left.\begin{array}{c} 2x - 4 \\ 3x^2 - 2x \\ 2x^2 + 2 \end{array}\right\} \text{binomials}$$

- A polynomial of three terms is called a **trinomial.**

$$\left.\begin{array}{c} x^2 + 2x - 4 \\ 2x^3 + 2x + 3 \\ -3x^3 - 4x - 1 \end{array}\right\} \text{trinomials}$$

A polynomial is usually described by the number of terms it contains and its degree. The degree of a polynomial is equal to the highest degree (of any term) in the polynomial.

The degree of a term is the exponent of the variable of that term. For example:

- The degree of $4x$ is 1.
- The degree of $4x^2$ is 2.
- The degree of $-2x^5$ is 5.
- The degree of 4 is 0.

NOTE:

> Since a variable is not present, the degree of a constant is zero.

If two or more variables are present in a term, the degree is the sum of the exponents of the variable of that term.

For example:

- The degree of $2x\,y$ is $1 + 1$ or   2.
- The degree of $2x^2y$ is $2 + 1$ or   3.
- The degree of $-3x^3y^2$ is $3 + 2$ or   5.
- The degree of $2x^2y^3z$ is $2 + 3 + 1$ or   6.
- The degree of $2^2x^2y^2$ is $2 + 2$ or   4.

The degree of polynomials is found by first finding the degree of each term, then choosing the largest.

For example:

- The degree of $2x^2 + 6x + 1$ is   2.
- The degree of $2x^2y + 3x\,y - 4y^2$ is   3.

Do the following practice set. Check your answers with the answers in the right-hand margin.

PRACTICE SET 6-1

For each polynomial give (a) the number of terms and (b) the degree.

1. $2x + 1$
2. 9
3. $4x + 3x^2 + 2x^3$
4. $2x^2y + 3xy - 4y + 4$

1.	(a) two
	(b) first
2.	(a) one
	(b) zero
3.	(a) three
	(b) third
4.	(a) four
	(b) third

EXERCISE 6-1

State the number of terms in the given algebraic expression and state the degree of the polynomial.

1. $x + 4$

2. $-2x^2$

3. $x^2 + 2x - 1$

4. $2(x + 1)$

5. $2ab - 4b^2$

6. $a^2 - \dfrac{b^2}{4}$

7. $a - 2b + c$

8. $5(a + b) + 4$

9. $\dfrac{t}{3} + \dfrac{8}{4}$

6-2 ADDITION OF POLYNOMIALS

In Section 3-3, we learned that there are two kinds of terms—like terms and unlike terms. Like terms are terms that contain the same variable factors to the same power. For example:

■ $3x$ and $-4x$ are like terms since they both contain the same variable factor (x) to the same power (x^1).

$3x$ and $-4x$

$x = x$

■ $4a$ and $4b$ are unlike terms since they contain different variable factors.

$4a$ and $4b$

$a \neq b$

■ $7ab$ and $-ab$ are like terms since they both contain the same variable factor (ab) to the same power. (Note: $-ab$ means $-1ab$.)

$7ab$ and $-1ab$

$ab = ab$

■ $2x$ and $2x^2$ are unlike terms. They contain the same variable factors (x) but not the same power $(x^2 \neq x^1)$.

$2x$ and $2x^2$

$x \neq x^2$

NOTE:

> *Like terms must contain the same variable(s) raised to exactly the same power(s).*

We also learned in Section 3-3 that we can simplify an expression by combining like terms. To do this, we first collect or group our like terms together before combining them. The same is true when adding polynomials. For example, to add $(2x + 3y) + (6x + 9y)$:

First, collect like terms.

$$(2x + 3y) + (6x + 9y)$$
$$= 2x + 3y + 6x + 9y$$
$$= (2x + 6x) + (3y + 9y)$$

Then, combine them accordingly.

$$(2x + 6x) + (3y + 9y)$$
$$= \quad 8x \quad + \quad 12y$$

Another method we can employ to add polynomials is to line up like terms under each other and then combine them, as opposed to writing them next to each other (as was done previously). Using the preceding example:

First, collect like terms by lining them up under each other.

$$2x + 3y + 6x + 9y$$
$$= 2x + 3y$$
$$+ 6x + 9y$$

Then, combine them accordingly.

$$\begin{array}{r} 2x + 3y \\ +6x + 9y \\ \hline 8x + 12y \end{array}$$

This approach to adding polynomials is often regarded as vertical addition, whereas the first approach is considered horizontal addition. In either case, addition of polynomials is accomplished by first grouping like terms together and then combining them. In the examples that follow, we use both approaches.

PROCEDURE:

To add polynomials:

1. Group all like terms together.
    a. In vertical addition, place like terms in the same column.
    b. In horizontal addition, place like terms adjacent to one another.

2. Combine like terms.

EXAMPLE 1:   Add $(12x^2 + 3x + 4)$ and $(2x^2 - 6x + 9)$.

VERTICAL FORM

SOLUTION: First, write like terms under each other.

$$12x^2 + 3x + 4$$
$$+ \ \ 2x^2 - 6x + 9$$

Now, combine accordingly.

$$12x^2 + 3x + 4$$
$$+ \ \ 2x^2 - 6x + 9$$
$$\overline{14x^2 - 3x + 13}$$

Final answer: $14x^2 - 3x + 13$

HORIZONTAL FORM

SOLUTION: First, collect like terms by grouping them together.

$$(12x^2 + 3x + 4) + (2x^2 - 6x + 9)$$
$$= 12x^2 + 3x + 4 + 2x^2 - 6x + 9$$
$$= 12x^2 + 2x^2 + 3x - 6x + 4 + 9$$

Combine accordingly.

$$\underbrace{12x^2 + 2x^2}_{14x^2} + \underbrace{3x - 6x}_{-3x} + \underbrace{4 + 9}_{+13}$$

EXAMPLE 2:   Add $(6ab + 3x - 2y)$ and $(-2ab + 4x + 9)$.

VERTICAL

SOLUTION: Line up like terms under each other.

$$6ab + 3x - 2y$$
$$- 2ab + 4x \ \ \ \ \ \ + 9$$

Combine.

$$6ab + 3x - 2y$$
$$- 2ab + 4x \ \ \ \ \ \ + 9$$
$$\overline{4ab + 7x - 2y + 9}$$

Final answer: $4ab + 7x - 2y + 9$

HORIZONTAL

SOLUTION: Group like terms together.

$$(6ab + 3x - 2y) + (-2ab + 4x + 9)$$
$$= 6ab + 3x - 2y - 2ab + 4x + 9$$
$$= 6ab - 2ab + 3x + 4x - 2y + 9$$

Combine.

$$\underbrace{6ab - 2ab}_{4ab} + \underbrace{3x + 4x}_{+7x} - \underbrace{2y + 9}_{-2y + 9}$$

NOTE:

In Example 2, since $-2y$ and 9 are unlike terms, we must keep them separate. When adding vertically, we do this by providing either a blank space or "dummy" term. For example:

$$6ab + 3x - 2y + 0$$
$$- 2ab + 4x + 0y + 9$$

dummy terms

EXAMPLE 3: Add $4 - 5x^2 + 3x^3$, $x^2 + x^3 - 3x$ and $10 - 8x^2 - 5x$.

Before adding these polynomials, we should first arrange the terms of each polynomial in descending order. That is, we begin with the term that has the largest degree and work our way down to the term with the smallest degree. Thus:

- $4 - 5x^2 + 3x^3 = 3x^3 - 5x^2 + 4$
- $x^2 + x^3 - 3x = x^3 + x^2 - 3x$
- $10 - 8x^2 - 5x = -8x^2 - 5x + 10$

Now, we can add.

VERTICAL FORM

SOLUTION: Line up like terms under each other. (Note the use of the dummy terms.)

$$3x^3 - 5x^2 + 0x + 4$$
$$x^3 + x^2 - 3x + 0$$
$$0x^3 - 8x^2 - 5x + 10$$

Combine like terms.

$$3x^3 - 5x^2 + 0x + 4$$
$$x^3 + x^2 - 3x + 0$$
$$0x^3 - 8x^2 - 5x + 10$$
$$\overline{4x^3 - 12x^2 - 8x + 14}$$

HORIZONTAL FORM

SOLUTION: Group like terms together.

$$(3x^3 - 5x^2 + 4) + (x^3 + x^2 - 3x)$$
$$+ (-8x^2 - 5x + 10)$$
$$= 3x^3 - 5x^2 + 4 + x^3 + x^2$$
$$- 3x - 8x^2 - 5x + 10$$
$$= 3x^3 + x^3 - 5x^2 + x^2 - 8x^2$$
$$- 3x - 5x + 4 + 10$$

Combine like terms.

$$\underbrace{3x^3 + x^3}_{4x^3} \underbrace{- 5x^2 + x^2 - 8x^2}_{- 12x^2} \underbrace{- 3x - 5x}_{- 8x} \underbrace{+ 4 + 10}_{+ 14}$$

Final answer: $4x^3 - 12x^2 - 8x + 14$

Do the following practice set. Check your answers with the answers in the right-hand margin.

PRACTICE SET 6-2

Add the following polynomials.

1. $3a$
    $7a$

2. $3ab$
    $-8ab$

3. $-2x^2 + (+6x^2)$

4. $-xy + (-2xy)$

5. $2x$
    $-4x$
    $-7x$

6. $2a + (+8a) + (-4a)$

1. $10a$
2. $-5ab$
3. $4x^2$
4. $-3xy$
5. $-9x$
6. $6a$

Add by using the vertical form (see example 1).

7. Add $(2a + 4)$ and $(a - 7)$.
8. Add $(x^2 + 6x)$ and $(-4x^2 + 3x)$.
9. Add $(6x^2 - 4)$ and $(2x^2 + 4)$.

7. $3a - 3$
8. $-3x^2 + 9x$
9. $8x^2$

Add by using the vertical form (see example 2 or 3).

10. Add $(x^2 + 3x)$ and $(x^2 - 2x - 4)$.
11. Add $(3 - 2x^3 - x)$ and $(2x^2 - 3x + 3x^3)$.
12. Add $(3x^3 + x - 7)$, $(5x^2 + 6x + 1)$ and $(-x^3 - 5x^2 - 4x - 1)$.

10. $2x^2 + x - 4$
11. $x^3 + 2x^2 - 4x + 3$
12. $2x^3 + 3x - 7$ or
     $2x^3 + 0x^2 + 3x - 7$

Add by using the horizontal form (see example 1).

13. Add $(3x - 4y)$ and $(x - 2y)$.
14. Add $(y^2 - 2y + 3)$ and $(y - 3y^2 + 4)$.
15. Add $(3x^2 + 6x - 5)$ and $(-2x^2 - x - 4)$ and $(7x^2 + 5x - 8)$.

13. $4x - 6y$
14. $-2y^2 - y + 7$
15. $8x^2 + 10x - 17$

Add by using the horizontal form (see example 2 or 3).

16. Add $(x^2 + x + 1)$ and $(x^2 - 2x)$ and $(x^2 - 4)$.
17. Add $(2z - 3z^3 + 4)$ and $(3z + 2z^3 + 3z^2)$.
18. Add $(-a + 2b)$ and $(5a - 9c)$ and $(8a + 3b + 6c)$.

16. $3x^2 - x - 3$
17. $-z^3 + 3z^2 + 5z + 4$
18. $12a + 5b - 3c$

EXERCISE 6-2

Add each of the following monomials by using the vertical form.

1. $10x$
    $7x$

2. $+8y^2$
    $-4y^2$

3. $-3ab$
    $-2ab$

4. $+2d$
    $-6d$

5. $-14xy$
    $+\ 7xy$

6. $+14z^3$
    $-42z^3$

7. $-28x^2y$
    $-\ 4x^2y$

8. $-18abc$
    $+29abc$

9. $+13y$
    $-32y$

10. $3(a+b)$
    $6(a+b)$

11. $13c$
    c

12. $+8d$
    $-3d$
    $+4d$

13. $-15x^2$
    $-\ x^2$
    $+13x^2$

14. $-6c$
    $+9c$
    $-3c$

15. $2(x+y)$
    $-4(x+y)$
    $-\ (x+y)$

Add the following monomials by using the horizontal form.

16. $(+7x) + (+8x)$

17. $(-18y) + (+6y)$

18. $(+14c) + (-c)$

19. $(-x^2) + (+7x^2)$

20. $(-6xy) + (-xy)$

21. $(-10x^2) + (+10x^2)$

22. $(+2c) + (+5c) + (-4c)$

23. $(-2xy) + (+5xy) + (-8xy)$

24. $(-13x^2y) + (+6x^2y) + (+7x^2y)$

25. $(-5c^2) + (+8c^2) + (+2c^2) + (-9c^2)$

Using the vertical form, add the following polynomials.

26. $(6a+7)$ and $(a+4)$

27. $(4x+9y)$ and $(3x+5y)$

28. $(-x+4y)$ and $(6x-2y)$

29. $(2a+2b)$ and $(3a-2b)$

30. $(3a+b-4c)$ and $(2a-4b+3c)$

31. $(4x^3+6)$ and (x^3-2) and $(-3x^3-5)$

32. $(4x+6y)$ and $(-4x-6y)$

33. $(x^2+3xy-y^2)$ and $(x^2-xy+2y^2)$

34. $(2a+b-3c)$ and $(4a-6b+4c)$

35. $(-a+8b-2c)$ and $(3a+9b+6c)$

36. $(4y^2-2y-5)$ and $(3y^2+2y-5)$

37. $(6x^2-3x-10)$ and $(2x^2+11)$

38. (y^2+3y-4) and (y^2+2) and $(-4y+3)$

39. $(2z^2-4z-5)$ and $(6z+2)$ and (z^2+3)

40. $(2a-9+3a^2)$ and (a^2-3-5a) and $(7a-a^2+12)$

Using the horizontal form, add the following polynomials.

41. $(2x+y)$ and $(x-7y)$

42. $(-x+8y)$ and $(4x-6y)$

43. $(6c+d)$ and $(c-4d)$

44. $(4x-3)$ and $(-4x-3)$

45. (x^2+4x) and $(2x^2-3x)$

46. $(6x^2-4x)$ and $(2x^2-6x)$ and $(-x^2+2x)$

47. $(4x^2-3x-8)$ and $(6x^2+9)$

48. $(4x^2 - 2x - 3)$ and $(x - 5)$ and $(2x^2 - 2)$
49. $(9 - 5y^2)$ and $(4y^2 - 3y + 6)$ and $(13y - 2)$
50. $(5a^2 - ab + b^2)$ and $(3b^2 + 2ab - a^2)$ and $(a^2 - b^2)$
51. $(x^3 - 3x^2 + 4)$ and $(2x^3 - 6x)$ and $(-3x^3 - 3x^2 + 2x + 4)$
52. $(z^2 + 4)$ and $(z^2 - 3)$ and $(z^2 + 2z)$ and $(z^2 - 1)$
53. $(4 - y)$ and $(6 - 2y + y^2)$ and $(-8 + 2y^2)$
54. $(2x^2 + 4x)$ and $(x^2 - 3x - 4)$ and $(-x^2)$ and $(5x - 1)$
55. $(4 - z^2)$ and $(-3 + z - z^2)$ and $(-1 - z + 2z^2)$

6-3 SUBTRACTION OF POLYNOMIALS

When subtracting polynomials, we must be careful to place the quantity being subtracted (called the subtrahend) within parentheses so that each term of the subtrahend gets subtracted. Then, using our distributive property, we remove the parentheses. Once this is done, we proceed in a similar manner as that for adding polynomials. Namely, group like terms together and combine them accordingly. As an example, consider the problem of subtracting $(2a - b)$ from $(6a + 2b)$.

First, write out the problem, placing parentheses around the subtrahend.

$$6a + 2b - (2a - b)$$

Next, remove the parentheses by applying our distributive property.

$$6a + 2b - (2a - b)$$
$$= 6a + 2b - 2a + b$$

Finally, group like terms and combine them accordingly.

$$6a + 2b - 2a + b$$
$$= 6a - 2a + 2b + b$$
$$= 4a + 3b$$

Thus,

$$6a + 2b - (2a - b) = 4a + 3b$$

NOTE:

Whenever we remove the parentheses around the subtrahend by using the distributive property, we are, in effect, changing the sign of each term of the subtrahend to its opposite. For example,

$$\left. \begin{array}{l} -(2a - b) \\ = -1(2a - b) \\ = -2a + 2b \end{array} \right\} \text{ Note that the signs of each term are now opposite}$$

Thus, subtraction of polynomials can be performed by first changing the sign of each term of the subtrahend and then combining like terms accordingly.

PROCEDURE:

> To subtract polynomials:
>
> 1. Place parentheses around the expression being subtracted.
>
> 2. Remove the parentheses by distributing the minus sign. This will change the sign of each term of the polynomial being subtracted.
>
> 3. Add the polynomials by combining like terms.

Subtraction of polynomials can also be performed either vertically or horizontally. We use both approaches in our examples.

EXAMPLE 1: Subtract $(3x - 2)$ from $(5x + 7)$.

VERTICAL

SOLUTION: Line up like terms under each other, and place parentheses around the subtrahend.

$$5x + 7$$
$$- (3x - 2)$$

Remove parentheses.

$$\begin{array}{lcl} 5x + 7 & & 5x + 7 \\ - (3x - 2) & = & - 3x + 2 \end{array}$$

Combine like terms.

$$\begin{array}{r} 5x + 7 \\ - 3x + 2 \\ \hline 2x + 9 \end{array}$$

Final answer: $2x + 9$

HORIZONTAL

SOLUTION: Write horizontally, placing parentheses around the subtrahend.

$$5x + 7 - (3x - 2)$$

Remove parentheses.

$$5x + 7 - (3x - 2)$$
$$= 5x + 7 - 3x + 2$$

Combine like terms.

$$5x + 7 - 3x + 2$$
$$= 5x - 3x + 7 + 2$$
$$= \quad 2x \quad + \quad 9$$

EXAMPLE 2: Subtract $18x - 6$ from $4x^2 - 12x$.

VERTICAL

SOLUTION: Line up like terms under each other. (Note the use of the parentheses and dummy terms.)

HORIZONTAL

SOLUTION: Write horizontally.

$$4x^2 - 12x - (18x - 6)$$

$$4x^2 - 12x + 0$$
$$- (0x^2 + 18x - 6)$$

Remove parentheses.   Remove parentheses.

$$4x^2 - 12x - (\overset{\frown}{18x - 6})$$
$$= 4x^2 - 12x - 18x + 6$$

$$4x^2 - 12x + 0 \qquad 4x^2 - 12x + 0$$
$$- (0x^2 + 18x - 6) = -0x^2 - 18x + 6$$

Combine.

Combine.

$$4x^2 \underbrace{- 12x - 18x}\ + 6$$
$$= 4x^2 \qquad -30x \qquad + 6$$

$$4x^2 - 12x + 0$$
$$- 0x^2 - 18x + 6$$
$$\overline{4x^2 - 30x + 6}$$

Final answer: $4x^2 - 30x + 6$

Do the following practice set. Check your answers with the answers in the right-hand margin.

PRACTICE SET 6-3

Subtract by using the vertical form (see example 1).

1. Subtract $6x - 3y$ from $9x - 5y$.
2. Subtract $-x + 4$ from $x - 6$.
3. Subtract $-4b + 3c$ from $5b - 2c$.

Subtract by using the vertical form (see example 2).

4. Subtract $10 + 3x^2 - 5x$ from $7x + 2x^2 - 8$.
5. Subtract $x^2 + 3xy - 4y^2$ from $4x^2 - 6xy + y^2$.
6. Subtract $7a - 4c - 3d$ from $a + 10c - 4d$.
7. Subtract $-2a^2 - 6a + 10$ from $9 + a^2$.
8. Subtract $x^2 + 3$ from $2x^2 + x - 4$.
9. Subtract $-7x + 4$ from 0.

Subtract by using the horizontal method (see example 1).

10. Subtract $2x^2 - 3x + 8$ from $x^2 + 6x - 12$.
11. Subtract $7x^2 - 3x - 6$ from $4x^2 - 7x + 5$.
12. Subtract $2x^2 - 6 - 9x$ from $-5x + 6x^2 + 3$.

1. $3x - 2y$
2. $2x - 10$
3. $9b - 5c$

4. $-x^2 + 12x - 18$
5. $3x^2 - 9xy + 5y^2$
6. $-6a + 14c - d$
7. $3a^2 + 6a - 1$
8. $x^2 + x - 7$
9. $+7x - 4$

10. $-x^2 + 9x - 20$
11. $-3x^2 - 4x + 11$
12. $4x^2 + 4x + 9$

Subtract by using the horizontal method (see example 2).

13. Subtract $10x - 11$ from $8x^2 - 7x - 4$.   13. $8x^2 - 17x + 7$

14. Subtract $3a - b + 3$ from $2b - 4$.   14. $-3a + 3b - 7$

15. Subtract $3y^2 + 4y + 7$ from $y^3 - y^2 + 20$.   15. $y^3 - 4y^2 - 4y + 13$

EXERCISE SET 6-3

Find the difference of the following monomials. Subtract the second term *from* the first term. (Use the vertical form.)

1. $9x$ and $2x$

2. $12x$ and $-4x$

3. $-4x$ and $-7x$

4. $-18y$ and $11y$

5. $-7z$ and $-7z$

6. 0 and $-4x$

7. $3y$ and 0

8. $8m$ and $-6n$

9. $1.5x^2$ and $0.5x^2$

10. $\dfrac{1}{2}A$ and $-\dfrac{3}{2}A$

Subtract the bottom term from the top term.

11. $\begin{array}{r} +2a \\ +9a \\ \hline \end{array}$

12. $\begin{array}{r} x^2 \\ 4x^2 \\ \hline \end{array}$

13. $\begin{array}{r} 10y^2 \\ 7y^2 \\ \hline \end{array}$

14. $\begin{array}{r} 4ab \\ 0 \\ \hline \end{array}$

15. $\begin{array}{r} 0 \\ 6ab \\ \hline \end{array}$

16. $\begin{array}{r} 0 \\ -2ab \\ \hline \end{array}$

17. $\begin{array}{r} 12y \\ -\ 7y \\ \hline \end{array}$

18. $\begin{array}{r} 3y \\ -6y \\ \hline \end{array}$

19. $\begin{array}{r} 6ab \\ 6ab \\ \hline \end{array}$

20. $\begin{array}{r} +4y^2 \\ -4y^2 \\ \hline \end{array}$

21. $\begin{array}{r} -2z \\ +2z \\ \hline \end{array}$

22. $\begin{array}{r} -8x^2 \\ +5x^2 \\ \hline \end{array}$

23. $\begin{array}{r} -4y^2 \\ +7y^2 \\ \hline \end{array}$

24. $\begin{array}{r} -3a \\ -3a \\ \hline \end{array}$

25. $\begin{array}{r} -6a \\ -2a \\ \hline \end{array}$

26. $\begin{array}{r} -\ 7a \\ -12a \\ \hline \end{array}$

27. $\begin{array}{r} +3(a+b) \\ -7(a+b) \\ \hline \end{array}$

28. $\begin{array}{r} -5(x+y) \\ -3(x+y) \\ \hline \end{array}$

29. $\begin{array}{r} -1.4x \\ +0.8x \\ \hline \end{array}$

30. $\begin{array}{r} 0.6cd \\ -0.8cd \\ \hline \end{array}$

31. $\begin{array}{r} 3.2d \\ -8.8d \\ \hline \end{array}$

Find the difference of the following monomials. (Use the horizontal form.)

32. $8x^2$ and $3x^2$

33. $4ab$ and $7ab$

34. $4a^2$ and $4a^2$

35. $-15x^2$ and $10x^2$

36. $-11abc$ and $-22abc$

37. $-6z$ and $-6z$

38. $2a$ and $-5a$

39. $6x$ and $-2x$

40. $9y$ and $-9y$

41. $-3x^2$ and $5x^2$

42. $-7xy$ and $4xy$

43. $-2(a+b)$ and $2(a+b)$

44. 0 and $-2x^2$

45. 0 and $7ab$

46. $3x$ and 0

Find the difference in each of the following cases.

47. $(+8xy) - (-5xy)$

48. $(+2a) - (-6a)$

49. $(-2x^2) - (4x^2)$

50. $(6mn) - (-2mn)$

51. $(-8cd) - (4cd)$

52. $(3x^2) - (10x^2)$

53. $\left(-5\frac{3}{4}x\right) - \left(-1\frac{1}{4}x\right)$

54. $(4.3y^2) - (-2.8y^2)$

55. $(13x^2) - (3x)$

56. $(-2xy) - (-x)$

Subtract the following polynomials. (Use the vertical form.)

57. Subtract $3a - 2$ from $6b + 5$.

58. Subtract $-3x - 4$ from $x - 3$.

59. Subtract $4a + 7b$ from $5a + 2b$.

60. Subtract $-4x^2 - 7$ from $6x^2 - 10$.

61. Subtract $12x - 3y$ from $-4x + 20y$.

62. Subtract $-6x^2 + 18x$ from $10x^2$.

63. Subtract $-y + 12$ from $3y - 5x + 9$.

64. Subtract $-a - 7$ from $7a + b - 5$.

65. Subtract $x^3 - 12$ from $10x^2 + 5x^3$.

66. Subtract $y^2 - 12y$ from $15 + 3y^2$.

Subtract in each of the following cases.

67. $10a + 7b$
     $3a + 5b$

68. $6x + 5y$
     $9x - 7y$

69. $4x - 7y$
     $7x - 7y$

70. 0
     $x^2 - 4$

71. $6x^2 - 4x$
     $9x^2 - 4x$

72. $x^2 - 8x + 6$
     $3x^2 - x - 2$

73. $3x^2 - 5x + 3$
     $- x^2 - 7x + 3$

74. $2x^2 + 10$
     $- x^2 + 5x - 4$

75. $y^2 + 6y - 7$
     $y^2 - 2y - 3$

76. $x^2 + 2x + 9$
     $-x^2 - 10$

77. $6x^2 - 10x$
     $8x - 5$

78. $7x^3 - 4x^2 + 5x$
     $6x^3 + 9x$

Find the difference of the following polynomials. (Use the horizontal form.)

79. $5a + 3b$ and $4a + b$

80. $7a - 3b$ and $-4a + 9b$

81. 0 and $3a - 2b$

82. $5x^2 - 9x$ and $-3x^2 + x$

83. $-2a + 10b$ and $15a - 3b$

84. $5x^2 + 2x$ and $9x^2 - 7x$

85. $x^2 - 4x$ and $-x^2 + 2x$

86. $3x + 9$ and 0

87. $12x^2 - 18$ and $9x^2 + 6$

88. $2x^2 - 3x - 1$ and $-4x^2 - 3x + 6$

89. $9x^2 + 6x - 8$ and $-5y - y^2$

90. $6 - 7y - y^2$ and $-5y - y^2$

91. $3y$ and $(6y - 2)$

92. $3x + 6$ and $8 - 9x$

93. $x^2 + 3$ and $3 + x^2$

Perform the indicated operations.

94. Subtract $8x - 6y + 3z$ from $9x + 11y - 8z$.

95. From $14r - 8s - 9t$ subtract $10s - 18t$.

96. Subtract $6a^2 + 4a - 1$ from 0.

97. From 0 subtract $-2x^2 + 3x - 1$.

98. Subtract $x^2 + 4x + 7$ from $x^3 - 2x^2 + 8$.

99. Subtract $2y + 6y^2 - 9$ from $3y^2 - 6y + 2$.

100. Subtract $y^2 - 5y + 9$ from $3y^2 + 5y$.

6-4   RULES OF EXPONENTS IN MULTIPLICATION

A.   Multiplying Factors with the Same Base

To multiply the powers of the same base such as $(a^2) \times (a^3)$:

First, express each power without exponents.

$$a^2 \times a^3$$
$$= (a \cdot a) \times (a \cdot a \cdot a)$$

Then, rewrite the product as a single power.

$$(a \cdot a) \times (a \cdot a \cdot a)$$
$$= a \cdot a \cdot a \cdot a \cdot a$$
$$= a^5$$

Notice that we can get this answer by writing the common base of the product and then adding the exponents of the factors to get the exponent of the product.

$$a^2 \cdot a^3$$
$$= a^{2+3}$$
$$= a^5$$

This will always be true as long as we have the same base. In general,

$$a^m \cdot a^n = a^{m+n}$$

PROCEDURE:

To multiply factors with the same base:

1. Write the common base as the base of the product.

2. Then, add the exponents, which is the exponent of the product. That is,

$$a^m \cdot a^n = a^{m+n}$$

EXAMPLE 1: Multiply $a^3 \cdot a^4$.

SOLUTION: The common base is a.

$$\underset{a}{a^3 \cdot a^4}$$

Add the exponents of the common base a.

$$a^3 \cdot a^4 = a^{3+4}$$
$$= a^7$$

Final answer: a^7

EXAMPLE 2: Multiply $a^3 \cdot a \cdot b^3 \cdot b^2$.

SOLUTION: The common bases are a and b.

Add the exponents of each common base.

$$a^3 \cdot a \cdot b^3 \cdot b^2 = a^{3+1} \cdot b^{3+2}$$
$$= a^4 \cdot b^5$$

Final answer: $a^4 \cdot b^5$

Do the following practice set. Check your answers with the answers in the right-hand margin.

PRACTICE SET 6-4A

Multiply (see example 1 or 2).

1. $x^7 \cdot x^3$

2. $3^5 \cdot 3^3$

3. $y^2 \cdot y \cdot y^4$

4. $x^3 \cdot y^2 \cdot x^5 \cdot y$

5. $a^2 \cdot b^3 \cdot a$

6. $x^2 y \cdot x \cdot y^2$

1. x^{10}
2. 3^8
3. y^7
4. $x^8 y^3$
5. $a^3 b^3$
6. $x^3 y^3$

B. Finding the Power of a Power of a Base

To evaluate an expression such as $(x^3)^4$:

First, write x^3 as a factor 4 times.

$$(x^3)^4$$
$$= x^3 \cdot x^3 \cdot x^3 \cdot x^3$$

Then, evaluate this by using the method we learned in the previous section.

$$x^3 \cdot x^3 \cdot x^3 \cdot x^3$$
$$= x^{3+3+3+3}$$
$$= x^{12}$$

Notice that we can obtain this answer by writing the base, x, as the base of our answer and multiplying the two exponents to obtain the exponent of our answer.

$$(x^3)^4$$
$$= x^{(3)(4)}$$
$$= x^{12}$$

In general,

$$(a^m)^n = a^{mn}$$

PROCEDURE:

To find the power of a power of a base:

1. Write the given base.
2. Multiply the exponents together to find the exponent of the base. That is,
$$(a^m)^n = a^{mn}$$

EXAMPLE 1: Evaluate $(a^4)^2$.

SOLUTION:

$$(a^4)^2 = a^{(4)(2)}$$
$$= a^8$$

Final answer: a^8

EXAMPLE 2: Evaluate $(x^3)^2(y^4)^2$.

SOLUTION:

$$(x^3)^2 \cdot (y^4)^2 = x^{(3)(2)}y^{(4)(2)}$$
$$= x^6 \cdot y^8$$

Final answer: $x^6 y^8$

Do the following practice set. Check your answers with the answers in the right-hand margin.

PRACTICE SET 6-4B

Evaluate each of the following (see example 1 or 2).

1. $(b^3)^5$

2. $(3^2)^3$

3. $(z^5)^5$

4. $(2^2)^2 \cdot (x^2)^3$

5. $(x^4)^3 \cdot (y)^3$

6. $(1)^3 \cdot (x^2)^3$

1. b^{15}
2. 3^6 or 729
3. z^{25}
4. $2^4 x^6$ or $16x^6$
5. $x^{12} y^3$
6. $1^3 x^6$ or x^6

C.   Finding the Power of an Indicated Product

When evaluating expressions containing exponents, we must remember that an exponent affects only the one factor that immediately precedes it. The exception is when parentheses are used. For example:

■   In the term, $2x^2$, the x is squared, not the 2.

$$2x^2$$
$$= 2 \cdot x \cdot x$$

■   In the term, $(2x)^2$, because of the parentheses, the entire quantity within parentheses, namely, $2x$, is squared.

$$(2x)^2$$
$$= (2x) \cdot (2x)$$
$$= 2 \cdot x \cdot 2 \cdot x$$
$$= 2 \cdot 2 \cdot x \cdot x$$
$$= 4 \cdot x^2$$
$$= 4x^2$$

In general,

$$(ab)^n = a^n b^n$$

PROCEDURE:

To find the power of an indicated product:

1.   Raise each factor of the product to the indicated power.

2.   Evaluate each factor. That is,

$$(a \cdot b)^n = a^n \cdot b^n$$

EXAMPLE 1: Evaluate $(ab^2)^3$.

SOLUTION: We must raise each factor within parentheses to the third power.

$$(ab^2)^3 = (a)^3 \cdot (b^2)^3$$

Now, we can evaluate by using our previous rules.

$$(a)^3 \cdot (b^2)^3 = a^3 \cdot b^{(2)(3)}$$
$$= a^3 b^6$$

Final answer: $a^3 b^6$

EXAMPLE 2: Evaluate $(2b)^3$.

SOLUTION: Raise each factor within parentheses to the third power.

$$(2b)^3 = (2)^3 \cdot (b)^3$$

Evaluate.

$$(2)^3 \cdot (b)^3 = 2^3 \cdot b^3$$
$$= 8b^3$$

Final answer: $8b^3$

EXAMPLE 3: Evaluate $(-x^2)^3$

$$-x^2 \quad \text{means} \quad (-1)(x^2)$$

SOLUTION:

$$(-x^2)^3 = (-1 \cdot x^2)^3$$
$$= (-1)^3 (x^2)^3$$
$$= (-1)^3 (x^6)$$
$$= -x^6$$

Final answer: $-x^6$

EXAMPLE 4: Evaluate $(-2a^2b)^2$

SOLUTION:

$$(-2a^2b)^2 = (-2)^2 \cdot (a^2)^2 \cdot (b^1)^2$$
$$= (4) \cdot (a^4) \cdot (b^2)$$
$$= 4a^4b^2$$

Final answer: $4a^4b^2$

Do the following practice set. Check your answers with the answers in the right-hand margin.

PRACTICE SET 6-4C

Evaluate (see examples 1 and 2).

1. $(x^2y^3)^2$ 2. $(2a^2)^3$ 3. $(ab^3)^4$

Evaluate (see examples 3 and 4).

4. $(-2x^2)^3$ 5. $(-y^3)^5$ 6. $(-z^2)^4$
7. $(2x^2y^2)^3$ 8. $(-z^3x^2y)^5$ 9. $(-3x^2y^5)^4$

1. x^4y^6
2. $8a^6$
3. a^4b^{12}
4. $-8x^6$
5. $-y^{15}$
6. z^8
7. $8x^6y^6$
8. $-z^{15}x^{10}y^5$
9. $81x^8y^{20}$

EXERCISE 6-4

Find each of the following indicated products.

1. $a^3 \cdot a^2$ 2. $x^2 \cdot x^3$ 3. $y^3 \cdot y^3$
4. $d^2 \cdot d$ 5. $a^2 \cdot a^2 \cdot a^2$ 6. $p^5 \cdot p \cdot p^2$
7. $a^3 \cdot a \cdot b \cdot b^4$ 8. $x \cdot x^3 \cdot y$ 9. $c^4 \cdot c^2 \cdot d \cdot d^3$
10. $x^3 \cdot y^2 \cdot x^2 \cdot y^2$ 11. $4^3 \cdot 4^2$ 12. $2^2 \cdot 2 \cdot 5$
13. $3^2 \cdot 3 \cdot x \cdot x^2$ 14. $3 \cdot x^2 \cdot 3 \cdot x$ 15. $x^2 \cdot x^b$

Evaluate the following.

16. $(x^3)^2$ 17. $(y^4)^2$ 18. $(z^4)^2$
19. $(y^5)^2$ 20. $(2^2)^3$ 21. $(2^5)^3$

22. $(3^5)^4$ 23. $(3^m)^n$ 24. $(a^3)^2 \cdot (b^2)^2$

25. $(z^3)^2 \cdot (z^4)^2$ 26. $(y^2)^4 \cdot y \cdot y^3$ 27. $(a^2)^b \cdot (a^b)^3$

Evaluate the following.

28. $(ab^2)^3$ 29. $(x^2y^3)^4$ 30. $(a^2b^3c^4)^2$ 31. $(7a)^2$

32. $(-3a)^2$ 33. $(2a^2)^3$ 34. $(-3c^2d)^3$ 35. $(2x^2)^2 \cdot (x)^3$

36. $(-a^2bc^3)^2$ 37. $(-3yz^2)^3$

6-5   MULTIPLICATION OF POLYNOMIALS

A.   Multiplying a Monomial by a Monomial

To multiply a monomial by a monomial, we rearrange the factors (using the commutative and associative properties of multiplication) so that we multiply only the numerical coefficients and multiply only the variables. For example, to multiply $(2x)(3y)$:

First, commute the factors, grouping the coefficients together and the variables together.

$$(2x)(3y)$$
$$= (2)(3)(x)(y)$$

Then, multiply the coefficients and multiply the variables.

$$= \underbrace{(2)(3)}\,\underbrace{(x)(y)}$$
$$= (2 \cdot 3) \quad (x \cdot y)$$
$$= \quad 6 \quad \cdot \quad xy$$

Thus,

$$\text{the product of } (2x)(3y) = 6xy$$

PROCEDURE:

> To multiply a monomial times a monomial:
>
> 1.   Multiply the numerical coefficients to find the coefficient of the product.
>
> 2.   Multiply the variables, simplifying powers with the same base.

EXAMPLE 1:   Multiply $(3x)$ by $(-2x^2)$.

SOLUTION: Multiply the numerical coefficients.

$$(\,3\,x)(\,-2\,x^2)$$
$$\underbrace{(3)(-2)}$$
$$-6$$

Multiply the variables.

$$(3\,x) \cdot (-2\,x^2) = (3)(-2) \cdot \underbrace{(x)(x^2)}_{x^3}$$
$$= \quad -6 \quad \cdot \quad x^3$$

Final answer: $-6x^3$

EXAMPLE 2: Multiply $(7a^3b)$ times $(-5a^2b^2)$.

SOLUTION: Multiply the coefficients.

$$(7\,a^3b)(-5\,a^2b^2)$$
$$\underbrace{(7)(-5)}$$
$$-35$$

Multiply the variables.

$$(7\,a^3b)(-5\,a^2b^2) = (7)(-5) \cdot \underbrace{a^3 \cdot a^2}_{a^5} \cdot \underbrace{b \cdot b^2}_{b^3}$$
$$= \quad -35 \quad \cdot \quad a^5 \qquad b^3$$

Final answer: $-35a^5b^3$

EXAMPLE 3: Multiply $(-4c^2d)^2$.

SOLUTION: First, express $(-4c^2d)^2$ without using exponents.

$$(-4c^2d)^2 = (-4c^2d)(-4c^2d)$$

Now, multiply.

$$(-4c^2d)(-4c^2d) = \underbrace{(-4)(-4)}\,\underbrace{(c^2)(c^2)}\,\underbrace{(d)(d)}$$
$$= \quad +16 \qquad c^4 \qquad d^2$$

Final answer: $16c^4d^2$

NOTE:

In Example 3, we can also find the value of $(-4c^2d)^2$ by raising each factor to the second power.

$$(-4c^2d)^2$$
$$= (-4)^2 \cdot (c^2)^2 \cdot (d)^2$$
$$= \quad 16 \quad \cdot \quad c^4 \quad \cdot \quad d^2$$

EXAMPLE 4: Multiply $(-5y)(3xy)(xy^2)$.

SOLUTION:

$$(-5y)(3xy)(xy^2) = \underbrace{(-5)(3)}\ \ \underbrace{(x)(x)}\ \ \underbrace{(y)(y)(y^2)}$$
$$= -15 \quad \cdot \quad x^2 \quad \cdot \quad y^4$$

Final answer: $-15x^2y^4$

Do the following practice set. Check your answers with the answers in the right-hand margin.

PRACTICE SET 6-5A

Multiply (see example 1).

1.	$(12y)(-3y^3)$	1.	$-36y^4$
2.	$(-4x^2)(-3x^3)$	2.	$+12x^5$
3.	$(2x^2)(2x^2)$	3.	$4x^4$

Multiply (see examples 2 and 3).

4.	$(2a^2b^3)(-4ab^4)$	4.	$-8a^3b^7$
5.	$(-2ab^2)(-3a^2c)$	5.	$+6a^3b^2c$
6.	$(-2x^2y)^2$	6.	$+4x^4y^2$

Multiply (see example 4).

7.	$(-3x^2y)(y^2)(4xy)$	7.	$-12x^3y^4$
8.	$(-x^2y^2z^2)(-3yz)(2xz)$	8.	$+6x^3y^3z^4$
9.	$(-2ab^4)^3$	9.	$-8a^3b^{12}$

B. Multiplying a Monomial by a Polynomial

To multiply a monomial by a polynomial, we apply the distributive law. For example, to multiply $3(2a + 4)$:

Multiply 3 by the first term, $2a$.

$$3(2a + 4)$$
$$3(2a)$$
$$6a$$

Multiply 3 by the second term, 4.

$$3(2a + 4)$$
$$= 3(2a) + 3(4)$$
$$= 6a + 12$$

Thus,

$$\text{the product of } 3(2a + 4) = 6a + 12$$

We can also perform this multiplication vertically. For example:

Multiply 3 by the first term.

$$\begin{array}{r} 2a + 4 \\ \times\ 3 \\ \hline 6a \end{array}$$

Multiply 3 by the second term.

$$\begin{array}{r} 2a + 4 \\ \times\ 3 \\ \hline 6a + 12 \end{array}$$

In the examples that follow, we use both the vertical and horizontal approach.

PROCEDURE:

To multiply a monomial times a polynomial:

Distribute the monomial over the polynomial by multiplying the monomial by every term in the polynomial and joining the products by their proper sign.

EXAMPLE 1: Multiply $(3x - 5)$ by 4.

VERTICAL

SOLUTION: Multiply the monomial by the first term.

$$\begin{array}{r} 3x - 5 \\ 4 \\ \hline 12x \end{array}$$

Multiply the monomial by the second term.

$$\begin{array}{r} 3x - 5 \\ 4 \\ \hline 12x - 20 \end{array}$$

Final answer: $12x - 20$

HORIZONTAL

SOLUTION: Multiply the monomial by the first term.

$$4(3x - 5)$$
$$= 12x$$

Multiply the monomial by the second term.

$$4(3x - 5)$$
$$= 12x - 20$$

EXAMPLE 2: Multiply $-5x$ by $(x^2 - 3x + 2)$.

VERTICAL

SOLUTION: Multiply the monomial by each term of the polynomial.

$$\begin{array}{r} x^2 - 3x + 2 \\ -5x \\ \hline -5x^3 + 15x^2 - 10x \end{array}$$

Final answer: $-5x^3 + 15x^2 - 10x$

HORIZONTAL

SOLUTION: Multiply the monomial by each term of the polynomial.

$$-5x(x^2 - 3x + 2)$$
$$= \underbrace{(-5x)(x^2)}_{} + \underbrace{(-5x)(-3x)}_{} + \underbrace{(-5x)(2)}_{}$$
$$= \quad -5x^3 \qquad\quad +15x^2 \qquad\quad -10x$$

Do the following practice set. Check your answers with the answers in the right-hand margin.

PRACTICE SET 6-5B

Multiply (see examples 1 and 2).

1.	$-4(2a + 3b - 4c)$	1.	$-8a - 12b + 16c$
2.	$-5x(x^2 - 3x + 4)$	2.	$-5x^3 + 15x^2 - 20x$
3.	$-2ab(a^2 - 2ab + 4b^2)$	3.	$-2a^3b + 4a^2b^2 - 8ab^3$
4.	$(x^2 - 2x + 4)(-3)$	4.	$-3x^2 + 6x - 12$
5.	$(-a^2 + 2ab - b^2)(2a)$	5.	$-2a^3 + 4a^2b - 2ab^2$
6.	$(x^2 - 3xy - 4y^2)(2xy)$	6.	$2x^3y - 6x^2y^2 - 8xy^3$
7.	$\begin{array}{r} -3a^2 + 4a - 6 \\ -2 \\ \hline \end{array}$	7.	$+6a^2 - 8a + 12$
8.	$\begin{array}{r} x^2 - 5x + 8 \\ -3x \\ \hline \end{array}$	8.	$-3x^3 + 15x^2 - 24x$
9.	$\begin{array}{r} -a^2 - 3ab + 4b^2 \\ -2ab \\ \hline \end{array}$	9.	$+2a^3b + 6a^2b^2 - 8ab^3$

C. Multiplying a Polynomial by a Polynomial

To multiply a polynomial times a polynomial, we again use the distributive property. Furthermore, we can perform our multiplication either vertically or horizontally. For example, to multiply $(x + 3)(x + 2)$:

Multiply the *first term* of the first factor by each term in the second factor.

$$(x + 3)(x + 2)$$
$$= (x)(x) + (x)(2)$$
$$= x^2 + 2x$$

Now, multiply the second term of the first factor by each term in the second factor.

$$(x + 3)(x + 2)$$
$$= (x)(x) + (x)(2) + (3)(x) + (3)(2)$$
$$= x^2 + 2x + 3x + 6$$

Then, combine like terms.

$$x^2 + 2x + 3x + 6$$
$$= x^2 + 5x + 6$$

Thus,

$$(x + 3)(x + 2) = x^2 + 5x + 6$$

Vertical multiplication would be performed in a similar manner.

Multiply each term of the binomial $x + 2$ by the first term in the multiplier.

$$x + 2$$
$$x + 3$$
$$\overline{x^2 + 2x}$$

Multiply each term of the binomial $x + 2$ by the second term in the multiplier.

$$x + 2$$
$$x + 3$$
$$\overline{x^2 + 2x}$$
$$+ 3x + 6$$

Notice how we line up like terms $2x$ and $3x$.

Add the partial products.

$$x + 2$$
$$x + 3$$
$$\overline{x^2 + 2x}$$
$$\underline{+ 3x + 6}$$
$$x^2 + 5x + 6$$

PROCEDURE:

To multiply a polynomial times a polynomial:

1. Multiply each term of the first polynomial by each term of the second polynomial.

2. Add the partial products by combining like terms.

EXAMPLE 1: Multiply $(3x - 4)(4x + 5)$.

VERTICAL

SOLUTION: Multiply each term of the binomial $3x - 4$ by the first term in the multiplier.

$$\begin{array}{r} 4x + 5 \\ 3x - 4 \\ \hline 12x^2 + 15x \end{array}$$

Multiply each term of the binomial $3x - 4$ by the second term in the multiplier.

$$\begin{array}{r} 4x + 5 \\ 3x - 4 \\ \hline 12x^2 + 15x \\ - 16x - 20 \end{array}$$

Add the partial products.

$$\begin{array}{r} 4x + 5 \\ 3x - 4 \\ \hline 12x^2 + 15x \\ - 16x - 20 \\ \hline 12x^2 - x - 20 \end{array}$$

Final answer: $12x^2 - x - 20$

HORIZONTAL

SOLUTION: Multiply each term of the second binomial by the first term of the first binomial.

$$(3x - 4)(4x + 5)$$
$$(3x)(4x) + (3x)(5)$$
$$= \quad 12x^2 \quad + \quad 15x$$

Multiply each term of the second binomial by the second term of the first binomial.

$$(3x - 4)(4x + 5)$$
$$= (3x)(4x) + (3x)(5) + (-4)(4x) + (-4)(5)$$
$$= \quad 12x^2 \quad + \quad 15x \quad\quad - 16x \quad\quad - 20$$

Combine like terms.

$$(3x - 4)(4x + 5)$$
$$= (3x)(4x) + (3x)(5) + (-4)(4x) + (-4)(5)$$
$$= \quad 12x^2 \quad + \quad 15x \quad - \quad 16x \quad - \quad 20$$
$$= \quad 12x^2 \quad\quad\quad - x \quad\quad\quad - \quad 20$$

EXAMPLE 2: Multiply $(x^2 + 2x + 3)$ by $(x - 2)$.

VERTICAL

SOLUTION: Multiply $x^2 + 2x + 3$ by x.

$$\begin{array}{r} x^2 + 2x + 3 \\ x - 2 \\ \hline x^3 + 2x^2 + 3x \end{array}$$

HORIZONTAL

SOLUTION: Distribute x over the trinomial.

$$(x - 2)(x^2 + 2x + 3)$$
$$= (x)(x^2) + (x)(2x) + (x)(3)$$
$$= \quad x^3 \quad + \quad 2x^2 \quad + \quad 3x$$

Multiply $x^2 + 2x + 3$ by -2.

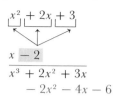

$$x - 2$$
$$x^3 + 2x^2 + 3x$$
$$\quad\quad - 2x^2 - 4x - 6$$

Add the partial products.

$$x^2 + 2x + 3$$
$$x \;\; - 2$$
$$\overline{}$$
$$x^3 + 2x^2 + 3x$$
$$\quad\quad - 2x^2 - 4x - 6$$
$$\overline{}$$
$$x^3 + 0x^2 - \;\; x - 6$$

Final answer: $x^2 - x - 6$

Distribute -2 over the trinomial.

$$(x - 2)(x^2 + 2x + 3)$$
$$(x)(x^2) + (x)(2x) + (x)(3)$$
$$\quad + (-2)(x^2) + (-2)(2x) + (-2)(3)$$
$$= x^3 + 2x^3 + 3x - 2x^2 - 4x - 6$$

Combine like terms.

$$(x - 2)(x^2 + 2x + 3)$$
$$= x^3 + 2x^2 + 3x - 2x^2 - 4x - 6$$
$$= x^3 + \underbrace{2x^2 - 2x^2} + \underbrace{3x - 4x} - 6$$
$$= x^3 \quad\quad + 0x^2 \quad\quad - x \;\; - 6$$
$$= x^3 - x - 6$$

Do the following practice set. Check your answers with the answers in the right-hand margin.

PRACTICE SET 6-5C

Multiply (see example 1 or 2).

1. $(2x - 5)(3x + 4)$
2. $(3x - 4)^2$
3. $(-x + 4)(x - 3)$
4. $(x^2 - 2x - 5)(2x - 3)$
5. $(z^4 + 3z^2 + 9)(z^2 - 3)$
6. $(4x - 5 + 2x^2)(4 + 3x)$

1. $6x^2 - 7x - 20$
2. $9x^2 - 24x + 16$
3. $-x^2 + 7x - 12$
4. $2x^3 - 7x^2 - 4x + 15$
5. $z^6 - 27$
6. $6x^3 + 20x^2 + x - 20$

EXERCISE 6-5

Find the product of the monomials.

1. $(6ab)(-4c)$
2. $(-2x^2)(-3x^3)$
3. $(+4x^2y^3)(+6xy^3)$
4. $(-3a^2b)(-7a)$
5. $(+8xy)\left(-\dfrac{1}{2}x^2y^2\right)$
6. $(-6a^2d)(2ad^3)$
7. $(2)(2a)$
8. $(-2)(6x^2)$

9. $(4)(-2y^2)$

10. $(z)(2z^3)$

11. $(-2)(4)(-6ab)$

12. $\left(\dfrac{1}{2}\right)(6)(3x^2)$

13. $(2x)(3y)(-z)$

14. $(-6y^2)(5y^6)(-2y^3)$

15. $\left(-\dfrac{1}{2}x^2\right)(8x)\left(-\dfrac{1}{4}x^3\right)$

16. $(-8z)(9z^4)(z^3)$

17. $(+6x^2y^3)(-3x^5y^2)$

18. $(2a^2b^2)(3ab^3)(-a^5b^4)$

19. $(-2y)(5xy)(2xy^2)$

20. $(6a)^2$

21. $(-2x)^2$

22. $(6x^3)^2$

23. $(-3a^2b^3)^2$

24. $(-2a^2b)^3$

25. $3(-3y)^2$

26. $(2x^2)(2x)^2$

27. $\left(\dfrac{1}{2}y^3\right)^3(-4y)^2$

28. $(-.4x)^2$

Multiply the following.

29. $4(a + b)$

30. $2x(x^2 - xy)$

31. $3(x + 2y)$

32. $xy(x + xy)$

33. $-3a(a^2 - a)$

34. $3x(2x + 3y - 4z)$

35. $x^2(2x^2 - 3x + 4)$

36. $-x^2(x^2 + 2x - 5)$

37. $ab(3a^2 + 4ab + 5b^2)$

38. $-7xy(-4xy^2 + 2x^2y)$

39. $3y(1 - 2y + 4y^2)$

40. $3x^2y^2(x^2 + xy + y^2)$

41. $5x^2y(2x^2 + 5xy - 4y^2)$

42. $16\left(\dfrac{1}{2}y^2 - \dfrac{3}{4}y + \dfrac{5}{8}\right)$

43. $\dfrac{1}{2}(4x^2 - 8x - 14)$

44. $-15xyz(5xz - 3xy - yz)$

45. $\begin{array}{r} 2x^2 - 4x + 5 \\ \underline{3x} \end{array}$

46. $\begin{array}{r} d^3 - d^2 + 20 \\ \underline{-d^2} \end{array}$

47. $\begin{array}{r} 4a - 2b - c + 4 \\ \underline{-3abc} \end{array}$

48. $\begin{array}{r} 2x^2 - 3x - 5 \\ \underline{-4c^2} \end{array}$

Multiply the following polynomials.

49. $(x + 3)(x + 4)$

50. $(y + 6)(y + 1)$

51. $(z + 2)(z + 2)$

52. $(x - 1)(x - 6)$

53. $(x - 7)(x - 2)$

54. $(x - 2)(x - 2)$

55. $(x + 7)(x - 3)$

56. $(y - 2)(y + 8)$

57. $(z + 2)(z - 9)$

58. $(x - 7)(x + 2)$

59. $(y - 2)(y + 2)$

60. $(z + 4)(z - 4)$

61. $(2a + 3)(a + 1)$

62. $(a - 1)(3a - 4)$

63. $(3x - 4)(x + 2)$

64. $(x - 7)(2x + 7)$

65. $(2y + 1)(y - 1)$

66. $(z + 1)(3z - 1)$

67. $(3 - z)(1 + z)$

68. $(2 - 2z)(1 - z)$

69. $(2x + 1)(x - 6)$

70. $(z + 2)(2z - 3)$

71. $(2a + 1)(3a + 9)$

72. $(4x + 3)(3x - 4)$

73. $(5x + 2)(2x - 2)$

74. $(2x + 3)(2x - 3)$

75. $(5x - 5)(2x - 3)$

76. $(3z - 2)(2z + 2)$

77. $(-x + 2)(x - 4)$

78. $(x - 3)(3x - 3)$

79. $(x + y)(2x + y)$

80. $(3a + b)(2a - b)$

81. $(a^2 + 2)(a^2 - 1)$

82. $(x^2 + 1)(x^2 - 1)$

83. $(x + y)(x + y)$

84. $(x + y)(x - y)$

85. $(x - y)(x - y)$

86. $(2x + y)(x - 2y)$

87. $(2x + 5y)(3x - 4y)$

88. $(x^2 + y^2)(x^2 + y^2)$

89. $(a + 1)^2$

90. $(x - 2)^2$

91. $(2a + b)^2$

92. $(2a - 3)^2$

93. $(-3a - 2b)^2$

94. $(x + 2)^3$

95. $(y - 3)^3$

96. $(a - 2b)^3$

97. $(a + 2)(2a + 3)(3a + 1)$

98. $(x + 3)(x - 1)(x)$

99. $(x + 3)(2x + 1)(3x - 4)$

100. $(x^2 + 2x + 1)(x + 1)$

101. $(x^2 - 3x - 3)(2x - 4)$

102. $(y - 6)(2y^2 - 3y + 4)$

103. $(2x^2 + 3xy - 2y^2)(2x - 3y)$

104. $(7 - 4x + 2x^2)(3 - 2x)$

105. $(-z^2 + 3z - 1)(4 - 3z)$

106. $(4 - x)(2x - 3 + x^2)$

107. $(5 - x)(2x - 3 + x^2)$

108. $(x^2 + x + 1)(x^2 + 2x - 1)$

109. $(x^2 + 2x + 3)(x^2 - 2x + 3)$

110. $(z^2 - z - 3)^2$

6-6   RULES OF EXPONENTS IN DIVIDING POWERS WITH LIKE BASES

A.   The Exponent of the Numerator Is Greater Than the Exponent of the Denominator

To divide $x^5 \div x^2$:

First, express the expression in fractional form.

$$\frac{x^5}{x^2}$$

Now, factor both the numerator and denominator.

$$\frac{x^5}{x^2} = \frac{xxxxx}{xx}$$

Reduce the fraction by canceling common factors.

$$\frac{x^5}{x^2} = \frac{{}^1\cancel{x}\,{}^1\cancel{x}xxx}{{}_1\cancel{x}\,{}_1\cancel{x}}$$

$$= \frac{(1)(1)(x)(x)(x)}{(1)(1)}$$

$$= x^3$$

Thus,

$$x^5 \div x^2 = x^3$$

Notice we could arrive at the same quotient if we subtracted exponents. $\qquad \dfrac{x^5}{x^2} = x^{5-2} = x^3$

This is always true provided the bases are the same. In general,

$$\frac{a^m}{a^n} = a^{m-n} \quad \text{if} \quad m > n.$$

PROCEDURE:

To divide powers of the same base when the exponent of the numerator is greater than the exponent of the denominator:

1. Write the common base as the base of the quotient.

2. Subtract the exponents; the difference of the exponents is the exponent of the quotient. That is,

$$\frac{a^m}{a^n} = a^{m-n}.$$

EXAMPLE 1: Find the value of $a^8 \div a^3$.

SOLUTION: The common base is a.

$$\frac{a^8}{a^3} \qquad a$$

Subtract the exponents.

$$\frac{a^8}{a^3} = a^{8-3}$$

$$= a^5$$

Final answer: a^5

EXAMPLE 2:   Find the value of $x^4 \div x$.

SOLUTION:

$$x^4 \div x = \frac{x^4}{x^1}$$

$$= x^{4-1}$$

$$= x^3$$

Final answer: x^3

EXAMPLE 3:   Find $x^3 y^2 \div xy$.

SOLUTION: x and y are common bases.

$$\frac{x^3 y^2}{xy} \qquad x \cdot y$$

Subtract the exponents for each common base.

$$\frac{x^3 y^2}{xy} = x^{3-1} \cdot y^{2-1}$$

$$= x^2 \cdot y^1$$

Final answer: $x^2 y$

Do the following practice set. Check your answers with the answers in the right-hand margin.

PRACTICE SET 6-6A

Divide the following (see example 1 or 2).

1. $x^8 \div x^2$ 2. $y^3 \div y$ 3. $z^7 \div z^6$

Divide the following (see example 3).

4. $a^4 b^3 \div ab^2$ 5. $x^2 y \div x$ 6. $x^2 y^4 z^3 \div xyz$

1. x^6
2. y^2
3. z

4. $a^3 b$
5. xy
6. $xy^3 z^2$

B. The Exponent of the Numerator Is Equal to the Exponent of the Denominator

To evaluate $x^2 \div x^2$, we proceed as before.

$$x^2 \div x^2 = \frac{x^2}{x^2}$$

$$= \frac{(x)(x)}{(x)(x)} = \frac{{}^1(x)\,{}^1(x)}{(x)_1\,(x)_1}$$

$$= \frac{(1)(1)}{(1)(1)} = 1$$

Thus,

$$x^2 \div x^2 = 1$$

If we were to apply our rule of exponents for dividing powers of the same base, we find that

$$\frac{x^2}{x^2} = x^{2-2} = x^0$$

Moreover, since

$$\frac{x^2}{x^2} = 1 \qquad \text{and} \qquad \frac{x^2}{x^2} = x^0$$

we conclude that $x^0 = 1$. In general,

$$\boxed{\frac{a^m}{a^m} = a^0 = 1}$$

NOTE:

Any nonzero factor or expression with an exponent of zero has a value of 1. For example:

$$a^0 = 1, \qquad (2a)^0 = 1, \qquad 4^0 = 1, \qquad (a+b)^0 = 1$$

Also, suffice it to say, 0^0 is not defined.

To divide powers of the same base whenever the exponents are equal, we apply the procedure that was previously outlined, noting that a nonzero base with an exponent of zero has a value of 1.

EXAMPLE 1:   Evaluate $\dfrac{x^6}{x^6}$.

SOLUTION: Since we are dividing like bases, we can subtract the exponents.

$$\frac{x^6}{x^6} = x^{6-6}$$
$$= x^0$$
$$= 1$$

Final answer: 1

EXAMPLE 2:   Evaluate $\dfrac{x^3 y^2}{x y^2}$.

SOLUTION:

$$\frac{x^3 y^2}{x y^2} = \frac{x^3}{x^1} \cdot \frac{y^2}{y^2}$$
$$= x^{3-1} y^{2-2}$$
$$= x^2 y^0$$
$$= x^2 \cdot 1$$

Final answer: x^2

NOTE:

When problems contain more than one base as in example 2, rewrite the problem so that only like bases are being divided.

Do the following practice set. Check your answers with the answers in the right-hand margin.

PRACTICE SET 6-6B

Divide (see example 1 or 2).

1. $\dfrac{y^4}{y^4}$

2. $\dfrac{x^2 y^3}{x^2 y}$

3. $\dfrac{x^3 y^2 z}{x y^2}$

1. 1
2. y^2
3. $x^2 z$

C. The Exponent of the Numerator Is Smaller Than the Exponent of the Denominator

Once again, to evaluate $x^2 \div x^5$, we proceed as before.

$$x^2 \div x^5 = \frac{x^2}{x^5}$$

$$= \frac{x \cdot x}{x \cdot x \cdot x \cdot x \cdot x} = \frac{{}^1x \cdot {}^1x}{\underset{1}{x} \cdot \underset{1}{x} \cdot x \cdot x \cdot x}$$

$$= \frac{(1)(1)}{(1)(1)(x)(x)(x)}$$

$$= \frac{1}{x^3}$$

Thus,

$$\frac{x^2}{x^5} = \frac{1}{x^3}$$

Notice that since the exponent of the numerator is smaller than the exponent of the denominator, the quotient will yield a fraction with a numerator of 1. Furthermore, the exponent of the denominator can be found by subtracting the smaller exponent from the larger exponent. Thus,

$$\frac{x^2}{x^5} = \frac{1}{x^{5-2}} = \frac{1}{x^3}$$

This will always be true provided the bases are equal. In general,

$$\boxed{\frac{a^n}{a^m} = \frac{1}{a^{m-n}} \qquad \text{where } m > n}$$

PROCEDURE:

To divide powers having the same base when the exponent of the denominator is greater than the exponent of the numerator, observe that:

1. The numerator of the quotient is 1.

2. The denominator of the quotient is found by keeping the common base and subtracting the smaller exponent from the larger exponent.

EXAMPLE 1: Evaluate $\dfrac{x^2}{x^7}$.

SOLUTION:

$$\frac{x^2}{x^7} = \frac{1}{x^{7-2}} \quad = (x^{-5}$$

$$= \frac{1}{x^5}$$

Final answer: $\dfrac{1}{x^5}$

EXAMPLE 2: Evaluate $(x^2y) \div (x^2y^3)$.

SOLUTION:

$$(x^2y) \div (x^2y^3) = \frac{x^2y}{x^2y^3} = \frac{x^2}{x^2} \cdot \frac{y}{y^3}$$

$$= (x^{2-2})\left(\frac{1}{y^{3-1}}\right)$$

$$= (x^0)\left(\frac{1}{y^2}\right)$$

$$= (1)\left(\frac{1}{y^2}\right)$$

$$= \frac{1}{y^2}$$

Final answer: $\dfrac{1}{y^2}$

EXAMPLE 3: Evaluate $(x^4y^2z^2) \div (xy^2z^6)$.

SOLUTION:

$$(x^4y^2z^2) \div (xy^2z^6) = \frac{x^4y^2z^2}{xy^2z^6} = \frac{x^4}{x} \cdot \frac{y^2}{y^2} \cdot \frac{z^2}{z^6}$$

$$= (x^{4-1})(y^{2-2})\left(\frac{1}{z^{6-2}}\right)$$

$$= (x^3)(y^0)\left(\frac{1}{z^4}\right)$$

$$= (x^3)(1)\left(\frac{1}{z^4}\right)$$

$$= (x^3)\left(\frac{1}{z^4}\right)$$

$$= \frac{x^3}{z^4}$$

Final answer: $\dfrac{x^3}{z^4}$

Do the following practice set. Check your answers with the answers in the right-hand margin.

PRACTICE SET 6-6C

Divide (see example 1, 2, or 3).

1. $\dfrac{y}{y^7}$ 2. $\dfrac{z^6}{z^7}$ 3. $\dfrac{y^3}{y^8}$

4. $\dfrac{y^4 z}{yz^4}$ 5. $\dfrac{x^3 y^3}{x^3 y^4}$ 6. $\dfrac{a^3 b^2 c}{a^3 b^4 c^5}$

1. $1/y^6$
2. $1/z$
3. $1/y^5$
4. y^3/z^3
5. $1/y$
6. $1/b^2 c^4$

EXERCISE 6-6

Divide the following.

1. $x^9 \div x^2$ 2. $y^4 \div y$ 3. $x^{10} \div x^5$

4. $y^6 \div y^5$ 5. $z^{3b} \div z^b$ 6. $w^9 \div w^8$

7. $10^4 \div 10$ 8. $3^5 \div 3^2$ 9. $y^2 \div y$

10. $\dfrac{x^8}{x^3}$ 11. $\dfrac{y^4}{y}$ 12. $\dfrac{z^6}{z^5}$

13. $\dfrac{10^9}{10^7}$ 14. $\dfrac{a^2 b^b}{ab}$ 15. $\dfrac{a^3 b}{a}$

16. $\dfrac{x^7 y^2}{x^5 y}$ 17. $\dfrac{x^2 y^3 z^4}{y^2}$ 18. $\dfrac{a^3 b^2 c^4}{a^2 bc}$

19. $\dfrac{a^5 b^3}{ba^3}$ 20. $\dfrac{a^4 b^8}{ab^7}$ 21. $\dfrac{x^4 y^3 z^7}{x^2 y^2}$

Divide the following.

22. $\dfrac{x^3}{x^3}$ 23. $\dfrac{y^5}{y^5}$ 24. $\dfrac{z}{z}$

25. $\dfrac{10^3}{10^3}$ 26. $\dfrac{5^2}{5^2}$ 27. $\dfrac{a^b}{a^b}$

28. $\dfrac{ab}{ab}$ 29. $\dfrac{x^2 y}{x^2 y}$ 30. $\dfrac{x^3 z^2}{x^3 z^2}$

31. $\dfrac{x^2 y}{xy}$ 32. $\dfrac{a^3 c^2}{a^3}$ 33. $\dfrac{ab}{a}$

34. $\dfrac{x^6 y^3}{x y^3}$

35. $\dfrac{a^2 b^2 c^2}{a b^2}$

36. $\dfrac{a^2 b^2 c^2}{a^2 b^2 c}$

Divide the following.

37. $y^2 \div y^8$

38. $8^2 \div 8^5$

39. $z \div z^2$

40. $a^3 \div a^3 b$

41. $b \div b^3$

42. $ab \div a^2 b^2$

43. $\dfrac{abc}{a^2 b^2 c^2}$

44. $\dfrac{a b^2 c^3}{a b^3 c^5}$

45. $\dfrac{ab}{b^3}$

46. $\dfrac{abc}{abc}$

47. $\dfrac{x^6 y z^2}{x^2 y z^6}$

48. $\dfrac{a^5 bc}{a b^5 c}$

6-7   DIVISION OF POLYNOMIALS

A.   Dividing a Monomial by a Monomial

A monomial usually consists of a numerical coefficient and variables. When dividing a monomial by a monomial, we should "break up" the problem so that we can use skills that we have already acquired. For example, to divide $12b^2/3b$:

First, find the quotient of the two numbers.

$$\dfrac{12\, b^2}{3\, b} \qquad 4$$

Then, divide the like base, b.

$$\dfrac{12\, b^2}{3\, b}$$

$$= 4b^{2-1}$$
$$= 4b^1$$
$$= 4b$$

Thus,

$$\dfrac{12b^2}{3b} = 4b$$

PROCEDURE:

To divide monomials:

1. Write the division problem in fractional form.

2. Divide the numerical coefficients.

3. Divide the variables, simplifying powers with the same base.

EXAMPLE 1:   Divide: $(21a^2) \div 7a$.

SOLUTION: Divide the coefficients.

$$21a^2 \div 7a = \frac{21\ a^2}{7\ a}$$

Divide the variable.

$$\frac{\overset{3}{21}\ a^2}{7\ a} = 3a^{2-1}$$
$$= 3a^1$$

Final answer: $3a$

EXAMPLE 2:   Divide: $\dfrac{20c^3d^4}{-4cd^3}$.

SOLUTION: Divide the coefficients.

$$\frac{20\ c^3d^4}{-4\ cd^3}$$

Divide the variables.

$$\frac{\overset{-5}{20}\ c^3\ d^4}{-4\ c\ d^3} = -5c^{3-1}d^{4-3}$$
$$= -5c^2d^1$$

Final answer: $-5c^2d$

EXAMPLE 3:   Divide $\dfrac{-18a^3c^3d^2}{6a^3c}$.

SOLUTION:

$$\frac{-18a^3c^3d^2}{6a^3c} = \frac{-18}{6} \cdot \frac{a^3}{a^3} \cdot \frac{c^3}{c} \cdot \frac{d^2}{1}$$
$$= (-3) \cdot (a^{3-3}) \cdot (c^{3-1}) \cdot (d^2)$$
$$= (-3)(a^0)(c^2)(d^2)$$
$$= (-3)(1)(c^2)(d^2)$$
$$= -3c^2d^2$$

Final answer: $-3c^2d^2$

EXAMPLE 4: Divide $\dfrac{12xy^6}{-3x^2y^2}$.

SOLUTION:

$$\dfrac{12xy^6}{-3x^2y^2} = \dfrac{12}{-3} \cdot \dfrac{x}{x^2} \cdot \dfrac{y^6}{y^2}$$

$$= (-4)\left(\dfrac{1}{x^{2-1}}\right)(y^{6-2})$$

$$= (-4)\left(\dfrac{1}{x}\right)(y^4)$$

$$= \dfrac{-4y^4}{x}$$

Final answer: $\dfrac{-4y^4}{x}$

Do the following practice set. Check your answers with the answers in the right-hand margin.

PRACTICE SET 6-7A

Divide (see example 1).

1. $\dfrac{9x^2}{3x}$

2. $\dfrac{6y^3}{2y^3}$

3. $\dfrac{-12z^6}{3z^2}$

1. $3x$
2. 3
3. $-4z^4$

Divide (see example 2 or 3).

4. $\dfrac{-10x^3y^4}{2xy^3}$

5. $\dfrac{-21xy^2z}{3xy}$

6. $\dfrac{4x^2y^3}{y^3}$

4. $-5x^2y$
5. $-7yz$
6. $4x^2$

Divide (see example 4).

7. $\dfrac{16x^2y^4}{4x^3y}$

8. $\dfrac{25x^2}{5x^2y}$

9. $\dfrac{-20ab^3c}{5abc^2}$

7. $4y^3/x$
8. $5/y$
9. $-4b^2/c$

B. Dividing a Polynomial by a Monomial

Dividing a polynomial by a monomial is very similar in technique to multiplying a polynomial by a monomial—both techniques require the application of the distributive law.

Dividing a polynomial $(a + b)$ by a monomial (2), $$\frac{a + b}{2}$$

Is the same as finding $\left(\frac{1}{2}\right)$ of $(a + b)$. $$\frac{a + b}{2} = \frac{1}{2}(a + b)$$

NOTE:

> Dividing by 2 is the same as multiplying by 1/2.
>
> Dividing by 3 is the same as multiplying by 1/3.
>
> Dividing by n is the same as multiplying by $1/n$.

To evaluate $\frac{1}{2}(a + b)$:

Apply the distributive property of multiplication, $\frac{1}{2}(a + b)$
$\frac{1}{2}(a) + \frac{1}{2}(b)$

Which is the same as $\dfrac{a}{2} + \dfrac{b}{2}$

If $(a + b)/2$ equals $a/2 + b/2$, then to divide a polynomial by a monomial, we divide every term in the polynomial by the monomial.

PROCEDURE:

> To divide a polynomial by a monomial:
>
> 1. Write as a fraction, placing the polynomial over the monomial.
>
> 2. Divide each term of the polynomial by the monomial.

EXAMPLE 1: Divide $(2a + 4b)$ by 2.

SOLUTION: First, write as a fraction.

$$\frac{2a + 4b}{2}$$

Divide the first term in the numerator by 2.

$$\frac{2a + 4b}{2}$$

$$\frac{2a}{2}$$

$$a$$

Divide the second term in the numerator by 2.

$$\frac{2a + 4b}{2}$$

$$= \frac{2a}{2} + \frac{4b}{2}$$

$$= a + 2b$$

Final answer: $a + 2b$

EXAMPLE 2:   Divide $(7x^2y - 21xy^2)$ by $(7xy)$.

SOLUTION: First, write as a fraction.

$$\frac{7x^2y - 21xy^2}{7xy}$$

Divide the first term by $7xy$.

$$\frac{7x^2y - 21xy^2}{7xy}$$

$$\frac{7x^2y}{7xy}$$

$$x$$

Divide the second term by $7xy$.

$$\frac{7x^2y - 21xy^2}{7xy}$$

$$= \frac{7x^2y}{7xy} + \frac{-21xy^2}{7xy}$$

$$= x - 3y$$

Final answer: $x - 3y$

Do the following practice set. Check your answers with the answers in the right-hand margin.

PRACTICE SET 6-7B

Divide (see example 1 or 2).

1. $(16x^2 + 8x + 4) \div 2$
2. $(25a - 5b + 15c) \div 5$
3. $(3a^2b - 9ab^2 + 6b^3) \div 3b$

1. $8x^2 + 4x + 2$
2. $5a - b + 3c$
3. $a^2 - 3ab + 2b^2$

C. Dividing a Polynomial by a Polynomial

Division is the inverse operation of multiplication and as such, we may compare expressions in the following manner. In the equation, $2(x + 3) = 2x + 6$, we know:

■ (2) multiplied by $(x + 3)$ is $(2x + 6)$.

$$2(x + 3)$$
$$= 2(x) + 2(3)$$
$$= 2x + 6$$

■ $(2x + 6)$ divided by (2) is $(x + 3)$.

$$\frac{2x + 6}{2} = \frac{2x}{2} + \frac{6}{2} = x + 3$$

■ $(2x + 6)$ divided by $(x + 3)$ is 2.

$$\frac{2x + 6}{x + 3} = \frac{2x}{x + 3} + \frac{6}{x + 3} = 2$$

Intuitively, we know division by a polynomial is possible. However, with our present techniques we are unable to evaluate

$$\left(\frac{2x + 6}{x + 3}\right) \qquad \text{or} \qquad \left(\frac{2x}{x + 3} + \frac{6}{x + 3}\right)$$

The following procedure, which parallels the method used in arithmetic for long division, should be employed when dividing by a polynomial.

PROCEDURE:

Dividing a polynomial by a polynomial:

1. Write the problem in long division form.
    a. Arrange the divisor and dividend in descending order.
    b. Account for missing terms.

2. To find the terms of the quotient:
    a. Divide the first term in the divisor into the first term in the dividend; write the answer as the first term in the quotient.
    b. Multiply the divisor by the first term in the quotient, writing the product under the dividend.
    c. Subtract like terms.
    d. Bring down the next term.

3. Repeat (2), divide, multiply, subtract, and bring down.
    Continue until all terms in the dividend have been used.

EXAMPLE 1: Divide $(x^2 + 5x + 6) \div (x + 2)$.

SOLUTION: 1. Write in long division form.

$$x + 2 \overline{\smash{)}\, x^2 + 5x + 6}$$

2. Find the first term in the quotient,

    a. Divide x into x^2.
$$x^2 \div x = x$$

$$x + 2 \overline{\smash{)}\, x^2 + 5x + 6} \quad \overset{x}{}$$

    b. Multiply (x) times $(x + 2)$.
$$x(x + 2) = x^2 + 2x.$$

$$\begin{array}{r} x \\ x + 2 \overline{\smash{)}\, x^2 + 5x + 6} \\ x^2 + 2x \end{array}$$

    c. Subtract like terms.
$$x^2 - (x^2) = 0$$
$$5x - (+2x) = 3x$$

$$\begin{array}{r} x \\ x + 2 \overline{\smash{)}\, x^2 + 5x + 6} \\ -(x^2 + 2x) \\ \hline 0 + 3x \end{array}$$

    d. Bring down the next term $(+6)$.

$$\begin{array}{r} x \\ x + 2 \overline{\smash{)}\, x^2 + 5x + 6} \\ -(x^2 + 2x) \quad\downarrow \\ \hline + 3x + 6 \end{array}$$

3. Find the second term in the quotient.
    a. Divide x into $+3x$.
$$3x \div x = +3$$

$$\begin{array}{r} x + 3 \\ x + 2 \overline{\smash{)}\, x^2 + 5x + 6} \\ -(x^2 + 2x) \\ \hline + 3x + 6 \end{array}$$

    b. Multiply $+3$ times $(x + 2)$.
$$+3(x + 2) = 3x + 6.$$

$$\begin{array}{r} x + 3 \\ x + 2 \overline{\smash{)}\, x^2 + 5x + 6} \\ -(x^2 + 2x) \\ \hline + 3x + 6 \\ + 3x + 6 \end{array}$$

    c. Subtract like terms.
$$(+3x) - (+3x) = 0$$
$$(+6) - (+6) = 0$$

$$\begin{array}{r} x + 3 \\ x + 2 \overline{\smash{)}\, x^2 + 5x + 6} \\ -(x^2 + 2x) \\ \hline + 3x + 6 \\ -(3x + 6) \\ \hline 0 + 0 \end{array}$$

Final answer: $(x + 3)$

EXAMPLE 2: Divide $(x^3 + 8) \div (x + 2)$.

SOLUTION: 1. Write in long division form and account for the missing terms.

$$x + 2 \overline{\smash{)}\, x^3 + 0x^2 + 0x + 8}$$

Example continued on next page.

2. Find the first term in the quotient.

$$\begin{array}{r} x^2 \\ x + 2 \overline{\smash{\big)}\, x^3 + 0x^2 + 0x + 8} \\ -(x^3 + 2x^2) \\ \hline - 2x^2 + 0x \end{array}$$

3. Repeat step (2). Find the second term in the quotient.

$$\begin{array}{r} x^2 - 2x \\ x + 2 \overline{\smash{\big)}\, x^3 + 0x^2 + 0x + 8} \\ -(x^3 + 2x^2) \\ \hline - 2x^2 + 0x \\ -(-2x^2 - 4x) \\ \hline + 4x + 8 \end{array}$$

4. Repeat step (2). Find the third term in the quotient.

$$\begin{array}{r} x^2 - 2x \; + 4 \\ x + 2 \overline{\smash{\big)}\, x^3 + 0x^2 + 0x + 8} \\ -(x^3 + 2x^2) \\ \hline - 2x^2 + 0x \\ -(-2x^2 - 4x) \\ \hline + 4x + 8 \\ -(+4x + 8) \\ \hline 0 \end{array}$$

Final answer: $x^2 - 2x + 4$

EXAMPLE 3: Divide: $(-5x + 4 + 4x^2) \div (x - 2)$.

SOLUTION:
1. Write in long division form.

$$x - 2 \overline{\smash{\big)}\, 4x^2 - 5x + 4}$$

2. Find the first term of the quotient.

$$\begin{array}{r} 4x \\ x - 2 \overline{\smash{\big)}\, 4x^2 - 5x + 4} \\ -(4x^2 - 8x) \\ \hline + 3x + 4 \end{array}$$

3. Repeat step (2). Find the second term of the quotient.

$$\begin{array}{r} 4x + 3 \\ x - 2 \overline{\smash{\big)}\, 4x^2 - 5x + 4} \\ -(4x^2 - 8x) \\ \hline + 3x + 4 \\ -(+3x - 6) \\ \hline + 10 \text{ (remainder)} \end{array}$$

Final answer: $4x + 3 + \dfrac{10}{x - 2}$

NOTE:

Remainders may be expressed in fractional form by writing the remainder over the divisor (e.g., $10/(x - 2)$). This fraction is added to the other terms of the quotient (e.g., $4x + 3 + 10/(x - 2)$).

Do the following practice set. Check your answers with the answers in the right-hand margin.

PRACTICE SET 6-7C

Divide (see example 1).

1. $(x^2 + 11x + 30) \div (x + 5)$
2. $(4x^3 + 4x^2 - 5x + 1) \div (2x - 1)$
3. $(3y^2 + 19y + 20) \div (3y + 4)$

1. $x + 6$
2. $2x^2 + 3x - 1$
3. $y + 5$

Divide (see example 2).

4. $(x^2 - 4) \div (x - 2)$
5. $(2x^3 + 7x^2 - 9) \div (2x + 3)$
6. $(3x^3 - 8x^2 - 9) \div (x - 3)$

4. $(x + 2)$
5. $(x^2 + 2x - 3)$
6. $3x^2 + x + 3$

Divide (see example 3).

7. $(4x^2 - 5x + 4) \div (x - 2)$
8. $(6x^2 + 5x - 5) \div (2x + 3)$
9. $(x^3 + 11) \div (x + 2)$

7. $4x + 3 + 10/(x - 2)$
8. $3x - 2 + 1/(2x + 3)$
9. $x^2 - 2x + 4 + 3/(x + 2)$

EXERCISE 6-7

Divide the following monomials.

1. $(+12x^5) \div (4x^2)$
2. $(-15x^5y^4) \div (3x^2y^3)$
3. $(-18a^3b^2) \div (a^2b)$
4. $(16a^3b^4c^2) \div (4a^3b)$
5. $12x \div 6$
6. $14x \div (-7)$
7. $-16y \div 8$
8. $-24z \div (-6)$
9. $2ab \div 2a$
10. $12x^2y \div 6x$
11. $-18x^3y^4 \div 6x^2y$
12. $20a^2 \div (-2a)$
13. $\dfrac{22b^2}{11b}$
14. $\dfrac{18b^2}{18b^2}$
15. $\dfrac{-6ab}{6ab}$
16. $\dfrac{14a^2b^2}{14a^2b}$
17. $\dfrac{16c^2d^2}{4c^2d}$
18. $\dfrac{26xyz}{13xyz}$
19. $\dfrac{-14x^8y^2z^3}{7x^4y}$
20. $\dfrac{-24x^2y^4}{-3xy^2}$
21. $\dfrac{-xyz}{xy}$
22. $\dfrac{-21rs^2}{-7rs}$
23. $\dfrac{-16rs^2}{-8r^3s^2}$
24. $\dfrac{42x^3y}{-7x^3y^2}$

25. $\dfrac{-18x^4y}{3xy^4}$

26. $\dfrac{3abc}{-3ab^3}$

27. $\dfrac{4x^2y}{x^2y^2}$

28. $\dfrac{xyz}{6xz}$

29. $\dfrac{25(a+b)^2}{-5(a+b)}$

30. $\dfrac{-12(x+y)^2}{-4(x+y)^2}$

Divide.

31. $(4x+6y) \div 2$

32. $(18x^2-12x) \div 3$

33. $(12x-4) \div 4$

34. $\dfrac{8a-4}{-2}$

35. $\dfrac{-6d^2-3d}{-3}$

36. $\dfrac{x^2-x}{x}$

37. $\dfrac{xy+xz}{x}$

38. $\dfrac{ax^2-ax}{a}$

39. $\dfrac{x^2+6x}{x}$

40. $\dfrac{z^2-6z}{-z}$

41. $\dfrac{12y^3-9y^2}{3y}$

42. $\dfrac{15y^2-10y}{-5y}$

43. $\dfrac{6y^3-12y^2}{6y^2}$

44. $\dfrac{14y^4+7y^2}{7y^2}$

45. $\dfrac{12a^3-4a^2}{-4a^2}$

46. $\dfrac{4a^2b-3ab^2}{ab}$

47. $\dfrac{12c^2d-4cd^2}{4cd}$

48. $\dfrac{-9a^4b^2-6a^3b^2}{3a^2b^2}$

49. $\dfrac{5x^3-15x^2-10}{-5}$

50. $\dfrac{24z^3-12z^2+6z}{6z}$

51. $\dfrac{9x^2y-3xy+6xy^2}{3xy}$

52. $\dfrac{\pi r^2+\pi rh}{\pi r}$

Divide the following polynomials.

53. $(x^2-5x+6) \div (x-3)$

54. $(x^2-9x+14) \div (x-7)$

55. $(x^2+5x+6) \div (x+2)$

56. $(2x^2+5x+2) \div (x+2)$

57. $(3x^2+5x-12) \div (3x-4)$

58. $(2x^2+3x-20) \div (2x-5)$

59. $(x^2-8x+15) \div (x-3)$

60. $(2x^2-13x+15) \div (2x-3)$

61. $(3x^2-23x+14) \div (3x-2)$

62. $(4x^2-12x+9) \div (2x-3)$

63. $(x^2-10x+25) \div (x-5)$

64. $(4x^2+12x+9) \div (2x+3)$

65. $(10x^2+x-21) \div (2x+3)$

66. $(15x^2+7x-2) \div (5x-1)$

67. $(x^3-3x^2+11x-6) \div (x-3)$

68. $(2y^3-3y^2-8y+12) \div (y+2)$

69. $(6y^3+y^2-16y+7) \div (2y-1)$

70. $(4y^3-17y+20) \div (2y+5)$

71. $(3y^3-8y^2-9) \div (y-3)$

72. $(27x^3-1) \div (3x-1)$

73. $(12x^3-17x^2+21x-16) \div (3x-2)$

74. $(20x^2+16x-21) \div (2x+3)$

75. $(6x^2+5x-15) \div (2x+3)$

76. $(x^6-8) \div (x^2-2)$

77. $(4y^4+1) \div (2y^2+2y+1)$

78. $(2y^3-y^2-4y+3) \div (y^2-2y+1)$

79. $(x^4-6x^2+8) \div (x^2-4)$

80. $(x^2+25) \div (x+5)$

REVIEW EXERCISES

Find the numerical coefficient of each term.

1. $4xy$

2. $-2xyz$

3. $2y^2 - 5y$

Find the number of terms in the following expressions.

4. $2x$

5. $2x^2 + x$

6. $\dfrac{1}{2}(x + y)$

Arrange the polynomials in descending powers of the indicated variable.

7. $x^2 - x^3 + x$ powers of x

8. $7y^2 - 2x^2 + 4xy$ powers of x;   powers of y

9. $s^4 - 3rs^3 - 3r^2s^2 + r^3s$ powers of r

Add the following monomials.

10. $5a^2$ and $7a^2$

11. $(2a) + (3a) + (-2a)$

12. $\begin{array}{r} 6a^3 \\ -4a^3 \\ \underline{-2a^3} \end{array}$

Add the following polynomials.

13. $3x^2 + 4x$ and $x^2 - 6x$

14. $3x^2 + 4x - 1$ and $4 + x - x^2$

15. $\begin{array}{l} x^3 + x^2 + 5x \\ 3x^3 + x^2 - 1 \\ 4x^2 - 6x + 5 \\ \underline{-2x^3 + 2x - 3} \end{array}$

16. $(3y^3 - 3y^2 - 3y - 9) + (4y^4 + 3y^3 - 6y^2 + 3y - 8)$

Subtract the following monomials.

17. $6x - (4x)$

18. $-2x^2 - (-6x^2)$

19. $-8ab - (-8ab)$

Subtract the following polynomials.

20. $\begin{array}{l} 6a + 4 \\ \underline{a - 6} \end{array}$

21. $\begin{array}{l} 3x^2 + 4x - 1 \\ \underline{2x^2 + 6x} \end{array}$

22. $\begin{array}{l} x^2 - 1 \\ \underline{3x^2 - x + 4} \end{array}$

23. $(2x^2 + 3x - 11) - (4x^2 - 4x + 6)$

Simplify and combine like terms.

24. $(+4x) + (-6x) - (+x)$
25. $x^2 + (2x^2 - 4x + 5)$
26. $(3x + 4) + (x - 4) - (2x - 3)$

Simplify the following.

27. $x^2 \cdot x^3$ 28. $2^2 \cdot 2^6 \cdot 2^3$ 29. $(a^2 b^3)^2 (c)^3$

Multiply the following monomials.

30. $(5a)(-2a)$ 31. $(3x^2)(4x^3)$ 32. $(3a^2 bc)(-2b^2)(2abc^3)$

Multiply the following.

33. $-4(2x - 3)$ 34. $2a(a^2 - 3a + 6)$ 35. $-3x^2(-x^2 + 3x - 4)$

36. $\begin{array}{r} 2x^2 - 4x \\ x - 4 \\ \hline \end{array}$ 37. $\begin{array}{r} z^2 - 3z + 5 \\ 5z - 2 \\ \hline \end{array}$ 38. $(a - 2)^3$

Simplify the following.

39. $\dfrac{y^{10}}{y^3}$ 40. $\dfrac{x^2 y^3}{xy}$ 41. $\dfrac{xy^2}{y^2 z}$ 42. $\dfrac{ab^2 c^6}{ab^5 c^3}$

Divide the following monomials.

43. $\dfrac{24a^4}{6a^2}$ 44. $\dfrac{-16a^2 b}{8a^2}$ 45. $\dfrac{125x^3 yz^2}{-5x^2 y}$

Divide the following.

46. $\dfrac{2x^2 + 6x}{2x}$ 47. $\dfrac{x^3 - 3x^2 - 4x}{-x}$ 48. $\dfrac{15x^2 y^2 - 10xy}{-5xy}$

49. $2x - 3 \, \overline{\smash{\big)}\, 4x^2 - 12x + 9}$ 50. $2a + b \, \overline{\smash{\big)}\, 6a^2 - 5ab - 4b^2}$

CHAPTER 7

Special Products and Factoring

7-1 INTRODUCTION: THE MEANING OF FACTORING

A product is found by multiplying numbers or terms together. For example:

2 multiplied by 3 gives a product of 6 $\qquad$ $2 \cdot 3 = \boxed{6}$

$\qquad\qquad\qquad\qquad\qquad\qquad\qquad\qquad\qquad$ product

and

2x multiplied by 3x gives a product of 6x^2. $\qquad$ $2x \cdot 3x = \boxed{6x^2}$

$\qquad\qquad\qquad\qquad\qquad\qquad\qquad\qquad\qquad$ product

The numbers or terms multiplied together are called the factors of the product. In the previous examples:

Factors of 6 are 2 and 3 $\qquad$ $\boxed{2} \cdot \boxed{3} = 6$

$\qquad\qquad\qquad\qquad\qquad\qquad\qquad\qquad$ factors

and

Factors of 6x^2 are 2x and 3x. $\qquad$ $\boxed{2x} \cdot \boxed{3x} = 6x^2$

$\qquad\qquad\qquad\qquad\qquad\qquad\qquad\qquad$ factors

A factor may consist of more than one term. For example, in the expression $5(x + y) = 5x + 5y$:

One factor is 5 $\qquad$ $\boxed{5}(x + y) = 5x + 5y$

and

The second factor is $(x + y)$. $\qquad$ $5\boxed{(x + y)} = 5x + 5y$

Notice that the value of an expression is not changed when expressing it in factored form.

A factor of an expression may also be thought of as a *divisor* of the expression. Therefore, if a factor of an expression is known, a second factor may be found by dividing the expression by the given factor. The quotient will be the second factor.

For example, if $2x$ is a factor of $6x$:

Find the missing factor by dividing $6x$ by $2x$.
$$\frac{6x}{2x} = (\quad)$$
missing factor

Thus,

the missing factor is 3.
$$\frac{6x}{2x} = 3$$

Another method that is commonly used to find a second factor is *multiplying back* (to find the missing *multiplier*). Using the same example, we find that

$2x$ times the missing factor is $6x$.
$$(2x)(\underset{\text{missing} \atop \text{factor}}{\blacksquare}) = 6x$$

The missing *multiplier* is 3.
$$(2x)(3) = 6x$$

Thus,

the missing factor is 3.

For a second example, if 5 is a factor of $(5x + 5y)$, we can find the other factor by:

Dividing the expression by 5.
$$\frac{5x + 5y}{5} = (\quad)$$

Thus,

the missing factor is the quotient $(x + y)$.
$$\frac{5x + 5y}{5}$$
$$\frac{5x}{5} + \frac{5y}{5} = (x + y)$$

In multiplication, we may use the given factor to multiply back.

5 times some expression equals $5x + 5y$.
$$5(\blacksquare) = 5x + 5y$$

The missing *multiplier* is $(x + y)$.
$$5\,(x + y) = 5x + 5y$$

Thus,

the missing factor is $(x + y)$.

From these two examples, we observe that if one factor of an expression is known, we may find the (second or missing) factor of the expression by thinking of the problem as either a division or a multiplication problem.

In practice, you will find that the multiplication method is intuitive in nature and usually faster than the division method. To find a missing factor, we use the multiplication method.

We now focus our attention on factoring algebraic expressions. Previously, we learned how to multiply to find the product form. This chapter reverses that procedure; that is, given the product, we express it in factored form.

7-2   FACTORING POLYNOMIALS WHOSE TERMS HAVE A COMMON MONOMIAL FACTOR

To factor a polynomial means to express it as the indicated product of other polynomials. For example, we know, according to the distributive property, that

$$5(x + y) = 5x + 5y$$

Notice that each term of our product contains a common factor of 5.

$$5x + 5y$$

5 is therefore called a *common monomial factor* of $5x + 5y$.

Using the distributive property in *reverse*, we can

Factor out 5 from each term.

$$5x + 5y$$
$$5(x + y)$$

Thus, the polynomial, $5x + 5y$, may be written in factored form as $5(x + y)$. Five (5) is the *common monomial factor* and the second factor, $(x + y)$, is obtained by factoring out 5.

NOTE:

Whenever we factor out a common factor, we are removing the common factor from each term and bringing it outside a set of parentheses. The set of parentheses will then contain the second factor of the polynomial. For example:

- In the polynomial $5x + 5y$,

  $5x + 5y$

- We factored out the common monomial factor from each term of the polynomial and brought it outside a set of parentheses.

  $5x + 5y$
  $5(\_\_\_)$

- The second factor is the remaining part after 5 is factored out.

  $5x + 5y$
  $= 5(x + y)$

In order to factor a polynomial whose terms contain a common monomial factor, we must be able to identify the *greatest common factor* (GCF) contained in each term of the given polynomial. For example:

The polynomial $4x^2 + 8x$ contains two terms, $4x^2$ and $8x$.

$$4x^2 + 8x$$

Notice that each term contains the common factor $4x$.

$$4x^2 \qquad 8x$$
$$= 4x \cdot x + 4x \cdot 2$$

It should be further observed that $4x$ is the *greatest common factor*.

If we factor out the GCF $(4x)$ from each term

$$4x^2 + 8x$$
$$= 4x(\_\_\_)$$

we can multiply back to find the terms of the missing second factor.

$$4x \quad \text{times} \quad x = 4x^2 \qquad\qquad 4x^2 + 8x$$
$$4x(x \quad)$$

$$4x \quad \text{times} \quad +2 = 8x \qquad\qquad 4x^2 + 8x$$
$$4x(x + 2)$$

Thus,

$$4x^2 + 8x \text{ may be factored as } 4x(x + 2).$$

NOTE:

The following observations should be made.

1. Whenever we factor out the GCF from a polynomial, the second factor will always have the same number of terms as the original expression. For example:

   Since the original expression $4x^2 + 8x$ contains two terms $\qquad 4x^2 + 8x$

   When we factor out the GCF, the second factor will also $\qquad 4x^2 + 8x$
   contain two terms. $\qquad\qquad\qquad\qquad\qquad\qquad\qquad\qquad 4x(x + 2)$

2. We find the missing factor one term at a time.

   $$4x \quad \text{times} \quad x = 4x^2 \qquad\qquad 4x^2 + 8x$$
   $$4x(x \quad)$$

   $$4x \quad \text{times} \quad +2 = 8x \qquad\qquad 4x^2 + 8x$$
   $$4x(x + 2)$$

Here's another example. Factor $3x^3 + 6x^2 + 12x$.

First, observe that the GCF of each term is $3x$. $\qquad\qquad 3x^3 + 6x^2 + 12x$
$$\text{GCF is } 3x$$

Next, factor out $3x$ from each term. Notice that the $\qquad 3x^3 + 6x^2 + 12x$
second (missing) factor will have three terms. $\qquad\qquad = 3x(x^2 + 2x + 4)$

Now, simply find each term of the second factor (one at a time) by using the GCF to multiply back.

$$3x \quad \text{times} \quad x^2 = 3x^3 \qquad\qquad 3x^3 + 6x^2 + 12x$$
$$3x(x^2 \_\_\_\_)$$

$$3x \quad \text{times} \quad +2x = 6x^2 \qquad\qquad 3x^3 + 6x^2 + 12x$$
$$3x(x^2 + 2x + \_\_)$$

$$3x \quad \text{times} \quad +4 = 12x \qquad\qquad 3x^3 + 6x^2 + 12x$$
$$3x(x^2 + 2x + 4)$$

Thus,

$$3x^3 + 6x^2 + 12x \text{ is factored as } 3x(x^2 + 2x + 4)$$

PROCEDURE:

To factor a polynomial whose terms contain a common monomial factor:

1. Identify the greatest common factor (GCF) of the polynomial. This will be the common monomial factor.
2. Factor out the common factor from each term of the polynomial.
3. Write the two factors as the indicated product.

EXAMPLE 1: Factor $3a + 6$.

$3(a + 2)$

SOLUTION: Notice that the GCF of $3a$ and 6 is 3. Factor out 3 from each term of the binomial.

$$3a + 6 = 3(a + 2)$$

Final answer: $3(a + 2)$

EXAMPLE 2: Factor $3b^2 + 6b$

$3b(b+2)$

SOLUTION: Notice that the GCF of $3b^2$ and $6b$ is $3b$. Factor out $3b$ from each term of the binomial.

$$3b^2 + 6b = 3b(b + 2)$$

Final answer: $3b(b + 2)$

EXAMPLE 3: Factor $6a^4 - 12a^3 + 3a^2$

$3a^2(2a^2 - 4a + 1)$

SOLUTION:

$$6a^4 - 12a^3 + 3a^2 = 3a^2(2a^2 - 4a + 1)$$

Final answer: $3a^2(2a^2 - 4a + 1)$

EXAMPLE 4: Factor $3ab^2 - 6a^2b$

SOLUTION:

$$3ab^2 - 6a^2b = 3ab(b - 2a)$$

Final answer: $3ab(b - 2a)$

EXAMPLE 5: Factor $a(x + 1) + b(x + 1)$

SOLUTION: Notice in this example that the GCF is the binomial $(x + 1)$. $a\,(x + 1)\; +\; b\,(x + 1)$

By treating $(x + 1)$ as a single value, we may factor out $(x + 1)$ from each term of the expression.

$$a(x + 1) + b(x + 1) = (x + 1)(a + b)$$

Final answer: $(x + 1)(a + b)$

why not $ab(x+1)$?

Do the following practice set. Check your answers with the answers in the right-hand margin.

PRACTICE SET 7-2

Factor each of the following polynomials whose terms contain a common monomial factor (see example 1, 2, or 3).

1. $6a + 12$	1. $6(a + 2)$
2. $18a^2 + 24a$	2. $6a(3a + 4)$
3. $2x^4 - 8x^3 - 2x^2$	3. $2x^2(x^2 - 4x - 1)$

Factor each of the following polynomials whose terms have a common monomial factor (see example 4).

4. $5ab^2 + 20a^2b$	4. $5ab(b + 4a)$
5. $14x^3y^3 - 21x^2y^2$	5. $7x^2y^2(2xy - 3)$
6. $2a^2b - 6ab^2 + 4ab$	6. $2ab(a - 3b + 2)$

Factor each of the following polynomials whose terms have a common *binomial* factor (see example 5).

7. $a(x - 1) + b(x - 1)$	7. $(x - 1)(a + b)$
8. $2x(y + 1) - 3(y + 1)$	8. $(y + 1)(2x - 3)$
9. $4(x + y) + 2(x + y)$	9. $(x + y)(4 + 2)$
	or $(x + y)(6)$
	or $6(x + y)$

EXERCISE 7-2

Find the missing factor of the polynomial.

1. $2x - 4x^3 = 2x($ $)$ 2. $6x + 12y = 6($ $)$

3. $ax + ay = a($ $)$ 4. $ax^2 + ax = ax($ $)$

5. $2x^3 + 4x^2 - 10x = 2x($ $)$ 6. $4x^2 + 4y^2 = 4($ $)$

7. $-5x + 10y = -5($ $)$ 8. $9x^2 + 12x = 3x($ $)$

9. $2x^3 - 10x^2 + 12x = 2x($ $)$ 10. $9a^2b^3 - 3a^2b^2 + 21a^2b = 3a^2b($ $)$

11. $-4xyz - 4x^2y + 4xy^2 = -4xy($ $)$ 12. $3x^3 - 6x^2 - 3x = 3x($ $)$

Find the greatest common factor.

13. $3x$ and $3y$ 14. $12a$ and 6

15. $7x$ and $14x$ 16. $6x^2$ and $9x^2$

17. $6xy$ and $8xz$ 18. $9x^2y$ and $12x^2z$

19. $10xy^2$ and $25xy^2$ 20. $4x^2y$ and $8x^2y$

21. $25x^2yz$ and $15xy^2z^2$ 22. $7a^3b^2$ and $-4a^2b$

23. $x^2yz, xy^2z,$ and xyz^2 24. $7ax^2, 14bx^2,$ and $-21cx^2$

Factor the polynomial whose terms have a common factor.

25. $7x + 7y$ 26. $4x - 4y$ 27. $4a + 6a$

28. $14x + 7x$ 29. $3ab - 6ac$ 30. $4ab + 6ab$

31. $9x^2 - 3x^2$ 32. $6x^2 + 6x$ 33. $6x^2 - 12xy + 6y^2$

34. $3ax^2 - 10ax - 7a$ 35. $3b^2 - 3ab^3 - 3b^4$ 36. $12a^3 + 16a^2 + 8a$

37. $a(x + 3) + b(x + 3)$ 38. $y(x - 3) - z(x - 3)$ 39. $x(x + 2) + 3(x + 2)$

40. $x(x - 3) - 3(x - 3)$

7-3   MULTIPLYING TWO BINOMIALS

A.   Multiplying by the FOIL Method

In Section 6-5B, we learned that by using the distributive law, we can multiply two polynomials by multiplying each term in one polynomial by each term in the other polynomial, and then combining like terms of the product.

In mathematics, we frequently encounter the product of two first-degree binomials. As a result, a shorter method for multiplying two binomials is presented. This method is commonly referred to as the **FOIL** method.

FOIL is an acronym for the order in which the terms are multiplied: F stands for First, O stands for Outer, I stands for Inner, and L stands for Last. Using $(x + 5)(x + 3)$ as an example, we illustrate the FOIL method of multiplication.

To multiply $(x + 5)(x + 3)$ by FOIL:

F—Multiply the first terms in each binomial.

$$\overset{\text{F}}{(x + 3)(x + 5)}$$
$$(x)(x)$$
$$x^2$$

O—Multiply the outer terms in each binomial.

$$\overset{\text{O}}{(x + 3)(x + 5)}$$
$$(x)(x) + (+5)(x)$$
$$x^2 \qquad +5x$$

I—Multiply the inner terms in each binomial.

$$(x + 3)(x + 5)$$
$$\qquad \text{I}$$
$$(x)(x) + (+5)(x) + (+3)(x)$$
$$x^2 \qquad +5x \qquad +3x$$

L—Multiply the last terms in each binomial.

$$(x + 3)(x + 5)$$
$$\qquad \text{L}$$
$$(x)(x) + (+5)(x) + (+3)(x) + (+3)(+5)$$
$$x^2 \qquad +5x \qquad +3x \qquad +15$$

Combine like terms.

$$x^2 + 5x + 3x + 15$$
$$= x^2 \qquad +8x \qquad = 15$$

Thus, to multiply two binomials, we multiply their first terms, their outer terms, their inner terms, and their last terms. Moreover, if the outer and inner products are combined mentally, the multiplication can be done in one step.

PROCEDURE:

To multiply binomials by using the FOIL method:

1. Write the multiplication problem in horizontal form and arrange the terms of the binomials in order.

2. Multiply the terms in the following order.

   First
   Outer
   Inner
   Last

3. Combine like terms.

EXAMPLE 1:   Multiply $(x + 4)$ by $(x - 7)$.

SOLUTION: Multiply the first terms.

$$(\overset{\lceil}{x} + 4)(\overset{\rceil}{x} - 7)$$
$$x^2$$

Multiply the outer terms.

$$(x + 4)(x - 7)$$
$$x^2 - 7x$$

Multiply the inner terms.

$$(x + 4)(x - 7)$$
$$x^2 - 7x + 4x$$

Multiply the last terms.

$$(x + 4)(x - 7)$$
$$x^2 - 7x + 4x - 28$$

Combine the outer and inner terms.

$$(x + 4)(x - 7)$$
$$= x^2 \underbrace{- 7x + 4x}_{} - 28$$
$$= x^2 - 3x - 28$$

Final answer: $x^2 - 3x - 28$

EXAMPLE 2:   Multiply $(2x - 7)(x + 5)$.

SOLUTION:

$$(2x - 7)(x + 5)$$

F: $(2x)(x) = 2x^2$
O: $(2x)(+5) = 10x$
I: $(-7)(x) = -7x$
L: $(-7)(+5) = -35$

$$= 2x^2 + \underbrace{10x - 7x}_{} - 35$$
$$= 2x^2 + 3x - 35$$

Final answer: $2x^2 + 3x - 35$

Do the following practice set. Check your answers with the answers in the right-hand margin.

PRACTICE SET 7-3A

Use the FOIL method to multiply the binomials (see example 1 or 2).

1.	$(x + 2)(x + 3)$	1.	$x^2 + 5x + 6$
2.	$(2x + 1)(x - 3)$	2.	$2x^2 - 5x - 3$
3.	$(x - 3)(x - 3)$	3.	$x^2 - 6x + 9$

B. Squaring a Binomial

To square a binomial such as $(x + 2)^2$:

Write this as the product of two binomial factors.	$(x + 2)^2$ $(x + 2)(x + 2)$
Then, find the product by applying the FOIL method.	$(x + 2)(x + 2)$ $x^2 + 2x + 2x + 4$
Combine like terms.	$x^2 + 2x + 2x + 4$ $x^2 + 4x + 4$

Thus,

the product is $x^2 + 4x + 4$

Therefore, the product or value of the binomial square can be found by multiplying the terms, using the FOIL method.

After working a number of these problems, we should recognize a relationship between the original binomial and the product. For example:

■ The value of $(x + 3)^2$ is $x^2 + 6x + 9$.

$$(x + 3)^2$$
$$x^2 + 3x + 3x + 9$$
$$x^2 + 6x + 9$$

■ The value of $(x - 3)^2$ is $x^2 - 6x + 9$.

$$(x - 3)^2$$
$$x^2 - 3x - 3x + 9$$
$$x^2 - 6x + 9$$

■ The value of $(2x - 3)^2$ is $4x^2 - 12x + 9$.

$$(2x - 3)^2$$
$$4x^2 - 6x - 6x + 9$$
$$4x^2 - 12x + 9$$

Notice that the first and last terms of our product are the square of the first and last terms of the binomial.

$(x)^2$ is x^2
$(+3)^2$ is 9

$(x +3)^2$
$x^2 + 6x + 9$

$(x)^2$ is x^2
$(-3)^2$ is $+9$

$(x -3)^2$
$x^2 - 6x + 9$

$(2x)^2$ is $4x^2$
$(-3)^2$ is $+9$

$(2x -3)^2$
$4x^2 - 12x + 9$

The middle term in our product is *twice* *the product of the first and last terms of* the binomial.

Twice $(x) \cdot (+3)$ is $2(3x) = 6x$.

$(x + 3)^2$

$x^2 + 6x + 9$

Twice $(x) \cdot (-3)$ is $2(-3x) = -6x$.

$(x - 3)^2$

$x^2 - 6x + 9$

Twice $(2x) \cdot (-3)$ is $2(-6x) = -12x$.

$(2x - 3)^2$

$4x^2 - 12x + 9$

Thus, to square a binomial, we square the first term of the binomial to get the first term of the product; multiply twice the product of the first and last terms to get the second term of the product; and finally, square the last term of the binomial to get the last term of the product. In general,

$$(A + B)^2 = A^2 + 2AB + B^2$$

PROCEDURE:

To square a binomial:

1. Square the first term of the binomial.

2. Multiply the first and last terms of the binomial and then multiply this product by two.

3. Square the last term of the binomial.

EXAMPLE 1: Find the product of $(2x + 1)^2$.

SOLUTION: Square the first term.

$$(2x + 1)^2$$

$$(2x)^2$$

$$4x^2$$

Multiply twice the product of the first and last terms.

$$(2x + 1)^2$$

$$(2x)^2 + 2(2x)(1)$$

$$4x^2 + 4x$$

Square the last term.

$$(2x + 1)^2$$

$$(2x)^2 + 2(2x)(1) + (+1)^2$$

$$4x^2 + 4x + 1$$

Final Answer: $4x^2 + 4x + 1$

EXAMPLE 2: Find the product of $(2x - 5)^2$.

SOLUTION:

$$(2x - 5)^2 = (2x)^2 + 2(2x)(-5) + (-5)^2$$
$$= 4x^2 - 20x + 25$$

Final answer: $3x^2 - 20x + 25$

NOTE:

When squaring a binomial, we find that the first and last terms of the product will always be positive.

Do the following practice set. Check your answers with the answers in the right-hand margin.

PRACTICE SET 7-3B

Find each of the following indicated squares (see example 1 or 2).

1. $(x + 5)^2$
2. $(6x - 1)^2$
3. $(2x - 3)^2$

1. $x^2 + 10x + 25$
2. $36x^2 - 12x + 1$
3. $4x^2 - 12x + 9$

EXERCISE 7-3

Find each of the following indicated products.

1. $(x + 3)(x + 5)$
2. $(y + 4)(y + 1)$
3. $(x - 7)(x + 2)$
4. $(y - 2)(y + 6)$
5. $(z + 2)(z - 7)$
6. $(z + 8)(z - 3)$
7. $(x - 1)(x - 4)$
8. $(x - 7)(x - 2)$
9. $(x + 1)(x - 1)$
10. $(y - 2)(y + 2)$
11. $(2x + 1)(x + 5)$
12. $(5x + 3)(x - 4)$
13. $(3x - 2)(x + 4)$
14. $(x - 6)(3x + 1)$
15. $(2x + 1)(2x - 1)$
16. $(2x - 3)(3x - 2)$
17. $(5x - 7)(3x + 5)$
18. $(2a + b)(3a - 2b)$
19. $(x^2 + 1)(x^2 + 3)$
20. $(3a^2 + b^2)(2a^2 - b^2)$
21. $(6 - x)(3 + x)$
22. $(2x + 7)(3x + 1)$
23. $(x + 6)^2$
24. $(2x + 3)^2$
25. $(y - 7)^2$
26. $(2y - 5)^2$
27. $(x^2 + 6)^2$
28. $(2a^2 - 3)^2$
29. $(2a - 3b)^2$
30. $(3a^2 - b^2)^2$

7-4 FACTORING TRINOMIALS BY THE FOIL METHOD

A. Factoring Trinomials of the Form $x^2 + bx + c$

The ability to factor a trinomial is one of the most important algebraic skills a student of algebra will possess. In this section we limit our discussion to factoring a trinomial of the form $x^2 + bx + c$. Notice that a trinomial of this form does not contain a common factor. Thus, we must develop a new technique to represent $x^2 + bx + c$ in factored form.

Before actually devising a method for factoring trinomials of this type, let us review the FOIL method for finding the product of two binomials. Consider the product $(x + 5)(x + 2)$.

$$(x + 5)(x + 2) =$$

$$
\begin{array}{cccc}
\text{F} & \text{O} & \text{I} & \text{L} \\
x^2 & +\, 5x & +\, 2x & +\, 10
\end{array}
$$

$$x^2 \qquad +\, 7x \qquad +\, 10$$

■ The first term in the product (x^2) is found by multiplying the first terms in each binomial factor.

$$(x + 5)(x + 2)$$
$$x^2 + 5x + 2x + 10$$
$$x^2 + 7x + 10$$

■ The last term in the product, $(+10)$, is found by multiplying the last terms in each binomial factor.

$$(x + 5)(x + 2)$$
$$x^2 + 5x + 2x + 10$$
$$x^2 + 7x + 10$$

■ The middle term of the product $(7x)$ is the algebraic sum of the outer product $(2x)$ and the inner product $(5x)$ of each binomial factor.

$$(x + 5)(x + 2)$$
$$x^2 + 5x + 2x + 10$$
$$x^2 + 7x + 10$$

Notice further that

■ The algebraic sum of the last terms in each binomial factor is equal to the coefficient of the middle term.

$$(x + 5)(x + 2)$$
$$x^2 + 7x + 10$$

In general, the product of two binomial factors equals a trinomial.

$$
\boxed{
\begin{aligned}
(x + A)(x + B) &= x^2 + Ax + Bx + A \cdot B \\
&= x^2 + (A + B)x + A \cdot B
\end{aligned}
}
$$

A trinomial may be expressed as the product of two binomial factors.

$$
\begin{array}{cc}
\text{trinomial} & \text{binomial factors}
\end{array}
$$
$$x^2 + (A + B)x + A \cdot B = (x + A)(x + B)$$

To reverse this procedure, that is, to factor a trinomial such as $x^2 + 7x + 10$, we must find two binomials that have the following characteristics.

■ The first term of each binomial factor must be x.

$$x^2 + 7x + 10$$
$$(x\quad)(x\quad)$$

■ The last term of the trinomial, $(+ 10)$, must be separated into two integer pairs such that their algebraic sum is equal to the coefficient of the middle term $(+ 7)$.

$$x^2 + 7x + 10$$
$$(5)(2) = 10$$
$$5 + 2 = 7$$
$$(x + 5)(x + 2)$$

To aid us in searching for these integer pairs, we consider the combinations of integers whose product is equal to the term of the trinomial being factored. For example, in $x^2 + 7x + 10$, the last term is 10. Factors of 10 are 1 and 10, and 2 and 5. However, only the combination of 2 and 5 will yield both a product of 10 and an algebraic sum of 7 (the coefficient of the middle term).

Here's an example. Factor $x^2 + 6x + 8$.

The first term of each factor must be x.

$$(x^2 + 6x + 8)$$
$$(x\quad)(x\quad)$$

Now, consider factors of 8 (since 8 is the last term in the trinomial).

$$(1)(8) = 8$$
$$(2)(4) = 8$$

Choose 2 and 4 since

$$2 \times 4 = 8 \qquad\qquad x^2 + 6x + 8$$
$$2 + 4 = 6 \qquad\qquad x^2 + 6x + 8$$

Thus, the factors are $(x + 2)$ and $(x + 4)$,

$$x^2 + 6x + 8$$
$$= (x + 2)(x + 4)$$

Confirmed by multiplying the factors.

$$(x + 2)(x + 4)$$
$$= x^2 + 2x + 4x + 8$$
$$= x^2 + 6x + 8$$

Thus,

the factors are $(x + 2)$ and $(x + 4)$

PROCEDURE:

To factor a trinomial of the form $x^2 + bx + c$:

1. Set the first term of each binomial factor as x.

2. Find two terms such that their product is equal to the last term of the trinomial and their algebraic sum is equal to the coefficient of the middle term of the trinomial.

3. Write the terms found in (2) as the last term of each binomial factor.

4. Check the factors by multiplying them and comparing the product with the given trinomial.

EXAMPLE 1: Factor $x^2 + 7x + 12$.

SOLUTION: The first term in each binomial factor is x.

$$x^2 + 7x + 12$$
$$(x \quad)(x \quad)$$

Search for two factors of 12 that have an algebraic sum of 7.

FACTORS OF 12

$$(1)(12)$$
$$(2)(6)$$
$$(3)(4)*$$

*Choose 3 and 4 since

$$3 \times 4 = 12 \quad \text{and} \quad 3 + 4 = 7$$

Factor by using the correct combination of terms.

$$x^2 + 7x + 12 = (x + 3)(x + 4)$$

Final answer: $(x + 3)(x + 4)$

(The check is left for the student.)

EXAMPLE 2: Factor $x^2 - 5x + 6$.

SOLUTION: The first terms are x.

$$x^2 - 5x + 6$$
$$(x \quad)(x \quad)$$

Search for two factors of 6 that have an algebraic sum of 7.

FACTORS OF 6

$$(1)(6)$$
$$(2)(3)*$$

*Choose 2 and 3 since

$$(-2)(-3) = +6 \quad \text{and} \quad (-2) + (-3) = -5$$

Factor by using the correct combination of terms.

$$x^2 - 5x + 6 = (x - 2)(x - 3)$$

Final answer: $(x - 2)(x - 3)$

(The check is left for the student.)

EXAMPLE 3:   Factor $x^2 - x - 6$.

SOLUTION: The first term in each factor is x.

$$x^2 - x - 6$$
$$(x \quad)(x \quad)$$

Search for two factors of -6 that have an algebraic sum of -1.

FACTORS OF -6

$$(-1)(6)$$
$$(1)(-6)$$
$$(-3)(2)*$$
$$(-2)(3)$$

*Choose -3 and 2 since

$$(-3)(2) = -6 \quad \text{and} \quad (-3) + 2 = -1$$

Factor accordingly.

$$x^2 - x - 6 = (x - 3)(x + 2)$$

Final answer: $(x - 3)(x + 2)$

(The check is left for the student.)

EXAMPLE 4:   Factor $y^2 + 3y - 18$.

SOLUTION: The first term is y.

$$y^2 + 3y - 18$$
$$(y \quad)(y \quad)$$

Search for two factors of -18 that have an algebraic sum of $+3$.

FACTORS OF -18

$$(-1)(18) \quad \text{or} \quad (1)(-18)$$
$$(-2)(9) \quad \text{or} \quad (2)(-9)$$
$$*(-3)(6) \quad \text{or} \quad (3)(-6)$$

*Choose -3 and 6 since

$$(-3)(6) = -18 \quad \text{and} \quad (-3) + 6 = +3$$

Factor accordingly.

$$y^2 + 3y - 18 = (y - 3)(y + 6)$$

Final answer: $(y - 3)(y + 6)$

(The check is left for the student.)

With sufficient practice the process of factoring becomes clearer and, ultimately, easier. Further, you will begin to do more of the work mentally, relying less on the need to write out all possible integer combinations of the last term of the trinomial in your quest for finding that seemingly elusive middle term.

NOTE:

This technique of factoring a trinomial in the form $x^2 + bx + c$ can only be applied to trinomials when the coefficient of the first term is 1. If we are given a trinomial such as $-x^2 - 5x - 6$, we can transform this trinomial by first factoring out the common monomial factor (-1). For example:

Factor (-1) from each term in the given trinomial.

$$-x^2 - 5x - 6$$
$$-1\,(x^2 + 5x + 6)$$

Now, with a positive first term, factor the resulting trinomial.

$$-1\,(x^2 + 5x + 6)$$
$$-1\,(x + 2)(x + 3)$$

Applying the associative law of multiplication, we express $-1(x + 2)(x + 3)$ in any one of the following ways.

The monomial (-1) may multiply the first factor

$$-1(x + 2)\,(x + 3)$$
$$(-x - 2)(x + 3)$$

or

multiply the second factor.

$$-1\,(x + 2)\,(x + 3)$$
$$(x + 2)\,(-x - 3)$$

Therefore,

$$-1(x + 2)(x + 3) = (-x - 2)(x + 3) = (x + 2)(-x - 3)$$

Do the following practice set. Check your answers with the answers in the right-hand margin.

PRACTICE SET 7-4A

Factor the trinomials (see example 1 or 2).

1. $x^2 + 8x + 12$
2. $x^2 + 4x + 3$
3. $x^2 - 7x + 12$

1. $(x + 2)(x + 6)$
2. $(x + 3)(x + 1)$
3. $(x - 3)(x - 4)$

Factor the trinomials (see example 3 or 4).

4.	$x^2 - 4x - 12$	4.	$(x + 2)(x - 6)$
5.	$x^2 - 2x - 3$	5.	$(x + 1)(x - 3)$
6.	$x^2 + x - 12$	6.	$(x - 3)(x + 4)$

B.   The Difference of Two Squares

If two binomials, such as $(x + 2)$ and $(x - 2)$ are multiplied, a special product is found. This product is called the *difference of two squares*. For example, multiply $(x + 2)$ times $(x - 2)$ by using FOIL.

$$(x + 2)(x - 2) = x^2 + 2x - 2x - 4$$
$$= x^2 + 0x - 4$$
$$= x^2 - 4$$

Notice that the product, $x^2 - 4$,

Is the difference $x^2 \boxed{-\ 4}$

Of two squares $x^2 - \boxed{2^2}$

NOTE:

> A perfect square (or square) is a number, expression, or term that can be expressed as the product of two equal factors.
>
> 4 is a perfect square, since it may be expressed as the product   4
> of two equal factors. $= \boxed{2 \cdot 2}$
> $= (2)^2$

Notice that since the last term in each binomial is opposite in sign, $(x \boxed{+}\ 2)(x \boxed{-}\ 2)$

The sum of the product of the outer and inner terms will equal zero.

$$\overset{\displaystyle -2x}{(x + 2)(x - 2)}$$
$$\underset{\displaystyle +2x}{}$$

Thus, the middle term of the product will always be 0.

$$(x + 2)(x - 2)$$
$$x^2 + \underline{2x - 2x} - 4$$
$$x^2 +\quad 0x\quad - 4$$
$$= x^2 - 4$$

The difference of two squares, $a^2 - b^2$, can always be immediately factored into the two binomials, $(a - b)$ and $(a + b)$. That is,

$$a^2 - b^2 = (a + b)(a - b)$$

EXAMPLE 1:   Factor $x^2 - 25$.

SOLUTION:

$$x^2 - 25 = (x)^2 - (5)^2$$
$$= (x + 5)(x - 5)$$

Final answer: $(x + 5)(x - 5)$

EXAMPLE 2:   Factor $4x^2 - 49$.

SOLUTION:

$$4x^2 - 49 = (2x)^2 - (7)^2$$
$$= (2x + 7)(2x - 7)$$

Final answer: $(2x + 7)(2x - 7)$

Do the following practice set. Check your answers with the answers in the right-hand margin.

PRACTICE SET 7-4B

Factor the difference of two squares (see example 1 or 2).

1. $y^2 - 9$	2. $z^2 - 1$	3. $x^2 - 16$
4. $4x^2 - 25$	5. $9y^2 - 100$	6. $49x^2 - 16y^2$

1. $(y + 3)(y - 3)$
2. $(z + 1)(z - 1)$
3. $(x + 4)(x - 4)$
4. $(2x + 5)(2x - 5)$
5. $(3y + 10)(3y - 10)$
6. $(7x + 4y)(7x - 4y)$

EXERCISE 7-4

Factor the following.

1. $x^2 + 3x + 2$	2. $y^2 + 4y + 3$	3. $x^2 + 5x + 6$	4. $y^2 + 6y + 5$
5. $z^2 + 4z + 4$	6. $z^2 + 5z + 4$	7. $x^2 + 10x + 9$	8. $y^2 + 6y + 9$
9. $x^2 + 9x + 18$	10. $y^2 + 12y + 35$	11. $x^2 - 6x + 9$	12. $y^2 - 10x + 9$
13. $x^2 - 7x + 10$	14. $z^2 - 11z + 10$	15. $x^2 - 8x + 7$	16. $y^2 - 5y + 6$
17. $x^2 - 6x + 8$	18. $y^2 - 4y + 3$	19. $y^2 - 7y + 6$	20. $z^2 - 13z + 36$

21. $y^2 + y - 2$ 22. $z^2 + 3z - 18$ 23. $x^2 + 5x - 24$ 24. $x^2 + 4x - 12$

25. $x^2 + 3x - 18$ 26. $x^2 + 4x - 21$ 27. $x^2 + 2x - 8$ 28. $x^2 + 7x - 60$

29. $y^2 + y - 6$ 30. $y^2 - y - 6$ 31. $y^2 - 3y - 10$ 32. $x^2 - 2x - 8$

33. $x^2 - 7x - 18$ 34. $y^2 - 3y - 4$ 35. $x^2 - 2x - 8$ 36. $x^2 - 6x - 7$

37. $z^2 - 5z - 24$ 38. $y^2 - 6y - 27$ 39. $x^2 - 2x - 15$ 40. $z^2 - 13z - 48$

Factor the difference of two squares.

41. $y^2 - 64$ 42. $z^2 - 25$ 43. $c^2 - 100$ 44. $x^2 - 81$

45. $9 - x^2$ 46. $16 - y^2$ 47. $4x^2 - 25$ 48. $25x^2 - y^2$

49. $121 - x^2$ 50. $x^2 - y^2$ 51. $36x^2 - 25y^2$ 52. $4y^2 - 9$

53. $x^2 - 1$ 54. $4z^2 - 1$ 55. $1 - x^2$ 56. $x^6 - 1$

57. $x^2 - \dfrac{1}{4}$ 58. $x^2 - \dfrac{1}{9}$

7-5   FACTORING TRINOMIALS OF THE FORM $ax^2 + bx + c$

In the previous section, we considered only trinomials of the form $x^2 + bx + c$. That is, the coefficient of the x^2 term was 1. In this section, we expand our discussion of factoring trinomials to include those trinomials in which the coefficient of the x^2 term is an integer greater than 1. For example, $2x^2 + 11x + 5$.

The procedure we use to factor these trinomials is the same. However, since the coefficient of the x^2 term is no longer 1, we will have more combinations of factors to try. We illustrate this with the following example. To factor $2x^2 + 11x + 5$:

Find two factors of 2	$\boxed{2}\,x^2 + 11x + 5$
And two factors of 5,	$2x^2 + 11x + \boxed{5}$
Such that the sum of the outer and inner products is 11. (Note: The sign in front of the last term indicates if the outer and inner products are to be added or subtracted.)	$2x^2 + 11x \boxed{+} 5$ $2x^2 \boxed{+} 11 x + 5$

To find the correct combination, we do the following.

First, write all possible factors of the coefficient of the x^2 term and the constant.	$\boxed{2}\,x^2 + 11x + \boxed{5}$ $1 \cdot 2 \mid 1 \cdot 5$ $5 \cdot 1$

NOTE:

When writing these factors we are not concerned with the signs. But the order of the factors is important. We must factor one of the terms from $1 \cdot N$ to $N \cdot 1$ (e.g., $1 \cdot 5$ to $5 \cdot 1$) to consider all possible combinations.

Set up our chart in this manner, so that we can now focus on the inner product and the outer product.

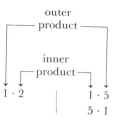

Since the coefficient of the x term represents the sum of the product of the outer factors and the product of the inner factors, we must try different combinations until a sum of 11 is found. To do this, we set up the following chart.

	Outer	Inner	Sum	
	1 · 5	2 · 1		
	5	2	7	
	1 · 1	2 · 5		*correct combination
	1	10	11	

Now, the factors of 2 (namely, 1 and 2) are the coefficients of the first term for each factor.

$2x^2 + 11x + 5$

$\quad 1 \cdot 2 \qquad 5 \cdot 1$

$(1x \quad) \quad (2x \quad)$

And the factors of 5 (namely, 5 and 1) are the last terms in each factor.

$2x^2 + 11x + 5$

$\quad 1 \cdot 2 \qquad 5 \cdot 1$

$(x \quad 5) \quad (2x \quad 1)$

The signs are inserted to show that the product of the last terms is $+5$ and the sum of the outer and inner product is $+11$.

$2x^2 + 11x + 5$

$(x + 5)(2x + 1)$

Thus, the factors are $(x + 5)$ and $(2x + 1)$. This can be checked by multiplying the binomials.

NOTE:

> The product
>
> $$(x + 5)(2x + 1)$$
>
> is different from the product
>
> $$(x + 1)(2x + 5).$$
>
> Thus, the placement of the constants is important and must be carefully considered.

Here's another example. Factor $4x^2 + 23x - 6$.

Find two factors of 4 and two factors of 6
such that the difference of the product
of outer terms and inner terms is 23.

$4x^2 + 23x - 6$
$4x^2 + 23x - 6$
$4x^2 + 23x - 6$
$4x^2 + 23x - 6$

To find the correct combinations:

Write all possible factors of 4 and 6.

$4x^2 + 23x - 6$

$1 \cdot 4$	$1 \cdot 6$
$2 \cdot 2$	$2 \cdot 3$
	$3 \cdot 2$
	$6 \cdot 1$

Now, set up our chart and look for
a difference (of the combinations) of 23.
Once we have found our combination,
our search is over.

$4x^2 + 23x - 6$
$4x^2 + 23x - 6$

Outer ↓ Inner ↓		**Outer**	**Inner**	**Difference** Larger—Smaller
$1 \cdot 4$	$1 \cdot 6$	$1 \cdot 6$	$4 \cdot 1$	$6 - 4 = 2$
$2 \cdot 2$	$2 \cdot 3$	$1 \cdot 3$	$4 \cdot 2$	$8 - 3 = 5$
	$3 \cdot 2$	$1 \cdot 2$	$4 \cdot 3$	$12 - 2 = 10$
	$6 \cdot 1$	$1 \cdot 1$	$4 \cdot 6$	$24 - 1 = 23$ * STOP

So, the factors of 4 (1 and 4)
are the coefficients of the
first term in each factor.

$4x^2 + 23x - 6$
$1 \cdot 4 \quad | \quad 6 \cdot 1$
$(1x \quad) \quad (4x \quad)$

And the factors of 6 (6 and 1)
are the last terms in each factor.

$4x^2 + 23x - 6$
$1 \cdot 4 \quad | \quad 6 \cdot 1$
$(1x \quad 6) \quad (4x \quad 1)$

The signs are inserted to
show that the product of the last
terms is (-6) and the algebraic
sum of the outer and inner
products is $(+23)$

$4x^2 + 23x - 6$
$(x + 6)(4x - 1)$

Thus,

the factors are $(x + 6)$ and $(4x - 1)$

Skill at factoring is the result of *Practice! Practice! Practice!* As you develop insight into the
process and your confidence grows, you will be doing most of the calculations mentally.

PROCEDURE:

> To factor a trinomial of the form $ax^2 + bx + c$:
>
> 1. Write the possible factor pairs of the coefficient of the second-degree term and the constant term.
>
> 2. Select the appropriate combination of pairs such that the sum or difference of the product of outer and inner terms is equal to the coefficient of the first-degree term.
>
> 3. Write the factors of the coefficient of the second-degree term as the coefficient of the first term in each binomial factor, and write the factor of the constant as the last term in each binomial factor.
>
> 4. Determine the correct signs such that the product of the last terms of the binomial factors equals the constant and the sum of the outer and inner product is the coefficient of x.
>
> 5. Check the result.

EXAMPLE 1:   Factor $5x^2 + 21x + 4$.

SOLUTION: Write all possible factors of 5 and 4.

$$5\,x^2 + 21x + 4$$

$$
\begin{array}{c|c}
1 \cdot 5 & 1 \cdot 4 \\
 & 2 \cdot 2 \\
 & 4 \cdot 1
\end{array}
$$

Search for the combination that will yield a sum of 21.

Outer	Inner	Sum
$1 \cdot 4$	$5 \cdot 1$	9
$1 \cdot 2$	$5 \cdot 2$	12
$1 \cdot 1$	$5 \cdot 4$	21 * STOP *

Since the correct combinations are 1 and 5, and 4 and 1, we can now factor accordingly.

$$5x^2 + 21x + 4$$

$$(x \quad 4)(5x \quad 1)$$

$$= (x + 4)(5x + 1)$$

Final answer: $(x + 4)(5x + 1)$

EXAMPLE 2:   Factor $4x^2 - 12x + 9$.

SOLUTION: Write all possible factors of 4 and 9.

$$\boxed{4}\,x^2 - 12x + \boxed{9}$$

$$
\begin{array}{cc}
1 \cdot 4 & 1 \cdot 9 \\
2 \cdot 2 & 3 \cdot 3 \\
 & 9 \cdot 1
\end{array}
$$

Search for the combination that will yield a sum of 12.

Outer	Inner	Sum
$1 \cdot 9$	$4 \cdot 1$	13
$1 \cdot 3$	$4 \cdot 3$	15
$1 \cdot 1$	$4 \cdot 9$	37
$2 \cdot 9$	$2 \cdot 1$	20
$2 \cdot 3$	$2 \cdot 3$	12 * <u>STOP</u> *

Since the correct combinations are 2 and 2, and 3 and 3, we can now factor accordingly.

$$4x^2 - 12x + 9$$

$$
\begin{array}{cc}
2 \cdot 2 & 3 \cdot 3 \\
\downarrow & \downarrow \\
(2x \quad 3x)(2x \quad 3) \\
\end{array}
$$

$$= (2x - 3)(2x - 3)$$

Final answer: $(2x - 3)(2x - 3)$

EXAMPLE 3:   Factor $3x^2 + 5x - 12$.

SOLUTION: Write all possible factors of 3 and 12.

$$\boxed{3}\,x^2 + 5x - \boxed{12}$$

$$
\begin{array}{c|c}
1 \cdot 3 & 1 \cdot 12 \\
 & 2 \cdot 6 \\
 & 3 \cdot 4 \\
 & 4 \cdot 3 \\
 & 6 \cdot 2 \\
 & 12 \cdot 1
\end{array}
$$

Search for a combination of pairs that will yield a difference of 5.

		Outer	Inner	Difference
1 · 3	1 · 12	1 · 12	3 · 1	9
	2 · 6	1 · 6	3 · 2	0
	3 · 4	1 · 4	3 · 3	5 * STOP *
	4 · 3			
	6 · 2			
	12 · 1			

Factor accordingly by using the pairs of 1 and 3, and 3 and 4.

$$3x^2 + 5x - 12$$

$$1 \cdot 3 \qquad 3 \cdot 4$$

$$(x \quad 3)(3x \quad 4)$$

$$= (x + 3)(3x - 4)$$

Final answer: $(x + 3)(3x - 4)$

EXAMPLE 4: Factor $16x^2 - 11x - 5$.

SOLUTION: Write all possible factors of 16 and 5.

$$16x^2 - 11x - 5$$

1 · 16	1 · 5
2 · 8	5 · 1
4 · 4	

Search for a combination of pairs that will yield a difference of 11.

	Outer	Inner	Difference
1 · 16 1 · 5			
2 · 8 5 · 1	1 · 5	16 · 1	11 * STOP *
4 · 4			

Factor accordingly.

$$16x^2 - 11x - 5$$

$$1 \cdot 16 \qquad 1 \cdot 5$$

$$(x \quad 1)(16x \quad 5)$$

$$= (x - 1)(16x + 5)$$

Final answer: $(x - 1)(16x + 5)$

Do the following practice set. Check your answers with the answers in the right-hand margin.

PRACTICE SET 7-5

Factor the following trinomials (see example 1 or 2).

1. $2x^2 + 5x + 3$
2. $6x^2 - 19x + 15$
3. $8x^2 - 10x + 3$

1. $(2x + 3)(x + 1)$
2. $(2x - 3)(3x - 5)$
3. $(2x - 1)(4x - 3)$

Factor the following trinomials (see example 3 or 4).

4. $2x^2 - 7x - 15$
5. $6x^2 + 5x - 6$
6. $6x^2 + 5x - 4$

4. $(2x + 3)(x - 5)$
5. $(2x + 3)(3x - 2)$
6. $(2x - 1)(3x + 4)$

EXERCISE 7-5

Factor the following trinomials.

1. $2x^2 + 5x + 2$
2. $2y^2 + 7y + 6$
3. $2x^2 + 5x + 3$
4. $3x^2 + 10x + 3$
5. $2x^2 + 11x + 12$
6. $3x^2 + 10x + 8$
7. $6x^2 + 25x + 4$
8. $4x^2 + 4x + 1$
9. $4x^2 - 8x + 3$
10. $7x^2 - 15x + 2$
11. $8x^2 - 14x + 3$
12. $3x^2 - 5x + 2$
13. $4y^2 - 12y + 5$
14. $10x^2 - 9x + 2$
15. $3y^2 - 13y + 14$
16. $9x^2 - 31x + 12$
17. $2x^2 + 7x - 15$
18. $4x^2 + 12x - 7$
19. $3y^2 + 5y - 12$
20. $4y^2 + 19y - 5$
21. $3x^2 + x - 10$
22. $6x^2 - x - 2$
23. $6x^2 - 5x - 4$
24. $15x^2 + x - 28$
25. $2x^2 - 7x - 15$
26. $3x^2 - 5x - 2$
27. $3x^2 - 5x - 12$
28. $18x^2 - 23x - 6$
29. $15x^2 - 2x - 1$
30. $6x^2 - 5x - 4$
31. $15y^2 - 14y - 8$
32. $18z^2 - 9z - 20$
33. $16x^2 - 25$
34. $4x^2 - 1$
35. $25x^2 - 49$
36. $9x^2 - 4$

7-6   FACTORING COMPLETELY

Factoring an expression once may not completely remove all factors from the expression. That is, some expressions may contain more than 2 factors, requiring additional factoring. For example, the difference of two squares is factored as follows.

The difference of two squares factors into the sum and difference of two expressions.

$a^4 - 16$

$(a^2 + 4)(a^2 - 4)$

Inspection of the two factors reveals the second factor $(a^2 - 4)$ is also the difference of two squares.

$(a^2 + 4)(a^2 - 4)$

$(a^2 + 4)\ (a^2 - 4)$

This factors into the sum and difference of two expressions.

$(a^2 + 4)(a^2 - 4)$

$(a^2 + 4)(a + 2)(a - 2)$

Thus,

$$a^4 - 16 \text{ factors completely into } (a^2 + 4)(a + 2)(a - 2)$$

NOTE:

$(a^2 + 4)$, the sum of two squares, cannot be factored.

As a second example, consider factoring $2x^2 + 4x - 6$.

First, factor the GCF 2 from each term.

$2x^2 + 4x - 6$
$= 2(x^2 + 2x - 3)$

Next, factor the remaining trinomial.

$2(x^2 + 2x - 3)$
$= 2(x + 3)(x - 1)$

Thus,

$$2x^2 + 4x - 6 \text{ is factored into } 2(x + 3)(x - 1)$$

In general, to factor a polynomial completely, first factor the GCF (if possible); next, factor the polynomial by using either the difference of two squares method or the FOIL method; and finally, check the factors for any additional factors.

PROCEDURE:

To factor completely:

1. Factor the greatest common factor (if possible).

2. Factor the polynomial into its binomial factors.
    a.   If it is a binomial, use the difference of two squares method.
    b.   If it is a trinomial, use the FOIL method.

3. Check the factors for additional factors.

EXAMPLE 1:   Factor: $3x^2 - 27y^2$.

Example continued on next page.

SOLUTION: Factor GCF of 3.

$$3x^2 - 27y^2 = 3(x^2 - 9y^2)$$

Factor the resulting binomial as a difference of two squares.

$$3(x^2 - 9y^2) = 3(x + 3y)(x - 3y)$$

Final answer: $3(x + 3y)(x - 3y)$

EXAMPLE 2: Factor: $2x^5 - 32x$.

SOLUTION: Factor GCF of $2x$.

$$2x^5 - 32x = 2x(x^4 - 16)$$

Factor $x^4 - 16$ as a difference of two squares.

$$2x(x^4 - 16) = 2x(x^2 + 4)(x^2 - 4)$$

Factor $x^2 - 4$ as a difference of two squares.

$$2x(x^2 + 4)(x^2 - 4) = 2x(x^2 + 4)(x + 2)(x - 2)$$

Final answer: $2x(x^2 + 4)(x + 2)(x - 2)$

EXAMPLE 3: Factor $h^2k - 4hk - 4k$.

SOLUTION:
$$h^2k - 4hk + 4k = k(h^2 - 4h + 4)$$
$$= k(h - 2)(h - 2)$$

Final answer: $k(h - 2)(h - 2)$

EXAMPLE 4: Factor $2x^3 + 2x^2 - 12x$.

SOLUTION:
$$2x^3 + 2x^2 - 12x = 2x(x^2 + x - 6)$$
$$= 2x(x + 3)(x - 2)$$

Final answer: $2x(x + 3)(x - 2)$

EXAMPLE 5: Factor $4x^2 + 6x - 4$.

SOLUTION:

$$4x^2 + 6x - 4 = 2(2x^2 + 3x - 2)$$
$$= 2(2x - 1)(x + 2)$$

Final answer: $2(2x - 1)(x + 2)$

Do the following practice set. Check your answers with the answers in the right-hand margin.

PRACTICE SET 7-6

Factor completely (see example 1 or 2).

1. $5x^2 - 20$

2. $28x^2 - 7$

3. $2y^2 - 162$

1. $5(x + 2)(x - 2)$

2. $7(2x + 1)(2x - 1)$

3. $2(y + 9)(y - 9)$

Factor completely (see example 3, 4, or 5).

4. $3x^2 + 6x + 3$

5. $6x^3 - 12x^2 + 6x$

6. $3xy^2 + 14xy - 5x$

4. $3(x + 1)(x + 1)$

5. $6x(x - 1)(x - 1)$

6. $x(3y - 1)(y + 5)$

EXERCISE 7-6

Factor completely.

1. $3x^2 - 27$

2. $8x^3 - 8x$

3. $z^4 - 16$

4. $2x^2 - 8y^2$

5. $ax^2 - ay^2$

6. $3x^2 - 27y^2$

7. $z^3 - z$

8. $x^4 - 1$

9. $4a^2 - 36$

10. $4x^2 - 16$

11. $x^3 - 25x$

12. $a^2b^2 - 121$

13. $2x^2 - 8x + 8$

14. $3y^2 + 18y + 27$

15. $ax^2 + 3ax + 2a$

16. $3ay^2 - 9ay + 6a$

17. $6y + 4x^2y - 10xy$

18. $4x^2 - 4x - 48$

19. $x^3 + 7x^2 + 10x$

20. $2ay^2 + 2ay - 12a$

21. $x^4 + x^2 - 2$

22. $x^4 - 10x^2 + 9$

23. $x^3y + 10x^2y^2 + 25xy^3$

24. $x^4 - 12x^3 + 36x^2$

7-7 SOLVING QUADRATIC EQUATIONS BY FACTORING

We have previously learned how to solve linear or first-degree equations. Now, we use our knowledge of solving equations and factoring to solve second-degree equations in one unknown.

The principle behind the use of factoring to solve second-degree equations is the fact that if the product of two or more factors is zero, then at least one of the factors must be zero. For example,

If x and $(x + 1)$ are two factors with a product of zero,	$x(x + 1) = 0$
Then either the first factor (x) equals zero, or	$\boxed{x}\,(x + 1) = 0$ $x = 0$
The second factor $(x + 1)$ equals zero.	$x\,\boxed{(x + 1)} = 0$ $x = 0 \quad\mid\quad (x + 1) = 0$
Assuming these statements are true, we can solve our two equations for x.	$x = 0 \quad\mid\quad x + 1 = 0$ $ x = 0 - 1$ $ x = -1$

If $x = 0$ or $x = -1$, our original statement $x(x + 1) = 0$ is true. This can be verified by a check.

REPLACE X WITH 0

$x(x + 1)$	0
$(0)(0 + 1)$	0
$(0)(1)$	
0	

REPLACE X WITH (-1)

$x(x + 1)$	0
$(-1)(-1 + 1)$	0
$(-1)(0)$	
0	

A second-degree equation such as $x^2 - 4x + 3 = 0$ is called a quadratic equation. Every quadratic equation will have at most two roots or values for x.

One method that we can use to solve a quadratic equation is factoring. (Other methods of solving quadratic equations will be discussed in Chapter 12.) To solve a quadratic equation by factoring, we must first make certain that the equation is expressed in standard form (i.e., $ax^2 + bx + c = 0$). Once this has been done, we next simply factor the quadratic expression completely; set each factor that contains the unknown equal to zero; and finally, solve each of these equations accordingly. Our results should then be checked back into the original equation. For example, to solve the equation $x^2 - 5x = -6$ by factoring,

First, write the equation in standard form.	$x^2 - 5x = 6$ $x^2 - 5x - 6 = 0$
Next, factor the quadratic expression.	$x^2 - 5x - 6 = 0$ $(x - 6)(x + 1) = 0$
Now, set each factor equal to zero and solve each equation for x.	$(x - 6)(x + 1) = 0$ $x - 6 = 0 \quad\mid\quad x + 1 = 0$ $ x = 6 \quad\mid\quad x = -1$

Thus,

$$x = 6 \text{ and } x = -1$$

Check these solutions now by substituting each solution into the original equation.

Check $x = 6$	
$x^2 - 5x$	6
$(6)^2 - 5(6)$	6
$36 - 30$	
6	

Check $x = -1$	
$x^2 - 5x$	6
$(-1)^2 - 5(-1)$	6
$1 + 5$	
6	

Thus,

the solutions are $x = 6$ and $x = -1$

PROCEDURE:

> To solve a quadratic equation by factoring:
>
> 1. Write the equation in standard form.
> 2. Factor the quadratic expression completely.
> 3. Set each factor found in (2) equal to zero and solve each equation for the variable.
> 4. Check all solutions by using the original equation.

EXAMPLE 1:   Solve $x^2 - 7x = 0$.

SOLUTION: Since the equation is in standard form, factor the quadratic expression.

$$x^2 - 7x = 0$$
$$x(x - 7) = 0$$

Set each factor equal to zero and solve accordingly.

$$x(x - 7) = 0$$
$$x = 0 \quad | \quad x - 7 = 0$$
$$x = 7$$

Final Answer: $x = 0, x = 7$

(The check is left to the student.)

EXAMPLE 2:   Solve $x^2 = 16$.

SOLUTION: Place in standard form and factor the quadratic expression.

$$x^2 = 16$$
$$x^2 - 16 = 0$$
$$(x + 4)(x - 4) = 0$$

Example continued on next page.

Set each factor equal to zero and solve accordingly.

$$(x + 4)(x - 4) = 0$$

$$\begin{array}{c|c} x + 4 = 0 & x - 4 = 0 \\ x = -4 & x = 4 \end{array}$$

Final answer: $x = -4, x = 4$

(The check is left to the student.)

EXAMPLE 3:   Solve $x^2 + 3x = -2$.

SOLUTION:

$$x^2 + 3x = -2$$
$$x^2 + 3x + 2 = 0$$
$$(x + 2)(x + 1) = 0$$

$$\begin{array}{c|c} x + 2 = 0 & x + 1 = 0 \\ x = -2 & x = -1 \end{array}$$

Final answer: $x = -2, x = -1$

(The check is left to the student.)

EXAMPLE 4:   Solve $4x^2 + 10x = 6$.

SOLUTION:

$$4x^2 + 10x = 6$$
$$4x^2 + 10x - 6 = 0$$
$$2(2x^2 + 5x - 3) = 0$$
$$2(2x - 1)(x + 3) = 0$$
$$(2x - 1)(x + 3) = 0$$

$$\begin{array}{c|c} 2x - 1 = 0 & x + 3 = 0 \\ 2x = 1 & x = -3 \\ x = \frac{1}{2} & \end{array}$$

Final answer: $x = \frac{1}{2}, x = -3$

(The check is left to the student.)

NOTE:

In Example 4, $2(2x - 1)(x + 3) = 0$ has the same solution as $(2x - 1)(x + 3) = 0$. Dividing each side of the first equation by 2 will produce the second equation.

$$2(2x - 1)(x + 3) = 0$$

$$\frac{2(2x - 1)(x + 3)}{2} = \frac{0}{2}$$

$$(2x - 1)(x + 3) = 0$$

(Do not divide by a variable.)

Do the following practice set. Check your answers with the answers in the right-hand margin.

PRACTICE SET 7-7

Solve and check the following equations (see example 1).

1. $x^2 - 4x = 0$ 　　　 2. $2x^2 - 6x = 0$ 　　　 3. $3x^2 + 9x = 0$

Solve and check the following equations (see example 2).

4. $x^2 - 9 = 0$ 　　　 5. $5x^2 = 125$ 　　　 6. $4x^2 = 16$

Solve and check the following equations (see example 3 or 4).

7. $x^2 - 4x - 5 = 0$
8. $2x^2 - 2x - 12 = 0$
9. $2x^2 - 11x - 21 = 0$

1. $x = 0, \quad x = 4$
2. $x = 0, \quad x = 3$
3. $x = 0, \quad x = -3$

4. $x = 3, \quad x = -3$
5. $x = 5, \quad x = -5$
6. $x = 2, \quad x = -2$

7. $x = 5, \quad x = -1$
8. $x = 3, \quad x = -2$
9. $x = 7, \quad x = -\frac{3}{2}$

EXERCISE 7-7

Solve and check the following equations.

1. $x^2 - 7x = 0$

2. $y^2 + 4y = 0$

3. $3x^2 = 12x$

4. $5y^2 = -15y$

5. $2x^2 - 3x = 0$

6. $3y^2 + 2y = 0$

7. $x^2 - 9 = 0$

8. $y^2 = 16$

9. $2z^2 = 18$

10. $9x^2 = 25$

11. $4x^2 - 9 = 0$

12. $36x^2 = 1$

13. $x^2 - 4x + 3 = 0$

14. $x^2 + 3x + 2 = 0$

15. $y^2 - 7y + 10 = 0$

16. $x^2 - 5x - 14 = 0$ 17. $x^2 + 5x = 6$ 18. $y^2 = 2y + 24$

19. $y^2 - 3y + 2 = 0$ 20. $x^2 - 4x = 21$ 21. $z^2 + 18 = 11z$

22. $y^2 = 15 + 2y$ 23. $y^2 - 7 = 6y$ 24. $x^2 - 2x - 35 = 0$

25. $2x^2 + 5x + 2 = 0$ 26. $3x^2 + 8x - 3 = 0$ 27. $3x^2 + x - 4 = 0$

28. $2x^2 + x - 3 = 0$ 29. $3x^2 + 10x + 3 = 0$ 30. $5x^2 + 11x = -2$

31. $2y^2 + 7y = -6$ 32. $3x^2 + 18x = -27$ 33. $4x^2 + 13x = 12$

34. $4x^2 - 8x + 4 = 0$ 35. $4y^2 - 4y + 1 = 0$ 36. $15x^2 - 2x = 8$

37. $6x^2 - 7x = 20$ 38. $2y^2 + 4y - 30 = 0$ 39. $2x^2 - 19x - 10 = 0$

40. $2y^2 + 2y = 24$

REVIEW EXERCISES

Find the missing factor of the polynomial.

1. $-24x^5 = -4x^2($ $)$ 2. $5x^2 - 20 = 5($ $)$

3. $2x^2 + 4x = 2x($ $)$ 4. $x(x + 1) - 3(x + 1) = (x + 1)($ $)$

Find the greatest common factor (GCF) of the following.

5. $5a$, $5b$ 6. $3x^3$, $6x^2$

7. $4x^2y$, $6xy^2$ 8. $10x^2y$, $15y^2z$, $25z^2x$

Factor each of the following polynomials for the common factor only.

9. $3x^2 + 6x$ 10. $ab^3 + a^2b$

11. $-12x^2 + 30x + 6$ 12. $8x^2 - 12x - 24x^3$

13. $x(x - 4) + 3(x - 4)$

Using the FOIL method, multiply the following binomials.

14. $(x + 3)(x - 7)$ 15. $(y + 9)(y - 2)$

16. $(2x + 3)(x - 4)$ 17. $(3y - 5)(2y + 3)$

18. $(x + 2)(x - 2)$ 19. $(3x + 4)(3x - 4)$

20. $(a + b)(a - b)$ 21. $(x + 3)(x - 3)$

Square the following binomials.

22. $(y + 5)^2$ 23. $(z - 3)^2$ 24. $(2x - 3)^2$

Factor the following trinomials.

25. $x^2 + 2x - 15$

26. $x^2 - 6x - 7$

27. $x^2 + 8x + 12$

28. $y^2 - 12x + 35$

29. $2x^2 - 7x - 4$

30. $6x^2 + 7x - 3$

31. $4x^2 - 4x + 1$

32. $6x^2 - 5x - 6$

33. $x^2 - 25$

34. $4y^2 - 25$

35. $16 - 9x^2$

Completely factor the following.

36. $3x^2 - 3$

37. $z^5 - z^3$

38. $y^4 - 16$

39. $3x^2 - 3x - 36$

40. $2x^3 - 4x^2 + 2x$

Solve and check the following equations.

41. $y^2 - 7y = 0$

42. $z^2 = 4z$

43. $y^2 = 16$

44. $3x^2 - 75 = 0$

45. $x^2 - 2x - 15 = 0$

46. $x^2 + 6x - 7 = 0$

47. $x^2 - 8x + 12 = 0$

48. $y^2 + 12x + 35 = 0$

49. $2x^2 + 7x - 4 = 0$

50. $6x^2 - 7x - 3 = 0$

CHAPTER 8

Algebraic Fractions

8-1 INTRODUCTION TO ALGEBRAIC FRACTIONS

An algebraic fraction represents the quotient of any two polynomials. For example:

■ $\dfrac{4}{x}$ is an algebraic fraction

  $\dfrac{4}{x}$

  and represents (4) divided by (x).

  $(4) \div (x)$

■ $\dfrac{x}{y}$ is also an algebraic fraction

  $\dfrac{x}{y}$

  and represents (x) divided by (y).

  $(x) \div (y)$

The terms of an algebraic fraction are also named:

■ Numerator, the top quantity in our fraction, or dividend

  $\dfrac{4}{x}$ or $\dfrac{x}{y}$

■ Denominator, the bottom quantity in our fraction, or divisor

  $\dfrac{4}{x}$ or $\dfrac{x}{y}$

NOTE:

> Remember, the denominator of a fraction cannot have a value of zero.

Algebraic fractions may have more than one term in the numerator or denominator. For example,

■ The algebraic fraction $\dfrac{a + b}{2}$

  $\dfrac{a + b}{2}$

  contains two terms in the numerator.

  $\dfrac{a + b}{2}$

■ The algebraic fraction $\dfrac{x^2 + 4x + 3}{x + 1}$

$$\dfrac{x^2 + 4x + 3}{x + 1}$$

contains three terms in the numerator

$$\dfrac{x^2 + 4x + 3}{x + 1}$$

and two terms in the denominator.

$$\dfrac{x^2 + 4x + 3}{x + 1}$$

Whenever more than one term appears in the numerator or denominator, parentheses are used to reinforce the concept of a "single value." For example:

■ The fraction $\dfrac{a + b}{2}$ expressed as a quotient is written as a numerator ÷ denominator; $(a + b)$ represents the numerator.

$$\dfrac{a + b}{2}$$

$$\underbrace{(a + b)}_{\text{numerator}} \div 2$$

The fraction $\dfrac{x^2 + 4x + 3}{x + 1}$ expressed as a quotient would again be written as a numerator ÷ denominator. Parentheses must be used to maintain the correct order of operations.

$$\dfrac{x^2 + 4x + 3}{x + 1}$$

$$\underbrace{(x^2 + 4x + 3)}_{\text{numerator}} \div \underbrace{(x + 1)}_{\text{denominator}}$$

NOTE:

> The numerator or denominator may consist of many terms but will be thought of as a single value—*the* numerator or *the* denominator. Parentheses help us remember this concept.

A. Factoring Algebraic Fractions

Although algebraic fractions represent division, we seldom find the quotient. Instead, we work with the *factors* of the numerator and factors of the denominator. In solving problems involving algebraic fractions, we find that our answers will be acceptable if the numerator and denominator are left in factored form.

The following examples demonstrate how to write algebraic fractions in factored form.

EXAMPLE 1: To express $\dfrac{2x}{3y}$ in factored form,

■ The numerator $2x$ is expressed as the product of (2) and (x).

$$\dfrac{2x}{3y} \qquad \dfrac{(2)(x)}{}$$

■ The denominator $3y$ is expressed as the product of (3) and (y).

$$\dfrac{2x}{3y} = \dfrac{(2)(x)}{(3)(y)}$$

EXAMPLE 2:   To express $\dfrac{2x^2 - 2}{x - 1}$ in factored form:

- The numerator factors into three factors 2, $(x + 1)$, and $(x - 1)$.

$$\dfrac{2x^2 - 2}{x - 1} \quad \dfrac{2(x^2 - 1)}{} \quad 2(x + 1)(x - 1)$$

- The denominator cannot be factored. $(x + 1)$ is a single factor.

$$\dfrac{2x^2 - 2}{x - 1} = \dfrac{2(x + 1)(x - 1)}{(x - 1)}$$

NOTE:

Parentheses are used to emphasize that the denominator is a single value; $(x - 1)$ is the factor of the denominator.

EXAMPLE 3:   To express $\dfrac{x^2 + 4x + 3}{3x - 3}$ in factored form:

- The numerator factors into the two factors $(x + 1)$ and $(x + 3)$.

$$\dfrac{x^2 + 4x + 3}{3x - 3} \rightarrow \dfrac{(x + 1)(x + 3)}{}$$

- The denominator factors into the two factors, (3) and $(x - 1)$.

$$\dfrac{x^2 + 4x + 3}{3x - 3} = \dfrac{(x + 1)(x + 3)}{3(x - 1)}$$

Do the following practice set. Check your answers with the answers in the right-hand margin.

PRACTICE SET 8-1A

Express the following fractions in factored form.

1. $\dfrac{3x}{3x + 6}$

2. $\dfrac{2x + 4}{3x + 6}$

1. $\dfrac{(3)(x)}{3(x + 2)}$

2. $\dfrac{2(x + 2)}{3(x + 2)}$

3. $\dfrac{x^2 + 4x + 4}{x^2 - x - 6}$

3. $\dfrac{(x + 2)(x + 2)}{(x + 2)(x - 3)}$

B.   Signs of Fractions

Every fraction has three signs associated with it.

- The sign in front of the fraction $\qquad + \dfrac{+a}{+b}$

- The sign of the numerator $\qquad + \dfrac{+a}{+b}$

- The sign of the denominator $\qquad + \dfrac{+a}{+b}$

For example, the following are all equivalent to $+2$.

$$+\dfrac{+10}{+5} = +(+2) = +2$$

$$+\dfrac{-10}{-5} = +(+2) = +2$$

$$-\dfrac{-10}{+5} = -(-2) = +2$$

$$-\dfrac{+10}{-5} = -(-2) = +2$$

Notice, in the preceding examples, by changing any *two* of the three signs of a fraction, the value of the fraction is not changed. This is indeed the case.

Thus, we can

- Change the sign of the fraction and the sign of the numerator. $\qquad -\dfrac{-7}{8} = +\dfrac{-7}{8}$

- Change the sign of the fraction and the sign of the denominator. $\qquad -\dfrac{6}{-7} = +\dfrac{6}{+7}$

    or

- Change the sign of the numerator and the sign of the denominator. $\qquad -\dfrac{-2}{-3} = -\dfrac{+2}{+3}$

The rule of signs for fractions is also true for algebraic fractions. For example, we can

- Change the sign of the fraction and the sign of the numerator. $\qquad -\dfrac{-x}{3} = -\dfrac{+x}{3}$

- Change the sign of the fraction and the sign of the denominator. $\qquad -\dfrac{x}{-3} = -\dfrac{x}{+3}$

    or

- Change the sign of the numerator and the sign of the denominator. $\qquad -\dfrac{x}{-3} = \dfrac{-x}{+3}$

Whenever our algebraic fraction contains more than one term in the numerator or denominator we apply the following "rule of signs" which applies to the factors of the fraction.

RULE OF SIGNS FOR ALGEBRAIC FRACTIONS

1. If the sign of the fraction and the sign(s) of *one* factor are changed, the value of the fraction will remain the same.

2. If the sign(s) of any *two* factors are changed, the value of the fraction will remain the same. (This step may be repeated so we may change two factors, four factors, six factors, etc.)

Using this rule, we can change $-3(x-1)/4$ in any one of the following ways, without changing the value of the fraction.

- Change the sign of the fraction and change the sign of the factor, 3.

$$-\frac{3(x-1)}{4} = +\frac{-3(x-1)}{4}$$

- Change the sign of the fraction and change the sign of the factor, 4.

$$-\frac{3(x-1)}{4} = +\frac{3(x-1)}{-4}$$

- Change the sign of the fraction and change the signs of the factor, $(x-1)$.

$$-\frac{3(x-1)}{4} = +\frac{3(-x+1)}{4}$$

- Change the signs of the two factors, $(x-1)$ and 4.

$$-\frac{3(x-1)}{4} = -\frac{3(-x+1)}{-4}$$

- Change the signs of the two factors, 3 and $(x-1)$.

$$-\frac{3(x-1)}{4} = -\frac{-3(-x+1)}{4}$$

This technique of changing signs of factors is very useful when reducing and combining fractions.

Do the following practice set. Check your answers with the answers in the right-hand margin.

PRACTICE SET 8-1B

Find the missing expression by supplying the correct sign without changing the value of the fraction.

1. $-\dfrac{-5}{6} = +\dfrac{}{6}$

2. $+\dfrac{x-1}{1-x} = -\dfrac{x-1}{}$

3. $-\dfrac{2(1-x)}{(x+1)(x-1)} = +\dfrac{}{(x+1)(x-1)}$

1. $+\dfrac{+5}{6}$

2. $-\dfrac{x-1}{-1+x}$

3. $+\dfrac{2(-1+x)}{(x+1)(x-1)}$

EXERCISE 8-1

Express the following fractions in factored form.

1. $\dfrac{3x}{4y}$

2. $\dfrac{2x+4}{4x+12}$

3. $\dfrac{x^2+2x}{2x+4}$

4. $\dfrac{2x^2+6x}{3x-9}$

5. $\dfrac{x^2-9}{x^2+6x+9}$

6. $\dfrac{x^2-x-6}{x^2+5x+6}$

7. $\dfrac{x^2+x-6}{x^2-4x+4}$

8. $\dfrac{2x^2-14x+24}{2x^2+2x-40}$

9. $\dfrac{2x^2+2x-4}{4x^2-4}$

10. $\dfrac{x^2-x}{3x^2-3}$

Find the missing expression by supplying the correct sign without changing the value of the fraction.

11. $-\dfrac{3x}{-4}=+\dfrac{3x}{\blacksquare}$

12. $\dfrac{2x}{-7}=\dfrac{2x}{\blacksquare}$

13. $\dfrac{x-2}{2-x}=\dfrac{x-2}{\blacksquare}$

14. $\dfrac{x(1-x)}{(x+1)(x-1)}=\dfrac{\blacksquare}{(x+1)(x-1)}$

15. $-\dfrac{2(1-y)}{3(y-1)}=\dfrac{\blacksquare}{3(y-1)}$

16. $-\dfrac{-2(x+1)}{3(x+1)}=\dfrac{\blacksquare}{3(x+1)}$

17. $-\dfrac{-2x(1-x)}{-3(x-1)}=\dfrac{\blacksquare}{-3(x-1)}$

8-2 REDUCING FRACTIONS TO LOWEST TERMS

The first fundamental law of fractions states that the numerator and denominator of a fraction may be divided by the same nonzero value without changing the value of the fraction. This technique is called reducing a fraction or simplifying a fraction.

In arithmetic, we reduced fractions to lowest terms by dividing (canceling) out the common factor(s). For example:

The fraction 12/18 can be reduced by canceling the common factor 6 from both the numerator and denominator.

$$\frac{12}{18}=\frac{^2\cancel{12}}{_3\cancel{18}}=\frac{2}{3}$$

Thus,

$$\frac{12}{18}\text{ is reduced to }\frac{2}{3}$$

In algebra, this same method is used. For example, to reduce the fraction $12x^2y/16xy^2$, we first identify the factors that are common to both the numerator and denominator, and then we simply cancel like factors. Observe.

The greatest common factor of 12 and 16 is 4.

$$\frac{12\ x^2y}{16\ x\ y^2}$$

Thus, we cancel 4 out of 12 and 16.

$$\frac{^3\cancel{12}x^2y}{_4\cancel{16}x\ y^2}$$

The greatest common factor of x^2 and x is x.

$$\frac{^3 12\ x^2\ y}{_4 16\ x\ y^2}$$

Thus, we cancel x out of x^2 and x.

$$\frac{^3 12\ ^x\cancel{x^2}\ y}{_4 16\ _1\cancel{x}y^2}$$

The greatest common factor of y and y^2 is y.

$$\frac{^3 12\ ^x x^2\ y}{_4 16\ _1 x\ y^2}$$

Thus, we cancel y out of y and y^2.

$$\frac{^3 12\ ^x x^2\ {}^1 y}{_4 16\ _1 \cancel{x}\ y\ y^2}$$

Multiplying the remaining factors of the numerator and denominator produces our answer.

$$\frac{^3\cancel{12}^x x^2\ {}^1\cancel{y}}{_4\cancel{16}_1\cancel{x}\ y\ y^2} = \frac{3 \cdot x \cdot 1}{4 \cdot 1 \cdot y} = \frac{3x}{4y}$$

Thus,

$$\frac{12x^2y}{16xy^2} \text{ is reduced to } \frac{3x}{4y}$$

As a second example, consider $(x^2 - y^2)/(4x - 4y)$. To reduce this fraction by canceling like factors, we should first factor the numerator and denominator completely. By doing this, we can then easily identify any common factors.

Factor the numerator.

$$\frac{x^2 - y^2}{4x - 4y} = \frac{(x + y)(x - y)}{4x - 4y}$$

Factor the denominator.

$$\frac{x^2 - y^2}{4x - 4y} = \frac{(x + y)(x - y)}{4(x - y)}$$

Clearly, the factor $(x - y)$ is common to both the numerator and denominator and hence can be canceled out.

$$\frac{(x + y)^1\cancel{(x - y)}}{4_1\cancel{(x - y)}} = \frac{(x + y)}{4}$$

Thus,

$$\frac{x^2 - y^2}{4x - 4y} \text{ is reduced to } \frac{(x + y)}{4}$$

In summary, to reduce an algebraic fraction to lowest terms, we first factor the numerator and denominator completely. Once this has been done, we can then cancel all like factors out of the numerator and denominator. Collecting the remaining factors in the numerator and denominator, respectively, yields the reduced form of the fraction.

PROCEDURE:

> To reduce an algebraic fraction to its lowest terms:
>
> 1. Factor the numerator and denominator completely.
>
> 2. Cancel all common factors between the numerator and denominator.
>
> 3. Multiply (collect) all remaining factors of the numerator and denominator, respectively.

EXAMPLE 1: Reduce $\dfrac{21x^3y}{35xy^2}$.

SOLUTION: Factor the numerator and denominator completely.

$$\frac{21x^3y}{35xy^2} = \frac{(21)(x^3)(y)}{(35)(x)(y^2)}$$

Cancel like factors out of the numerator and denominator.

$$\frac{(21)(x^3)(y)}{(35)(x)(y^2)} = \frac{^3\,(21)\;^{x^2}(x^3)\;^1\,(y)}{_5\,(35)\;_1\,(x)\;_y\,(y^2)}$$

Collect the remaining factors.

$$\frac{^3\,(21)\;^{x^2}(x^3)\;^1\,(y)}{_5\,(35)\;_1\,(x)\;_y\,(y^2)} = \frac{(3)(x^2)(1)}{(5)(1)(y)} = \frac{3x^2}{5y}$$

Final answer: $\dfrac{3x^2}{5y}$

EXAMPLE 2: Reduce $\dfrac{5x-10}{15x}$.

SOLUTION: Factor the numerator and denominator completely.

$$\frac{5x-10}{15x} = \frac{5(x-2)}{(15)(x)}$$

Cancel like factors out of the numerator and denominator.

$$\frac{5(x-2)}{(15)(x)} = \frac{^1\,5(x-2)}{_3\,(15)(x)}$$

Example continued on next page.

Collect the remaining factors.

$$\frac{^1\,5(x-2)}{_3\,(15)(x)} = \frac{(1)(x-2)}{(3)(x)} = \frac{x-2}{3x}$$

Final answer: $\dfrac{x-2}{3x}$

EXAMPLE 3: Reduce $\dfrac{4x^2 - 1}{4x + 2}$.

SOLUTION: Factor the numerator and denominator completely.

$$\frac{4x^2 - 1}{4x + 2} = \frac{(2x+1)(2x-1)}{2(2x+1)}$$

Cancel like factors out of the numerator and denominator.

$$\frac{(2x+1)(2x-1)}{2(2x+1)} = \frac{^1\,(2x+1)(2x-1)}{2\,_1\,(2x+1)}$$

Collect the remaining factors.

$$\frac{^1\,(2x+1)(2x-1)}{2\,_1\,(2x+1)} = \frac{(1)(2x-1)}{(2)(1)} = \frac{(2x-1)}{2}$$

Final answer: $\dfrac{2x-1}{2}$

EXAMPLE 4: Reduce $\dfrac{z^2 + 3z}{z^2 + 10z + 21}$

SOLUTION: Factor completely.

$$\frac{z^2 + 3z}{z^2 + 10z + 21} = \frac{z(z+3)}{(z+3)(z+7)}$$

Cancel like factors.

$$\frac{z(z+3)}{(z+3)(z+7)} = \frac{z^1\,(z+3)}{_1\,(z+3)(z+7)}$$

Collect all remaining factors.

$$\frac{z^1\,(z+3)}{_1\,(z+3)(z+7)} = \frac{(z)(1)}{(1)(z+7)} = \frac{z}{z+7}$$

Final answer: $\dfrac{z}{z+7}$

NOTE:

In Example 4, it is very tempting to cancel out z from the numerator and denominator. However, we *cannot* do this! z in the denominator is not a factor; it is a term that is part of the factor $(z + 7)$.

$$\frac{^1\cancel{z}}{_1\cancel{z} + 7} = \frac{1}{7} \quad \text{(wrong)} \qquad \frac{^1\cancel{z}}{_1\,\cancel{(z)}(7)} = \frac{1}{7} \quad \text{(correct)}$$

A similar situation exists in examples 2 and 3.

EXAMPLE 5: Reduce $\dfrac{x^2 + 2x - 15}{2x^2 - 12x + 18}$.

SOLUTION:

$$\frac{x^2 + 2x - 15}{2x^2 - 12x + 18} = \frac{(x + 5)(x - 3)}{2(x^2 - 6x + 9)}$$

$$= \frac{(x + 5)(x - 3)}{2(x - 3)(x - 3)}$$

$$= \frac{(x + 5)\,^1\,\cancel{(x - 3)}}{2(x - 3)\,_1\,\cancel{(x - 3)}} = \frac{x + 5}{2(x - 3)}$$

Final answer: $\dfrac{x + 5}{2(x - 3)}$

EXAMPLE 6: Reduce $\dfrac{x^2 + x - 6}{4 - 2x}$.

SOLUTION:

$$\frac{x^2 + x - 6}{4 - 2x} = \frac{(x + 3)(x - 2)}{2(2 - x)} \qquad \text{or}$$

$$= \frac{(x + 3)(x - 2)}{-2(x - 2)} \qquad \left(\begin{matrix}\text{Recall the three signs}\\\text{of a fraction}\end{matrix}\right.$$

$$= \frac{(x + 3)\,^1\,\cancel{(x - 2)}}{-2\,_1\,\cancel{(x - 2)}}$$

$$= \frac{(x + 3)}{-2} = -\frac{(x + 3)}{2}$$

Final answer: $-\dfrac{(x + 3)}{2}$

Do the following practice set. Check your answers with the answers in the right-hand margin.

PRACTICE SET 8-2

Reduce the following fractions to lowest terms (see example 1 or 2).

1. $\dfrac{5x^2}{10x}$ 2. $\dfrac{2x^2 + 4x}{4x^2 - 8x}$ 3. $\dfrac{6x - 6}{2x - 4}$

1. $x/2$
2. $(x + 2)/2(x - 2)$
3. $3(x - 1)/(x - 2)$

Reduce the following fractions (see example 3, 4, 5, or 6).

4. $\dfrac{2x - 2}{3x - 3}$ 5. $\dfrac{x^2 - 1}{x^2 + 3x + 2}$

6. $\dfrac{4x^2 - 4x - 8}{2x^2 - 4x - 6}$ 7. $\dfrac{3 - 3x}{3x^2 - 3}$

4. $\frac{2}{3}$
5. $(x - 1)/(x + 2)$
6. $2(x - 2)/(x - 3)$
7. $-1/(x + 1)$

EXERCISE 8-2

Reduce the following fractions.

1. $\dfrac{3x}{3y}$ 2. $\dfrac{5}{15x}$ 3. $\dfrac{4x}{16}$ 4. $\dfrac{6x}{3x}$

5. $\dfrac{2x^2}{4x^2}$ 6. $\dfrac{3b^2}{12b}$ 7. $\dfrac{3x^3y^2}{12x^2y^4}$ 8. $\dfrac{-14x^2y^2}{21x^3y^3}$

9. $-\dfrac{6x^2y}{9xy^2}$ 10. $\dfrac{30x^3y}{-36x^2y^3}$ 11. $\dfrac{3(x + 2)}{4(x + 2)}$ 12. $\dfrac{4x^2(x + 1)}{2x(x + 1)}$

13. $\dfrac{6(x - 2)}{12x(x - 2)}$ 14. $\dfrac{2(1 - x)}{4(x - 1)}$ 15. $\dfrac{3x - 3y}{5x - 5y}$ 16. $\dfrac{3x + 3}{3x - 3}$

17. $\dfrac{5x - 5y}{20x - 20y}$ 18. $\dfrac{3(x + y)(x + y)}{6(x + y)}$ 19. $\dfrac{4(x - y)^2}{6x - 6y}$ 20. $\dfrac{x^2 - y^2}{2(x - y)}$

21. $\dfrac{2x + 4}{x^2 - 4}$ 22. $\dfrac{(x + 2)^2}{x^2 - 4}$ 23. $\dfrac{x}{x^2 + x}$ 24. $\dfrac{x^2 - 9}{x^2 - 6x + 9}$

25. $\dfrac{7x + 7}{x^2 + 2x + 1}$ 26. $\dfrac{2y - 2}{y^2 - 2y + 1}$ 27. $\dfrac{x^2 - 4}{x^2 + 4x + 4}$ 28. $\dfrac{x^2 - 4x + 4}{x^2 - 4}$

29. $\dfrac{4x + 8}{x^2 + 5x + 6}$ 30. $\dfrac{3x - 3}{x - 1}$ 31. $\dfrac{3x^2 - 3}{x + 1}$ 32. $\dfrac{x^2 - x}{x^2 - 4x + 3}$

33. $\dfrac{x^2 - 3x + 2}{x^2 + 2x - 8}$

34. $\dfrac{x^2 + 4x + 3}{x^2 - 2x - 3}$

35. $\dfrac{y^2 - 8y + 15}{y^2 - y - 6}$

36. $\dfrac{2x^2 + 5x + 2}{2x^2 - 5x - 3}$

37. $\dfrac{2x - 2}{3 - 3x}$

38. $\dfrac{y^2 - 3y + 2}{-y^2 + 3y - 2}$

39. $\dfrac{x^2 + 3x - 10}{x^2 - 10x + 16}$

40. $\dfrac{3x^2 - 5x + 2}{9x^2 - 4}$

8-3 MULTIPLICATION OF FRACTIONS

In arithmetic, we found the product of fractions such as $\frac{2}{3} \times \frac{1}{4}$ by

First, finding the product of the numerators

$$\frac{2}{3} \times \frac{1}{4} = \frac{(2)(1)}{} = \frac{2}{}$$

And the product of the denominators.

$$\frac{2}{3} \times \frac{1}{4} = \frac{(2)(1)}{(3)(4)} = \frac{2}{12}$$

Then, reducing the resulting fraction.

$$\frac{^1 2}{_6 12} = \frac{1}{6}$$

This problem is simplified by first reducing the fraction while in factored form. For example, before we find the product of the numerator and the product of the denominator, cancel factors that are common to both the numerator and denominator.

Cancel 2 from the numerator and the denominator before multiplying.

$$\frac{2}{3} \times \frac{1}{4} = \frac{2 \cdot 1}{3 \cdot 4}$$

$$\frac{^1 2 \cdot 1}{3 \cdot {_2} 4} = \frac{1}{6}$$

Thus,

$$\text{the product is } \frac{1}{6}$$

The product of algebraic fractions is found the same way as in arithmetic.

To multiply two or more fractions together, multiply the numerators together to find the numerator of the product and multiply the denominators to find the denominator of the product. For example, to multiply $2/x \cdot 3/y$:

Multiply the numerators.

$$\frac{2}{x} \cdot \frac{3}{y} = \frac{2 \cdot 3}{} = 6$$

Multiply the denominators.

$$\frac{2}{x} \cdot \frac{3}{y} = \frac{2 \cdot 3}{x \cdot y} = \frac{6}{xy}$$

Thus,

$$\text{the product of } \frac{2}{x} \text{ and } \frac{3}{y} \text{ is } \frac{6}{xy}$$

In most of our problems, we do not actually multiply to find the product of the numerator or denominator. Instead, we indicate the product by writing the numerator and denominator in factored form. Our answer is acceptable in factored form, and the common factors are used to reduce our product. Consider the following example.

To multiply $(5x - 5y)/9x^2$ by $3x^2/(x^2 - y^2)$, we do not find the product of the numerators and the product of the denominators. Rather, we indicate the product by expressing them in factored form. For example:

The numerator is the product of $(5x - 5y)$ and $(3x^2)$.

$$\frac{5x - 5y}{9x^2} \cdot \frac{3x^2}{x^2 - y^2} = (5x - 5y)(3x^2)$$

The denominator is the product of $(9x^2)$ and $(x^2 - y^2)$.

$$\frac{5x - 5y}{9x^2} \cdot \frac{3x^2}{x^2 - y^2} = \frac{(5x - 5y)(3x^2)}{(9x^2)(x^2 - y^2)}$$

Now, factor the numerator completely.

$$\begin{bmatrix} 5x - 5y = 5(x - y) \\ 3x^2 = 3 \cdot x^2 \end{bmatrix}$$

$$\frac{(5x - 5y)(3x^2)}{(9x^2)(x^2 - y^2)} = 5(x - y) \cdot 3 \cdot x^2$$

Factor the denominator completely.

$$\begin{bmatrix} 9x^2 = 9 \cdot x^2 \\ x^2 - y^2 = (x + y)(x - y) \end{bmatrix}$$

$$\frac{(5x - 5y)(3x^2)}{(9x^2)(x^2 - y^2)} = \frac{5(x - y) \cdot 3 \cdot x^2}{9 \cdot x^2(x + y)(x - y)}$$

The common factors of $(x - y)$, x^2, and 3 are canceled.

$$\frac{5(x - y) \cdot 3 \cdot x^2}{9 \cdot x^2(x + y)(x - y)}$$

$$\frac{5 \; ^1(x - y) \cdot \; ^1 3 \cdot \; ^1 x^2}{_3 9 \cdot \; _1 x^2(x + y) \; _1 (x - y)}$$

And the remaining factors are collected.

$$\frac{5 \cdot 1 \cdot 1 \cdot 1}{3 \cdot 1(x + y) \cdot 1} = \frac{5}{3(x + y)}$$

Thus,

$$\text{the product is } \frac{5}{3(x + y)}$$

NOTE:

Answers will be kept in factored form.

So, to multiply fractions, we, first, write the product of the numerators over the product of the denominators; then, reduce the resulting fraction by completely factoring the numerator and denominator, canceling common factors and collecting the remaining factors; and, finally, leave our answer in factored form.

PROCEDURE:

To multiply fractions:

1. Write the product of the numerators over the product of the denominators.

2. Reduce the resulting fraction.
    a. Factor the numerator and denominator completely.
    b. Cancel all common factors.
    c. Collect all remaining factors.

EXAMPLE 1:   Multiply $\dfrac{7x}{12y}$ by $\dfrac{48y^2}{35y^2}$.

SOLUTION: Write the product of the numerators over the product of the denominators.

$$\frac{7x}{12y} \cdot \frac{48y^2}{35x^2} = \frac{(7x)(48y^2)}{(12y)(35x^2)}$$

Reduce the fraction.

$$\text{Factor} \quad \frac{(7x)(48y^2)}{(12y)(35x^2)}$$

$$= \frac{(7)(48)(x)(y^2)}{(12)(35)(x^2)(y)}$$

$$\text{Cancel common factors} = \frac{^1\cancel{(7)}\ ^4\cancel{(48)}\ ^1(x)\ ^y\cancel{(y^2)}}{_1\cancel{(12)}\ _5\cancel{(35)}\ _x\cancel{(x^2)}\ _1\cancel{(y)}}$$

$$\text{Collect remaining factors} = \frac{(1)(4)(1)(y)}{(1)(5)(x)(1)}$$

$$= \frac{4y}{5x}$$

Final answer: $\dfrac{4y}{5x}$

EXAMPLE 2:   Multiply $\dfrac{2a}{b+c}$ by $\dfrac{3b+3c}{2a^2+2a}$.

Example continued on next page.

SOLUTION: Write the product of the numerators over the product of the denominators.

$$\frac{2a}{b+c} \cdot \frac{3b+3c}{2a^2+2a} = \frac{(2a)(3b+3c)}{(b+c)(2a^2+2a)}$$

Reduce the fraction.

$$\text{Factor} \quad \frac{(2a)(3b+3c)}{(b+c)(2a^2+2a)}$$

$$= \frac{(2a)(3)(b+c)}{(b+c)(2a)(a+1)}$$

$$\text{Cancel common factors} = \frac{^1\cancel{(2a)}(3)^{\ 1}\cancel{(b+c)}}{_1\cancel{(b+c)}_1\cancel{(2a)}(a+1)}$$

$$\text{Collect remaining factors} = \frac{(1)(3)(1)}{(1)(1)(a+1)}$$

$$= \frac{3}{(a+1)}$$

Final answer: $\dfrac{3}{a+1}$

EXAMPLE 3: Multiply $\dfrac{4}{a-b}$ by (a^2-b^2).

SOLUTION: Write the product of the numerators over the product of the denominators.

$$\frac{4}{a-b} \cdot \frac{a^2-b^2}{1} = \frac{(4)(a^2-b^2)}{(a-b)(1)}$$

Reduce the fraction.

$$\text{Factor} \quad \frac{(4)(a^2-b^2)}{(a-b)(1)}$$

$$= \frac{(4)(a+b)(a-b)}{(a-b)(1)}$$

$$\text{Cancel common factors} = \frac{(4)(a+b)^{\ 1}\cancel{(a-b)}}{_1\cancel{(a-b)}(1)}$$

$$\text{Collect remaining factors} = \frac{(4)(a+b)(1)}{(1)(1)}$$

$$= (4)(a+b)$$

Final answer: $4(a+b)$

EXAMPLE 4:   Multiply $\dfrac{(x-3)^2}{x^2-x-6}$ by $\dfrac{x+2}{x-3}$.

SOLUTION: Write the product of the numerators over the product of the denominators.

$$\frac{(x-3)^2}{x^2-x-6} \cdot \frac{x+2}{x-3} = \frac{(x-3)^2(x+2)}{(x^2-x-6)(x-3)}$$

Reduce the fraction.

$$\frac{(x-3)^2(x+2)}{(x^2-x-6)(x-3)} = \frac{(x-3)(x-3)(x+2)}{(x+2)(x-3)(x-3)}$$

$$= \frac{^1(x-3)\,^1(x-3)\,^1(x+2)}{_1(x+2)\,_1(x-3)\,_1(x-3)}$$

$$= \frac{(1)(1)(1)}{(1)(1)(1)}$$

$$= 1$$

Final answer: 1

EXAMPLE 5:   Multiply $\dfrac{x^2+6x+5}{7x^2-63}$ by $\dfrac{7x+21}{(5+x)^2}$.

SOLUTION:

$$\frac{x^2+6x+5}{7x^2-63} \cdot \frac{7x+21}{(5+x)^2} = \frac{(x^2+6x+5)(7x+21)}{(7x^2-63)(5+x)^2}$$

$$= \frac{(x+1)(x+5)(7)(x+3)}{7(x^2-9)(5+x)(5+x)}$$

$$= \frac{(x+1)(x+5)(7)(x+3)}{7(x+3)(x-3)(5+x)(5+x)}$$

$$= \frac{(x+1)\,^1(x+5)\,^1(7)\,^1(x+3)}{_1 7\,_1(x+3)(x-3)\,_1(5+x)(5+x)}$$

$$= \frac{(x+1)(1)(1)(1)}{(1)(1)(x-3)(1)(5+x)}$$

$$= \frac{x+1}{(x-3)(5+x)}$$

Final answer: $\dfrac{x+1}{(x-3)(5+x)}$ or $\dfrac{x+1}{(x-3)(x+5)}$

Do the following practice set. Check your answers with the answers in the right-hand margin.

PRACTICE SET 8-3

Multiply and express the product in reduced form (see example 1 or 2).

1. $\dfrac{7x}{9y} \cdot \dfrac{36y^2}{21x^2}$

2. $\dfrac{x+y}{x-y} \cdot \dfrac{x-y}{2}$

3. $\dfrac{2x^2-2x}{3} \cdot \dfrac{1}{x-1}$

1. $4y/3x$
2. $(x+y)/2$
3. $2x/3$

Multiply and express the product in reduced form (see example 3).

4. $6 \cdot \dfrac{a}{2}$

5. $(a+b) \cdot \dfrac{2}{(a+b)^2}$

6. $\dfrac{3}{(x-2)} \cdot (x^2-4)$

4. $3a$
5. $2/(a+b)$
6. $3(x+2)$

Multiply and express the product in reduced form (see example 4 or 5).

7. $\dfrac{2x+4}{4x} \cdot \dfrac{x^2}{x^2-4}$

8. $\dfrac{x-3}{x+3} \cdot \dfrac{x^2+6x+9}{3x^2-27}$

9. $\dfrac{x^2+4x+4}{2x^2-8} \cdot \dfrac{4x-8}{x^2+2x}$

7. $x/2(x-2)$

8. $\frac{1}{3}$

9. $2/x$

EXERCISE 8-3

Multiply and express the product in reduced form.

1. $\dfrac{2}{3} \cdot \dfrac{9}{10}$

2. $\dfrac{4x^3}{5} \cdot \dfrac{3}{2x}$

3. $\dfrac{5a}{3} \cdot \dfrac{1}{a^2}$

4. $9x^2y \cdot \dfrac{2}{3xy}$

5. $\dfrac{1}{x^2y^2} \cdot 3xy$

6. $\dfrac{3x}{4} \cdot \dfrac{4x}{3}$

7. $\left(\dfrac{-3x^3}{4}\right) \cdot \left(-\dfrac{2}{9x^2}\right)$

8. $\left(\dfrac{-4ab}{9c}\right) \cdot \left(\dfrac{-3ac}{8b}\right)$

9. $\dfrac{6x^2}{5y^2} \cdot \dfrac{10xy}{6y^3}$

10. $4\pi r^2 \cdot \dfrac{1}{2\pi r}$

11. $\dfrac{2}{x+1} \cdot \dfrac{x+1}{3}$

12. $\dfrac{x+2}{x+3} \cdot \dfrac{x-1}{x+2}$

13. $\dfrac{2x+2y}{3x-3y} \cdot \dfrac{9x-9y}{4x+4y}$

14. $\dfrac{2x+10}{6} \cdot \dfrac{3x-15}{x^2-25}$

15. $\dfrac{3x-3y}{(x-y)^2} \cdot \dfrac{x^2-y^2}{4x+4y}$

16. $\dfrac{2x^2 + x}{4x^2 - 1} \cdot \dfrac{4x - 2}{6x}$

17. $\dfrac{x + 9}{4x - 8} \cdot \dfrac{30x}{x^2 + 6x - 27}$

18. $\dfrac{x - 2y}{3y - x} \cdot \dfrac{6y - 2x}{(x - 2y)^2}$

19. $\dfrac{x^2 - 9}{2x - 6} \cdot \dfrac{4x - 4}{x^2 + 2x - 3}$

20. $\dfrac{(x - 4)^2}{16x} \cdot \dfrac{4x + 16}{2x^2 - 32}$

21. $\dfrac{1 - 9x^2}{x^2 - 4} \cdot \dfrac{3x + 6}{6 - 18x}$

22. $\dfrac{x^2 - x - 6}{x + 3} \cdot \dfrac{3x + 3}{x - 3}$

23. $\dfrac{x^2 - 1}{x^2 - 16} \cdot \dfrac{x^2 - 4x}{x - 1}$

24. $\dfrac{x^2 + x - 6}{2x^2 + 6x} \cdot \dfrac{6x^2}{x^2 - 5x + 6}$

25. $\dfrac{2x^2 - 9x - 5}{2x + 1} \cdot \dfrac{3x - 2}{3x^2 + 13x - 10} \cdot \dfrac{x^2 + 6x + 5}{x^2 - 4x - 5}$

8-4 DIVISION OF FRACTIONS

In arithmetic, we learned to divide fractions by changing the problem to a multiplication problem and multiplying by the *reciprocal* of the divisor. For example, to divide $\frac{2}{3}$ by $\frac{8}{15}$,

Change the sign of the operation ($\div$ to $\times$), and the reciprocal of $\dfrac{8}{15}$ is $\dfrac{15}{8}$.

$$\dfrac{2}{3} \div \dfrac{8}{15}$$

$$\dfrac{2}{3} \times \dfrac{15}{8}$$

Now, find the product of the numerator and the product of the denominator.

$$\dfrac{2}{3} \times \dfrac{15}{8} = \dfrac{(2)(15)}{(3)(8)}$$

Then, reduce our fraction by canceling like factors.

$$\dfrac{^1(2)\,^5(15)}{_1(3)\,_4(8)} = \dfrac{1 \cdot 5}{1 \cdot 4} = \dfrac{5}{4}$$

Thus,

$$\text{the quotient of } \dfrac{2}{3} \div \dfrac{8}{15} = \dfrac{5}{4}$$

We use the same technique to divide algebraic fractions—that is, changing the division problem to a multiplication problem by multiplying by the reciprocal of the divisor. For example, to divide $(2x - 4)/(2x + 6)$ by $(x - 2)/(3x + 9)$:

First, change the problem to multiplication by multiplying the first fraction by the reciprocal of the second fraction.

$$\dfrac{2x - 4}{2x + 6} \div \dfrac{x - 2}{3x + 9}$$

$$= \dfrac{2x - 4}{2x + 6} \cdot \dfrac{3x + 9}{x - 2}$$

Now, proceed as in multiplication, factoring the numerator and denominator, and then canceling like factors.

$$= \dfrac{(2x - 4)(3x + 9)}{(2x + 6)(x - 2)}$$

$$= \dfrac{2(x - 2)(3)(x + 3)}{2(x + 3)(x - 2)}$$

The product is found by collecting the remaining terms.

$$= \dfrac{^1(2)\,^1(x - 2)(3)\,^1(x + 3)}{_1 2\,_1(x + 3)\,_1(x - 2)}$$

$$= \dfrac{3}{1}$$

Thus,

$$\frac{2x-4}{2x+6} \text{ divided by } \frac{x-2}{3x+9} = 3$$

PROCEDURE:

To divide fractions:

1. Invert the divisor and proceed as in multiplication.

2. Multiply the fractions together.
   a. Factor the numerator and denominator completely.
   b. Cancel all common factors.
   c. Multiply the remaining factors.

EXAMPLE 1: Divide $\dfrac{9a^2}{4b}$ by $\dfrac{3a}{16b^2}$.

SOLUTION: Change to a multiplication problem.

$$\frac{9a^2}{4b} \div \frac{3a}{16b^2} = \frac{9a^2}{4b} \cdot \frac{16b^2}{3a}$$

Multiply.

$$\frac{9a^2}{4b} \cdot \frac{16b^2}{3a} = \frac{(9)(16)(a^2)(b^2)}{(4)(3)(a)(b)}$$

$$= \frac{^3\,(9)\,^4\,(16)\,^a\,(a^2)\,^b\,(b^2)}{_1\,(4)\,_1\,(3)\,_1\,(a)\,_1\,(b)}$$

$$= \frac{(3)(4)(a)(b)}{(1)(1)(1)(1)}$$

$$= 12ab$$

Final answer: $12ab$

EXAMPLE 2: Divide $\dfrac{5x^2}{y^2-36}$ by $\dfrac{25xy-25x}{y^2-7y+6}$.

SOLUTION: Change to a multiplication problem.

$$\frac{5x^2}{y^2-36} \div \frac{25xy-25x}{y^2-7y+6} = \frac{5x^2}{y^2-36} \cdot \frac{y^2-7y+6}{25xy-25x}$$

Multiply.

$$\frac{5x^2}{y^2 - 36} \cdot \frac{y^2 - 7y + 6}{25xy - 25x} = \frac{(5x^2)(y^2 - 7y + 6)}{(y^2 - 36)(25xy - 25x)}$$

$$= \frac{(5)(x^2)(y - 6)(y - 1)}{(y + 6)(y - 6)(25x)(y - 1)}$$

$$= \frac{(5)(x^2)(y - 6)(y - 1)}{(y + 6)(y - 6)(25)(x)(y - 1)}$$

$$= \frac{\overset{1}{\cancel{(5)}}{}^x \overset{}{\cancel{(x^2)}}{}^1 \cancel{(y - 6)}{}^1 \cancel{(y - 1)}}{(y + 6)_1 \cancel{(y - 6)}_5 (25)_1 \cancel{(x)}_1 \cancel{(y - 1)}}$$

$$= \frac{(1)(x)(1)(1)}{(y + 6)(1)(5)(1)(1)}$$

$$= \frac{x}{(y + 6)5}$$

$$= \frac{x}{5(y + 6)}$$

Final answer: $\dfrac{x}{5(y + 6)}$

Do the following practice set. Check your answers with the answers in the right-hand margin.

PRACTICE SET 8-4

Divide the following fractions (see example 1 or 2).

1. $\dfrac{6x^2y}{8xy} \div \dfrac{3xy}{4x^2y}$ 1. x^2

2. $\dfrac{x - 3}{x + 4} \div \dfrac{x^2 - 9}{2x + 8}$ 2. $2/(x + 3)$

3. $\dfrac{x^2 + 2x + 1}{x - 3} \div \dfrac{x + 1}{2}$ 3. $2(x + 1)/(x - 3)$

EXERCISE 8-4

Divide and express the quotient in lowest terms.

1. $\dfrac{2}{x^2} \div \dfrac{2}{x}$ 2. $\dfrac{x}{24} \div \dfrac{x^2}{8}$ 3. $\dfrac{x^2}{y} \div \dfrac{x}{y}$

4. $\dfrac{3a}{4b^2} \div \dfrac{15a^2}{16b^3}$ 5. $6x \div \dfrac{x}{2}$ 6. $3x \div \dfrac{1}{5}$

7. $18 \div \dfrac{3}{x}$

8. $8xy \div \dfrac{24x}{y}$

9. $\dfrac{5}{7}xy \div 15x^2$

10. $\dfrac{3x^2y^2}{8z} \div 3xy$

11. $\dfrac{4x^2}{7} \div 8x$

12. $5x^2 \div 2\dfrac{1}{2}x$

13. $\dfrac{x-4}{2(x-1)} \div \dfrac{x-4}{3(x-1)}$

14. $\dfrac{3(x+2)}{6(x-2)} \div \dfrac{9(x+2)}{12(x-2)}$

15. $\dfrac{(x-y)^2}{x+y} \div \dfrac{3(x-y)}{2(x+y)}$

16. $\dfrac{15}{x^2-1} \div \dfrac{5}{x+1}$

17. $\dfrac{2x-2y}{x^2y^2} \div \dfrac{x^2-y^2}{x^2y^2}$

18. $\dfrac{2x+4}{3x+9} \div \dfrac{4x+8}{5x+15}$

19. $\dfrac{x^2-4x-5}{x^2+4x+3} \div \dfrac{x-5}{x+3}$

20. $\dfrac{x^2-4x+4}{2x-4} \div (x-2)$

21. $(x^2-9) \div \dfrac{x^2+2x-15}{3x+15}$

22. $\dfrac{x^2-1}{x^2-3x+2} \div \dfrac{1-x}{2-x}$

8-5 ADDITION AND SUBTRACTION OF FRACTIONS HAVING THE SAME DENOMINATOR

In arithmetic, we learned to add or subtract fractions with common denominators by finding the sum or difference of the numerators and dividing this value by the common denominator. For example, to add $\frac{3}{9} + \frac{2}{9}$,

First, add the numerators. $\qquad\qquad\qquad \dfrac{3}{9} + \dfrac{2}{9} = 5$

Then, divide by the common denominator. $\qquad \dfrac{3}{9} + \dfrac{2}{9} = \dfrac{5}{9}$

The answer is already reduced.
Thus,

$$\text{the sum of } \dfrac{3}{9} + \dfrac{2}{9} = \dfrac{5}{9}$$

In algebra, we use the same technique. That is, to add or subtract fractions having the same denominator, we find the sum or difference of the numerators and divide this value by the common denominator.

A more appropriate description of this procedure would be to combine the numerators according to their sign and place this result over the common denominator. The resulting fraction should then be reduced. For example, to combine $3x/(2x+10)$ and $15/(2x+10)$,

Combine the numerators according to their sign. $\qquad \dfrac{3x}{2x+10} \boxed{+} \dfrac{15}{2x+10} \quad 3x+15$

And place this result over the common denominator. $\qquad \dfrac{3x}{2x+10} + \dfrac{15}{2x+10} = \dfrac{3x+15}{2x+10}$

The resulting fraction is then reduced. $\qquad \dfrac{3x+15}{2x+10} = \dfrac{3\,^1\cancel{(x+5)}}{2\,_1\cancel{(x+5)}} = \dfrac{3}{2}$

Thus,

$$\frac{3x}{2x + 10} + \frac{15}{2x + 10} = \frac{3}{2}$$

PROCEDURE:

To add or subtract fractions with the same denominator:

1.   Combine the numerators according to their sign to get the numerator of the answer.

2.   Place the result found in (1) over the common denominator.

3.   Reduce the fraction.
  a.   Factor completely.
  b.   Cancel the common factors.
  c.   Multiply the remaining factors.

EXAMPLE 1:   Combine $\dfrac{2x}{15} + \dfrac{7x}{15}$.

SOLUTION: Combine numerators according to their sign.

$$\frac{2x}{15} + \frac{7x}{15}$$

$$\frac{(2x) + (7x)}{} $$
$$= 9x$$

Place the answer over the common denominator.

$$\frac{2x}{15} + \frac{7x}{15} = \frac{9x}{15}$$

Reduce the fraction.

$$\frac{9x}{15} = \frac{^3 9x}{_5 15} = \frac{3x}{5}$$

Final answer: $\dfrac{3x}{5}$

EXAMPLE 2:   Combine $\dfrac{x + 5}{3x} - \dfrac{1 - x}{3x} - \dfrac{7x + 4}{3x}$

Example continued on next page.

SOLUTION: Combine numerators according to their sign.

$$\frac{x+5}{3x} - \frac{1-x}{3x} - \frac{7x+4}{3x}$$

$$(x+5) - (1-x) - (7x+4)$$
$$= x + 5 - 1 + x - 7x - 4$$
$$= x + x - 7x + 5 - 1 - 4$$
$$= -5x + 0$$
$$= -5x$$

Place this answer over the common denominator.

$$\frac{x+5}{3x} - \frac{1-x}{3x} - \frac{7x+4}{3x} = \frac{-5x}{3x}$$

Reduce the fraction.

$$\frac{-5x}{3x} = \frac{-5\,^1(x)}{3\,_1(x)} = \frac{-5}{3}$$

Final answer: $\dfrac{-5}{3}$

NOTE:

When combining numerators according to their sign, we are careful to place each numerator within parentheses. We then remove parentheses and combine like terms.

EXAMPLE 3:   Combine $\dfrac{5}{x^2 + 3x - 4} + \dfrac{7x - 8}{x^2 + 3x - 4} - \dfrac{3x + 1}{x^2 + 3x - 4}$.

SOLUTION: Combine numerators according to their sign.

$$\frac{5}{x^2 + 3x - 4} + \frac{7x - 8}{x^2 + 3x - 4} - \frac{3x + 1}{x^2 + 3x - 4}$$

$$(5) + (7x - 8) - (3x + 1)$$
$$= 5 + 7x - 8 - 3x - 1$$
$$= 7x - 3x + 5 - 8 - 1$$
$$= 4x - 4$$

Place this answer over the common denominator.

$$\frac{5}{x^2 + 3x - 4} + \frac{7x - 8}{x^2 + 3x - 4} - \frac{3x + 1}{x^2 + 3x - 4} = \frac{4x - 4}{x^2 + 3x - 4}$$

Reduce the fraction.

$$\frac{4x - 4}{x^2 + 3x - 4} = \frac{4^1 (x - 1)}{(x + 4)_1 (x - 1)}$$

$$= \frac{4}{x + 4}$$

Final answer: $\dfrac{4}{x + 4}$

Do the following practice set. Check your answers with the answers in the right-hand margin.

PRACTICE SET 8-5

Combine the fractions and simplify (see example 1).

1. $\dfrac{4x}{6} + \dfrac{9x}{6} - \dfrac{15x}{6}$

2. $\dfrac{7}{x - 2} - \dfrac{3}{x - 2}$

3. $\dfrac{x}{2(x - 2)} - \dfrac{2}{2(x - 2)}$

Combine the fractions and simplify (see example 2 or 3).

4. $\dfrac{x + 2}{x - 1} + \dfrac{3x - 4}{x - 1}$

5. $\dfrac{3x}{7} - \dfrac{x + 4}{7}$

6. $\dfrac{x + 2}{3x} + \dfrac{4x - 1}{3x} - \dfrac{x + 1}{3x}$

7. $\dfrac{2x}{x^2 - x - 2} - \dfrac{x + 2}{x^2 - x - 2}$

1. $-\frac{x}{3}$

2. $4/(x - 2)$

3. $\frac{1}{2}$

4. $2(2x - 1)/(x - 1)$

5. $2(x - 2)/7$

6. $\frac{4}{3}$

7. $1/(x + 1)$

EXERCISE 8-5

Combine the fractions and simplify.

1. $\dfrac{7x}{8} + \dfrac{1x}{8}$

2. $\dfrac{7}{x} - \dfrac{2}{x}$

3. $\dfrac{7x}{12} + \dfrac{5x}{12} - \dfrac{7x}{12}$

4. $\dfrac{8}{15x} + \dfrac{4}{15x} - \dfrac{3}{15x}$

5. $\dfrac{2}{x+1} + \dfrac{3}{x+1}$

6. $\dfrac{6}{x-1} - \dfrac{4}{x-1}$

7. $\dfrac{x}{x+3} + \dfrac{3}{x+3}$

8. $\dfrac{2y}{y-4} - \dfrac{8}{y-4}$

9. $\dfrac{x^2}{x-y} - \dfrac{y^2}{x-y}$

10. $\dfrac{x^2}{x+4} - \dfrac{16}{x+4}$

11. $\dfrac{9x-3}{6} + \dfrac{3x+5}{6}$

12. $\dfrac{4x+1}{3} - \dfrac{x-2}{3}$

13. $\dfrac{3x-3y}{x-y} - \dfrac{2x+5y}{x-y} + \dfrac{8x-y}{x-y}$

14. $\dfrac{6x-5}{x^2-1} - \dfrac{5x-6}{x^2-1}$

15. $\dfrac{6x-4}{2x+6} - \dfrac{2x-6}{2x+6}$

16. $\dfrac{x-2}{x^2-3x-4} + \dfrac{3}{x^2-3x-4}$

8-6   COMBINING UNLIKE FRACTIONS

To add or subtract fractions with different denominators, we must form fractions that are equivalent to the original fractions that have common denominators. We then proceed to combine the fractions with the same denominator (as was done in Section 8-5). Two skills are needed to do this: We must be able (1) to find equivalent fractions and (2) to find the lowest common denominator.

A.   Finding Equivalent Fractions

The second fundamental law of fractions states that if the numerator and denominator of a fraction are multiplied by the same nonzero value, the value of the fraction is not changed. Thus, the resulting fraction is equivalent to the given fraction. (Equivalent fractions or equal fractions are fractions that have the same value.)

For example, to find fractions that are equivalent to $\frac{2}{3}$:

Multiply the numerator and denominator by 2—thus, $\dfrac{2}{3} = \dfrac{4}{6}$. $\dfrac{2 \cdot 2}{3 \cdot 2} = \dfrac{4}{6}$

or

Multiply the numerator and denominator by 3—thus, $\dfrac{2}{3} = \dfrac{6}{9}$. $\dfrac{2 \cdot 3}{3 \cdot 3} = \dfrac{6}{9}$

or

Multiply the numerator and denominator by x—thus, $\dfrac{2}{3} = \dfrac{2x}{3x}$.

$\dfrac{2 \cdot \boxed{x}}{3 \cdot \boxed{x}} = \dfrac{2x}{3x}$

or

Multiply the numerator and denominator by $(x + 3)$—thus, $\dfrac{2}{3} = \dfrac{2(x + 3)}{3(x + 3)}$.

$\dfrac{2 \boxed{(x + 3)}}{3 \boxed{(x + 3)}} = \dfrac{2(x + 3)}{3(x + 3)}$

Evidently, there are many fractions that are equivalent to $\frac{2}{3}$. In fact, every fraction has an unlimited number of equivalent fractions.

Our task in algebra will be to find an equivalent fraction with a *given* denominator. For example, to find a fraction that is equivalent to $\frac{2}{3}$ with a denominator of 12:

First, determine the multiplier of the denominator; that is, what the original denominator must be multiplied by to get the new denominator.

$\dfrac{2}{3} = \dfrac{?}{12}$

In this example, the multiplier is 4, since $3 \cdot 4 = 12$.

$3 \cdot 4 = 12$

Next, multiply the numerator by the multiplier.

$\dfrac{2 \cdot \boxed{4}}{3 \cdot \boxed{4}} \times \dfrac{8}{12}$

Thus,

$$\dfrac{2}{3} = \dfrac{8}{12}$$

When working with algebraic fractions, we often find it easier to identify the multiplier of the denominator if the denominators are in factored form. As an example, consider finding the missing numerator for the following problem.

$$\dfrac{2x - 5}{x + 3} = \dfrac{?}{3x + 9}$$

First, factor the denominators.

$\dfrac{2x - 5}{x + 3} = \dfrac{?}{\boxed{3x + 9}}$

$\dfrac{2x - 5}{(x + 3)} = \dfrac{?}{3(x + 3)}$

Clearly, the multiplier of the denominator is 3, so multiply the numerator by 3.

$\dfrac{2x - 5}{x + 3} = \dfrac{?}{3(x + 3)}$

$\dfrac{2x - 5}{x + 3} = \dfrac{\boxed{3(2x - 5)}}{3(x + 3)}$

Cleaning up the numerator yields our solution.

$\dfrac{3(2x - 5)}{3(x + 3)} = \dfrac{6x - 15}{3(x + 3)}$

Thus,

$$\dfrac{2x - 5}{x + 3} = \dfrac{6x - 15}{3(x + 3)}$$

So, to find equivalent fractions with a given denominator, we first factor the denominator completely; then, identify the multiplier of the denominator by comparing factors; and finally, multiply the numerator of the original fraction by the same multiplier. All like terms of the new numerator should be combined.

PROCEDURE:

To find equivalent fractions with a given denominator:

1. Factor the denominator completely.

2. Identify the denominator's multiplier by comparing factors.

3. Multiply the numerator by the same multiplier.

EXAMPLE 1:   Find the missing numerator.

$$\frac{2x}{6} = \frac{?}{18}$$

SOLUTION: The multiplier is easily recognized as 3, so multiply the numerator by 3.

$$\frac{2x}{6} = \frac{3 \cdot 2x}{3(6)} = \frac{6x}{18}$$

Final answer: $\dfrac{6x}{18}$

EXAMPLE 2:   Find the missing numerator.

$$\frac{2}{x} = \frac{?}{x^2 + x}$$

SOLUTION: Factor the denominator.

$$\frac{2}{x} = \frac{?}{x^2 + x}$$

$$\frac{2}{x} = \frac{?}{x(x + 1)}$$

Multiply the numerator by the multiplier $(x + 1)$

$$\frac{2}{x} = \frac{2 \cdot (x + 1)}{x \cdot (x + 1)}$$

$$= \frac{2x + 2}{x(x + 1)}$$

Final Answer: $\dfrac{2x + 2}{x(x + 1)}$

Do the following practice set. Check your answers with the answers in the right-hand margin.

PRACTICE SET 8-6A

Find the missing numerator that will make the second fraction equal to the first (see example 1 or 2).

1. $\dfrac{3}{2x} = \dfrac{}{4x^2}$

2. $\dfrac{2}{x + 1} = \dfrac{}{x^2 - 1}$

3. $\dfrac{x - 1}{x - 2} = \dfrac{}{x^2 - 5x + 6}$

1. $6x$

2. $2x - 2$

3. $x^2 - 4x + 3$

B. Finding the Least Common Denominator

The least common denominator (LCD) is an expression that is a multiple of each denominator. For example, the least common denominator of $\frac{3}{4}$ and $\frac{5}{6}$ is found by:

■ Finding the multiples of 4

$\dfrac{3}{4}$ and $\dfrac{5}{6}$

4, 8, 12, 16, 20, 24, 28, . . .

■ Finding the multiples of 6

$\dfrac{3}{4}$ and $\dfrac{5}{6}$

6, 12, 18, 24, 30, 36, 42, . . .

■ Finding the smallest multiple common to both.

4, 8, 12 , 16, 20, 24, 28, . . .

6, 12 , 18, 24, 30, 36, 42, . . .

Thus,

the least common denominator of $\dfrac{3}{4}$ and $\dfrac{5}{6}$ is 12

We apply this principle to algebraic expressions in the following manner. To find the least common denominator (LCD) of $5/2x^2$ and $3/4x^3$:

Separate each denominator into its factors.

$$\frac{5}{2x^2} \quad \text{and} \quad \frac{3}{4x^3}$$

$$= \frac{5}{(2) \cdot (x^2)} \quad \text{and} \quad \frac{3}{(4) \cdot (x^3)}$$

The least common multiple of the coefficients 2 and 4 is 4.

$$\frac{5}{\boxed{2} \cdot x^2} \quad \text{and} \quad \frac{3}{\boxed{4} \cdot x^3}$$

$$\text{LCM} = 4$$

The least common multiple of the variables x^2 and x^3 is x^3.

$$\frac{5}{2 \cdot \boxed{x^2}} \quad \text{and} \quad \frac{3}{4 \cdot \boxed{x^3}}$$

$$\text{LCM} = x^3$$

Thus,

$$\text{the LCD of } \frac{5}{2x^2} \text{ and } \frac{3}{4x^3} \text{ is } 4x^3$$

NOTE:

> The LCD includes each factor the *greatest number of times* that it occurs in any single denominator.

For fractions with polynomial denominators (two or more terms) such as $5/(2x - 6)$ and $2/(3x - 9)$, we find the LCD by:

First, factor the denominators of each fraction.

$$\frac{5}{2x - 6} \quad \text{and} \quad \frac{2}{3x - 9}$$

$$\frac{5}{2(x - 3)} \quad \text{and} \quad \frac{2}{3(x - 3)}$$

Then, find the least common multiple of the monomials. (The LCM of 2 and 3 is 6.)

$$\frac{5}{\boxed{2} (x - 3)} \quad \text{and} \quad \frac{2}{\boxed{3} (x - 3)}$$

$$\text{LCM} = 6$$

The LCD as a factor contains each different factor occurring in any of the denominators. $(x - 3)$ is a factor in both denominators.

$$\frac{5}{2 \boxed{(x - 3)}} \quad \text{and} \quad \frac{2}{3 \boxed{(x - 3)}}$$

$$\text{LCM} = (x - 3)$$

The LCD is then the product of these factors.

$$6(x - 3)$$

Thus,

$$\text{the LCD of } \frac{5}{2x - 6} \text{ and } \frac{2}{3x - 9} \text{ is } 6(x - 3)$$

In summary, to find the LCD of algebraic fractions, we factor each denominator. The LCD is then the *product* of each different factor that occurs in any of the denominators, with each factor occurring the *greatest* number of times that it occurs in any single denominator.

PROCEDURE:

To find the least common denominator (LCD) of algebraic fractions:

1. Factor each denominator completely.

2. Determine the LCD, which is the product of each different factor that occurs in any of the denominators, with each factor occurring the greatest number of times that it occurs in any single denominator.

EXAMPLE 1: Find the LCD of $\dfrac{1}{a^2}$ and $\dfrac{2}{3a}$.

SOLUTION: Separate the denominators into factors.

$$\frac{1}{a^2} \qquad \text{and} \qquad \frac{2}{3a}$$

$$= \frac{1}{(1)(a^2)} \qquad \text{and} \qquad \frac{2}{(3)(a)}$$

The LCD is $3a^2$ since the LCM of 1 and 3 is 3, and the LCM of a and a^2 is a^2.

Final answer: $3a^2$

EXAMPLE 2: Find the LCD of $\dfrac{4}{x^2 + 2x}$ and $\dfrac{2}{x}$.

SOLUTION: Factor the denominators.

$$\frac{4}{x^2 + 2x} \qquad \text{and} \qquad \frac{2}{x}$$

$$= \frac{4}{x(x + 2)} \qquad \text{and} \qquad \frac{2}{x}$$

The LCD is the product of the common factor x and the factor $(x + 2)$.

Final answer: $x(x + 2)$

EXAMPLE 3: Find the LCD of $\dfrac{3}{x^2 + x - 2}$ and $\dfrac{4}{x^2 - x - 6}$.

SOLUTION: Factor the denominators.

$$\dfrac{3}{x^2 + x - 2} \quad \text{and} \quad \dfrac{4}{x^2 - x - 6}$$

$$= \dfrac{3}{(x + 2)(x - 1)} \quad \text{and} \quad \dfrac{4}{(x + 2)(x - 3)}$$

The LCD is the product of the common factor $(x + 2)$ and the factors $(x - 1)$ and $(x - 3)$.

Final answer: $(x + 2)(x - 1)(x - 3)$

EXAMPLE 4: Find the LCD of $\dfrac{2}{x^2 + 4x + 4}$ and $\dfrac{7}{x^2 + x - 2}$.

SOLUTION: Factor the denominators.

$$\dfrac{2}{x^2 + 4x + 4} \quad \text{and} \quad \dfrac{7}{x^2 + x - 2}$$

$$= \dfrac{2}{(x + 2)(x + 2)} \quad \text{and} \quad \dfrac{7}{(x + 2)(x - 1)}$$

The LCD is the product of the common factor $(x + 2)$ and the factor $(x - 1)$. However, since the factor $(x + 2)$ is used twice in the first fraction, it must be used twice in the LCD.

Final answer: $(x + 2)(x + 2)(x - 1)$ or $(x + 2)^2(x - 1)$

Do the following practice set. Check your answers with the answers in the right-hand margin.

PRACTICE SET 8-6B

Find the LCD of the following fractions (see example 1 or 2).

1. $\dfrac{1}{8x}$ and $\dfrac{3}{12x}$

2. $\dfrac{2}{x}$ and $\dfrac{5}{x + 2}$

3. $\dfrac{1}{2x + 6}$ and $\dfrac{4}{x + 3}$

4. $\dfrac{1}{x + 1}$ and $\dfrac{1}{x - 1}$

1. $24x$
2. $(x)(x + 2)$
3. $(2)(x + 3)$
4. $(x + 1)(x - 1)$

Find the LCD of the following fractions (see example 3 or 4).

5. $\dfrac{1}{x^2 - 1}$ and $\dfrac{3}{x + 1}$

6. $\dfrac{1}{2x + 18}$ and $\dfrac{9}{x^2 - 81}$

7. $\dfrac{1}{x^2 - 2x - 35}$ and $\dfrac{1}{x + 7}$

8. $\dfrac{1}{x^2 + 6x + 9}$ and $\dfrac{1}{x + 3}$ and $\dfrac{1}{3x + 9}$

5. $(x + 1)(x - 1)$

6. $2(x + 9)(x - 9)$

7. $(x - 7)(x + 5)(x + 7)$

8. $3(x + 3)^2$

C. Addition and Subtraction of Fractions with Different Denominators

Previously, we learned how to add or subtract algebraic fractions that contain common denominators. We now focus our attention on adding or subtracting algebraic fractions that contain different denominators. To do this, we must use the skills we have just acquired in parts A and B of this section. Specifically, before combining unlike algebraic fractions, we first find a common denominator of the fractions, and then find fractions that are equivalent to the original fractions that have this LCD. Next, we combine fractions as we did before.

For example, to combine the unlike fractions $1/2x + 2/3x$, we proceed as follows.

First, find the LCD.

$$\dfrac{1}{2x} + \dfrac{2}{3x}$$

$$\text{LCD} = 6x$$

Then, find fractions that are equivalent to the original fractions having the LCD as the denominator.

$$\dfrac{1}{2x} = \dfrac{?}{6x}$$

$$\dfrac{2}{3x} = \dfrac{?}{6x}$$

■ The common multiplier for the first fraction is 3.

$$\dfrac{1}{2x} = \dfrac{3 \cdot 1}{3 \,(2x)} = \dfrac{3}{6x}$$

■ The common multiplier for the second fraction is 2.

$$\dfrac{2}{3x} = \dfrac{(2)\,(2)}{(2)\,(3x)} = \dfrac{4}{6x}$$

Now, combine like fractions, $\dfrac{3}{6x}$ and $\dfrac{4}{6x}$, accordingly.

$$\dfrac{3}{6x} + \dfrac{4}{6x}$$

$$= \dfrac{3 + 4}{6x}$$

$$= \dfrac{7}{6x}$$

Thus,

the sum of $\dfrac{1}{2x}$ and $\dfrac{2}{3x}$ is $\dfrac{7}{6x}$

As a second example, consider $3x/(x^2 - 4) - 4/(2x - 4)$.

First, find the LCD by factoring the denominators and selecting each different factor with its highest exponent.

$$\frac{3x}{x^2 - 4} - \frac{4}{2x - 4}$$

$$= \frac{3x}{(x + 2)(x - 2)} - \frac{4}{2(x - 2)}$$

$$\text{LCD} = 2(x + 2)(x - 2)$$

With the denominators in factored form, identify the common multiplier (in shaded box) for each fraction and thus find the equivalent fractions having the LCD as the denominator.

$$2 \left(\frac{3x}{(x+2)(x-2)} - \frac{4}{2(x-2)} \right) \quad x + 2$$

$$\frac{3x}{2(x+2)(x-2)} - \frac{4}{2(x+2)(x-2)}$$

$$= \frac{(2)(3x)}{2(x+2)(x-2)} - \frac{(4)(x+2)}{2(x+2)(x-2)}$$

Now, combine the fractions accordingly.

$$\frac{(2)(3x)}{2(x+2)(x-2)} - \frac{4(x+2)}{2(x+2)(x-2)}$$

$$= \frac{6x - 4(x+2)}{2(x+2)(x-2)} = \frac{6x - 4x - 8}{2(x+2)(x-2)}$$

$$= \frac{2x - 8}{2(x+2)(x-2)} = \frac{2(x-4)}{2(x+2)(x-2)}$$

$$= \frac{x-4}{(x+2)(x-2)}$$

Thus,

$$\frac{3x}{x^2 - 4} - \frac{4}{2x - 4} = \frac{x - 4}{(x + 2)(x - 2)}$$

PROCEDURE:

To combine algebraic fractions with different denominators:

1. Find the least common denominator.
   a. Factor each denominator.
   b. Determine the LCD, which will contain each different factor to the highest power it occurs in any single denominator.

2. Change each fraction to an equivalent fraction that has the LCD.
   a. Find the common multiplier of each fraction by comparing the factors of the denominators.
   b. Multiply the numerator by the multiplier of that fraction.

3. Combine the fractions with the common denominator.
   a. Combine the numerators according to their sign and place over the common denominator.
   b. Factor and reduce the fraction.

EXAMPLE 1: Combine $\dfrac{x-3}{2x^2} + \dfrac{3x}{10}$.

SOLUTION: Find the LCD.

By inspection, the LCD is $10x^2$.

$$\dfrac{x-3}{2x^2} + \dfrac{3x}{10}$$

Find equivalent fractions.

■ The common multiplier for the first fraction is 5 since $(2x^2)(5) = 10x^2$.

$$\dfrac{x-3}{2x^2} + \dfrac{3x}{10}$$

$$= \dfrac{5\,(x-3)}{5\,(2x^2)} + \dfrac{3x\,(x^2)}{10\,(x^2)}$$

■ The common multiplier for the second fraction is x^2 since $(10)(x^2) = 10x^2$.

$$= \dfrac{5(x-3)}{10x^2} + \dfrac{(3x)(x^2)}{10x^2}$$

Combine the fractions accordingly.

$$\dfrac{5(x-3)}{10x^2} + \dfrac{3x(x^2)}{10x^2} = \dfrac{5(x-3) + 3x(x^2)}{10x^2}$$

$$= \dfrac{5x - 15 + 3x^3}{10x^2}$$

$$= \dfrac{3x^3 + 5x - 15}{10x^2}$$

Final answer: $\dfrac{3x^3 + 5x - 15}{10x^2}$

EXAMPLE 2: Combine $\dfrac{7y-1}{y-9} - \dfrac{y+2}{2y-18}$.

SOLUTION: Find the LCD.

Factor the denominator.

$$\dfrac{7y-1}{y-9} - \dfrac{y+2}{2y-18}$$

$$= \dfrac{7y-1}{y-9} - \dfrac{y+2}{2(y-9)}$$

The LCD is $2(y-9)$.

Find the equivalent fractions.

■ The common multiplier for the first fraction is 2 since $(y-9)(2) = 2(y-9)$.

$$\dfrac{7y-1}{y-9} - \dfrac{y+2}{2y-18}$$

Example continued on next page.

■ The common multiplier for the second fraction is 1
since $(1)(2)(y - 9) = 2(y - 9)$.

$$= \frac{2(7y - 1)}{2(y - 9)} - \frac{y + 2}{2(y - 9)}$$

Combine the fractions accordingly.

$$\frac{2(7y - 1)}{2(y - 9)} - \frac{y + 2}{2(y - 9)} = \frac{2(7y - 1) - (y + 2)}{2(y - 9)}$$

$$= \frac{14y - 2 - y - 2}{2(y - 9)}$$

$$= \frac{13y - 4}{2(y - 9)}$$

Final answer: $\dfrac{13y - 4}{2(y - 9)}$

EXAMPLE 3: Combine $\dfrac{5x + 1}{2x^2 - 2} + \dfrac{7}{6x + 6}$.

SOLUTION: Find the LCD.

Factor the denominators.

$$\frac{5x + 1}{2x^2 - 2} + \frac{7}{6x + 6}$$

$$= \frac{5x + 1}{2(x^2 - 1)} + \frac{7}{6(x + 1)}$$

$$= \frac{5x + 1}{2(x + 1)(x - 1)} + \frac{7}{6(x + 1)}$$

$$\text{LCD} = 6(x + 1)(x - 1).$$

Find the equivalent fractions.

■ The common multiplier of the
first fraction is 3.

$$\frac{5x + 1}{2(x + 1)(x - 1)} + \frac{7}{6(x + 1)}$$

$$= \frac{3(5x + 1)}{(3)(2)(x + 1)(x - 1)} + \frac{7(x - 1)}{6(x + 1)(x - 1)}$$

■ The common multiplier of the
second fraction is $(x - 1)$.

$$= \frac{3(5x + 1)}{6(x + 1)(x - 1)} + \frac{7(x - 1)}{6(x + 1)(x - 1)}$$

Combine the fractions accordingly.

$$\frac{3(5x + 1)}{6(x + 1)(x - 1)} + \frac{7(x - 1)}{6(x + 1)(x - 1)} = \frac{3(5x + 1) + 7(x - 1)}{6(x + 1)(x - 1)}$$

$$= \frac{15x + 3 + 7x - 7}{6(x + 1)(x - 1)}$$

$$= \frac{22x - 4}{6(x + 1)(x - 1)}$$

$$= \frac{2(11x - 2)}{6(x + 1)(x - 1)}$$

$$= \frac{11x - 2}{3(x + 1)(x - 1)}$$

Final answer: $\dfrac{11x - 2}{3(x + 1)(x - 1)}$

EXAMPLE 4: Combine $y - \dfrac{y - 3}{2}$.

SOLUTION: By inspection, the LCD is 2.

$$y - \frac{y - 3}{2} = \frac{y}{1} - \frac{y - 3}{2}$$

$$\text{LCD} = 2.$$

Find equivalent fractions and combine accordingly.

$$\frac{y}{1} - \frac{y - 3}{2} = \frac{2(y)}{2} - \frac{y - 3}{2}$$

$$= \frac{2y - (y - 3)}{2}$$

$$= \frac{2y - y + 3}{2}$$

$$= \frac{y + 3}{2}$$

Final answer: $\dfrac{y + 3}{2}$

EXAMPLE 5: Combine $\dfrac{3a^2}{a^2 - 1} + \dfrac{2a + 1}{2a - 2} - \dfrac{2a - 1}{2a + 2}$.

SOLUTION: Factoring the denominators, we find the LCD to be $2(a + 1)(a - 1)$.

$$\frac{3a^2}{a^2 - 1} + \frac{2a + 1}{2a - 2} - \frac{2a - 1}{2a + 2} = \frac{3a^2}{(a + 1)(a - 1)} + \frac{2a + 1}{2(a - 1)} - \frac{2a - 1}{2(a + 1)}$$

$$\text{LCD} = 2(a + 1)(a - 1)$$

Example continued on next page.

Find equivalent fractions and combine accordingly.

$$\frac{3a^2}{(a+1)(a-1)} + \frac{2a+1}{2(a-1)} - \frac{2a-1}{2(a+1)}$$

$$= \frac{2(3a^2)}{2(a+1)(a-1)} + \frac{(2a+1)(a+1)}{2(a-1)(a+1)} - \frac{(2a-1)(a-1)}{2(a+1)(a-1)}$$

$$= \frac{2(3a^2) + (2a+1)(a+1) - (2a-1)(a-1)}{2(a-1)(a+1)}$$

$$= \frac{6a^2 + 2a^2 + a + 2a + 1 - (2a^2 - a - 2a + 1)}{2(a-1)(a+1)}$$

$$= \frac{6a^2 + 2a^2 + 3a + 1 - (2a^2 - 3a + 1)}{2(a-1)(a+1)}$$

$$= \frac{8a^2 + 3a + 1 - 2a^2 + 3a - 1}{2(a-1)(a+1)}$$

$$= \frac{6a^2 + 6a}{2(a-1)(a+1)}$$

$$= \frac{6a(a+1)}{2(a-1)(a+1)}$$

$$= \frac{\overset{3}{6a}\,\overset{1}{(a+1)}}{\underset{1}{2}(a-1)\,\underset{1}{(a+1)}}$$

$$= \frac{3a}{a-1}$$

Final answer: $\dfrac{3a}{a-1}$

Do the following practice set. Check your answers with the answers in the right-hand margin.

PRACTICE SET 8-6C

Combine the fractions as indicated. Reduce (see example 1 or 2).

1. $\dfrac{12x}{35} + \dfrac{5x}{7}$

2. $\dfrac{x+5}{2x} + \dfrac{2x-1}{4x}$

3. $\dfrac{5}{2x-4} - \dfrac{1}{3x-6}$

4. $\dfrac{2x}{x+5} - \dfrac{3x+1}{2x+10}$

1. $37x/35$
2. $(4x+9)/4x$
3. $13/6(x-2)$
4. $(x-1)/2(x+5)$

Combine the fractions as indicated. Reduce (see example 3 or 4).

5. $\dfrac{9}{x+8} + \dfrac{2}{x}$

6. $\dfrac{x+1}{x^2-9} + \dfrac{3}{2x-6}$

7. $4 + \dfrac{2}{y}$

8. $2 - \dfrac{3}{x+1}$

Combine the fractions as indicated. Reduce (see example 5).

9. $\dfrac{2x}{x^2-4} + \dfrac{x}{x+2} - \dfrac{4}{5x-10}$

10. $\dfrac{3}{x-3} + \dfrac{6}{x^2-9} + \dfrac{2}{x+3}$

11. $\dfrac{5x}{x+7} + \dfrac{2x}{2x-10} - \dfrac{9}{x^2+2x-35}$

5. $(11x+16)/x(x+8)$

6. $\dfrac{(5x+11)}{2(x+3)(x-3)}$

7. $2(2y+1)/y$

8. $(2x-1)/(x+1)$

9. $\dfrac{5x^2-4x-8}{5(x+2)(x-2)}$

10. $\dfrac{5x+9}{(x+3)(x-3)}$

11. $\dfrac{6x^2-18x-9}{(x+7)(x-5)}$

EXERCISE 8-6

Find the missing numerator that will make the second fraction equal to the first.

1. $\dfrac{3}{4} = \dfrac{\rule{1.5em}{0.8em}}{8x}$

2. $\dfrac{-5}{7} = \dfrac{\rule{1.5em}{0.8em}}{35x}$

3. $\dfrac{2}{x^2} = \dfrac{\rule{1.5em}{0.8em}}{x^3}$

4. $\dfrac{x-1}{3x} = \dfrac{\rule{1.5em}{0.8em}}{6x}$

5. $\dfrac{3}{4} = \dfrac{\rule{1.5em}{0.8em}}{8x}$

6. $\dfrac{x-2}{xy^2} = \dfrac{\rule{2.5em}{0.8em}}{x^2y^2}$

7. $\dfrac{4}{x+2} = \dfrac{\rule{4em}{0.8em}}{x^2+3x+2}$

8. $\dfrac{x-1}{x+2} = \dfrac{\rule{4em}{0.8em}}{x^2+3x+2}$

9. $\dfrac{x-2}{x+1} = \dfrac{\rule{4em}{0.8em}}{x^2-x-2}$

10. $\dfrac{x-3}{x+2} = \dfrac{\rule{4em}{0.8em}}{2x^2-8}$

Find the least common denominator (LCD) of the following fractions.

11. $\dfrac{2}{3x}$ and $\dfrac{3}{4x}$

12. $\dfrac{1}{4x}$ and $\dfrac{7}{12y}$

13. $\dfrac{2}{x^2}$ and $\dfrac{1}{x}$

14. $\dfrac{1}{x^2}$ and $\dfrac{1}{4x}$

15. $\dfrac{2}{10x}$ and $\dfrac{5}{25x^2}$

16. $\dfrac{4}{x+1}$ and $\dfrac{3}{4}$

17. $\dfrac{1}{x}$ and $\dfrac{4}{x+1}$

18. $\dfrac{2}{3x}$ and $\dfrac{2}{3x-3}$

19. $\dfrac{2}{x+y}$ and $\dfrac{3}{x-y}$

20. $\dfrac{1}{x^2-25}$ and $\dfrac{2}{3x-15}$

21. $\dfrac{1}{x^2-5x+6}$ and $\dfrac{5}{x-3}$

22. $\dfrac{2}{x^2-5x+6}$ and $\dfrac{3}{x^2-9}$

Combine the fractions as indicated and reduce the results to lowest terms.

23. $\dfrac{4x}{15} + \dfrac{11x}{30}$

24. $\dfrac{2}{x} - \dfrac{3}{4x}$

25. $\dfrac{5}{2x} - \dfrac{3}{5x}$

26. $\dfrac{8}{25x^2} + \dfrac{3}{5x}$

27. $\dfrac{1}{6x^2} - \dfrac{1}{9y^2}$

28. $\dfrac{x-3}{2x^2} + \dfrac{3x}{10}$

29. $\dfrac{3-4x}{2} + \dfrac{2x-3}{4}$

30. $\dfrac{3}{2(3x-y)} + \dfrac{4}{3(3x-y)}$

31. $\dfrac{y}{(y+4)(y-4)} - \dfrac{4}{3(y-4)}$

32. $\dfrac{3x}{x-2} + \dfrac{5x}{x+2}$

33. $\dfrac{4x}{3y+6} - \dfrac{x}{y+2}$

34. $\dfrac{5}{x^2+x} - \dfrac{2}{3x^2+3x}$

35. $\dfrac{4x}{x^2-36} - \dfrac{4}{x+6}$

36. $\dfrac{3x}{2x^2-x} - \dfrac{4}{6x-3}$

37. $\dfrac{3}{x-1} + \dfrac{4}{x^2-3x+2}$

38. $\dfrac{4}{x^2-x-6} + \dfrac{3}{x^2+x-2}$

39. $\dfrac{2x}{x^2+7x+6} - \dfrac{x-2}{x+6}$

40. $\dfrac{x^2-3x}{x^2+x-6} - \dfrac{x+1}{x+3}$

41. $\dfrac{7x}{x^2+2x-3} + \dfrac{2x-5}{x^2+5x+6}$

42. $\dfrac{1}{x^2+3x+2} + \dfrac{1}{x^2-x-2} + \dfrac{2x}{x^2-4}$

43. $x + \dfrac{3}{x}$

44. $3 - \dfrac{4}{t+2}$

45. $x + \dfrac{1}{x+1}$

46. $x - \dfrac{1}{x-1}$

47. $\dfrac{4}{x-2} - 4$

48. $x+1 + \dfrac{1}{x-1}$

49. $3 + \dfrac{5}{x+1}$

50. $x^2 - 2x + 1 - \dfrac{1}{x}$

8-7 COMPLEX FRACTIONS

A complex fraction is a fraction that has a fraction in its numerator or in the denominator or in both. The following are complex fractions.

- $\dfrac{a}{2}$ is a fraction in the numerator

$$\dfrac{\frac{a}{2}}{3}$$

- $\dfrac{a}{2}$ is a fraction in the denominator

$$\dfrac{3}{\frac{a}{2}}$$

■ $\dfrac{a}{2}$ is a fraction in the numerator and

$$\dfrac{\frac{a}{2}}{\frac{a}{3}}$$

$\dfrac{a}{3}$ is a fraction in the denominator.

■ $\dfrac{a}{2}$ is a fraction in the numerator.

$$\dfrac{1 + \frac{a}{2}}{3}$$

To simplify a complex fraction, we simply perform the indicated division. For example, to simplify $\frac{3}{4}/2$:

Divide the numerator by the denominator. (Recall division of fractions is performed by multiplying the first fraction by the reciprocal of the second fraction.)

$$\dfrac{3}{4} \div 2$$

$$\dfrac{3}{4} \times \dfrac{1}{2}$$

$$\dfrac{3 \cdot 1}{4 \cdot 2} = \dfrac{3}{8}$$

Thus,

$$\dfrac{\frac{3}{4}}{2} \text{ simplifies to } \dfrac{3}{8}$$

Whenever the numerator or denominator contains one or more fractional terms, we must combine the terms of the fraction before dividing. For example, to simplify $(y + \frac{1}{3})/2$,

First, combine the terms of the numerator and express it as a single fraction.

$$\dfrac{y + \frac{1}{3}}{2}$$

$$y + \dfrac{1}{3}$$
$$= \dfrac{y}{1} + \dfrac{1}{3}$$
$$= \dfrac{3y + 1}{3}$$

Replacing our "old" numerator with the "new" numerator, we are now able to divide.

$$\dfrac{y + \frac{1}{3}}{2} = \dfrac{\frac{3y + 1}{3}}{\frac{2}{1}}$$

$$= \dfrac{3y + 1}{3} \div \dfrac{2}{1}$$

$$= \dfrac{3y + 1}{3} \cdot \dfrac{1}{2}$$

$$= \dfrac{(3y + 1)(1)}{(3)(2)}$$

$$= \dfrac{3y + 1}{6}$$

Thus,

$$\dfrac{y + \frac{1}{3}}{2} \text{ is equal to } \dfrac{3y + 1}{6}$$

In summary, in order to simplify a complex fraction, we first express the numerator and denominator as a single fraction. Once this has been done, we then divide the fraction as before.

PROCEDURE:

To simplify complex fractions:

1. Express the numerator and denominator as a single fraction.
2. Divide the new numerator by the new denominator.

EXAMPLE 1: Simplify $\dfrac{\dfrac{x}{3} - \dfrac{x}{5}}{\dfrac{1}{2}}$.

SOLUTION: Express the numerator as a single fraction.

$$\frac{x}{3} - \frac{x}{5} = \frac{5x}{15} - \frac{3x}{15}$$

$$= \frac{5x - 3x}{15}$$

$$= \frac{2x}{15}$$

Divide the new numerator by the denominator.

$$\frac{2x}{15} \Big/ \frac{1}{2} = \frac{2x}{15} \div \frac{1}{2}$$

$$= \frac{2x}{15} \times \frac{2}{1}$$

$$= \frac{(2x)(2)}{(15)(1)}$$

$$= \frac{4x}{15}$$

Final answer: $\dfrac{4x}{15}$

EXAMPLE 2: Simplify $\dfrac{\dfrac{x}{2} + \dfrac{x}{5}}{2x - \dfrac{3x}{10}}$.

SOLUTION: Express both the numerator and the denominator as a single fraction.

NUMERATOR	**DENOMINATOR**
$\dfrac{x}{2} + \dfrac{x}{5} = \dfrac{5x}{10} + \dfrac{2x}{10}$	$2x - \dfrac{3x}{10} = \dfrac{2x}{1} - \dfrac{3x}{10}$
$= \dfrac{5x + 2x}{10}$	$= \dfrac{10(2x)}{10} - \dfrac{3x}{10}$
$= \dfrac{7x}{10}$	$= \dfrac{20x}{10} - \dfrac{3x}{10}$
	$= \dfrac{20x - 3x}{10}$
	$= \dfrac{17x}{10}$

Divide the new numerator by the new denominator.

$$\dfrac{7x}{10} \bigg/ \dfrac{17x}{10} = \dfrac{7x}{10} \div \dfrac{17x}{10}$$

$$= \dfrac{7x}{10} \times \dfrac{10}{17x}$$

$$= \dfrac{(7x)(10)}{(10)(17x)}$$

$$= \dfrac{7}{17}$$

Final answer: $\dfrac{7}{17}$

Do the following practice set. Check your answers with the answers in the right-hand margin.

PRACTICE SET 8-7

Simplify each of the following complex fractions (see example 1 or 2).

1. $$\dfrac{x + \dfrac{1}{3}}{2}$$

2. $$\dfrac{x}{1 + \dfrac{1}{x}}$$

3. $$\dfrac{\dfrac{x}{2} - \dfrac{2}{x}}{\dfrac{1}{x} + \dfrac{1}{2}}$$

1. $(3x + 1)/6$
2. $x^2/(x + 1)$
3. $x - 2$

EXERCISE 8-7

Simplify each of the following complex fractions.

1. $$\dfrac{\dfrac{3}{x}}{x - \dfrac{1}{x}}$$

2. $$\dfrac{\dfrac{1}{x} + \dfrac{1}{2x}}{\dfrac{3}{x} - \dfrac{2}{3x^2}}$$

3. $$\dfrac{\dfrac{3}{5} + x}{\dfrac{3}{5} - x}$$

4. $$\dfrac{x + \dfrac{1}{x}}{x - \dfrac{1}{x}}$$

5. $$\dfrac{x - \dfrac{4}{x}}{3 + \dfrac{x^2 + 12}{x^2 - 4}}$$

6. $$\dfrac{1 - \dfrac{5}{x + 3}}{3 + \dfrac{15}{x - 2}}$$

7. $$\dfrac{x + 2 + \dfrac{1}{x}}{x - \dfrac{1}{x}}$$

8. $$\dfrac{\dfrac{1}{2y} - \dfrac{1}{x}}{\dfrac{1}{y} + \dfrac{1}{2x}}$$

8-8 EQUATIONS HAVING FRACTIONS

The general rule of thumb in solving equations that contain fractions is first to eliminate the fractions. We do this by multiplying each term of the equation by the least common denominator (LCD) of the fractions. For example, to solve the equation $(4y - 3)/3 = (3 + y)/2$:

First, "clear the fractions" by multiplying both sides of the equation by the LCD, 6.

$$\frac{4y - 3}{3} = \frac{3 + y}{2}$$

$$6\left(\frac{4y - 3}{3}\right) = 6\left(\frac{3 + y}{2}\right)$$

$$2(4y - 3) = 3(3 + y)$$

The result is an equation that is free of fractions. Now, solve this equation as before.

$$2(4y - 3) = 3(3 + y)$$
$$8y - 6 = 9 + 3y$$
$$8y - 3y = 9 + 6$$
$$5y = 15$$
$$y = \frac{15}{5}$$
$$y = 3$$

Thus,

$$\frac{4y - 3}{3} = \frac{3 + y}{2} \text{ is true when } y = 3$$

This result can be easily checked.

Check:

$$
\begin{array}{c|c}
\dfrac{4y - 3}{3} & \dfrac{3 + y}{2} \\[2ex]
\dfrac{4(3) - 3}{3} & \dfrac{3 + (3)}{2} \\[2ex]
\dfrac{12 - 3}{3} & \dfrac{6}{2} \\[2ex]
\dfrac{9}{3} & 3 \\[2ex]
3 &
\end{array}
$$

PROCEDURE:

To solve equations containing fractions:

1. Find the common denominator of all the terms in the equation, factoring the denominators if necessary.

2. Clear the equation of fractions by multiplying each term of the equation by the common denominator.

3. Solve the resulting equation.

4. Check the solution.

EXAMPLE 1: Solve $\dfrac{x}{3} - \dfrac{x}{6} = 2$.

SOLUTION: Multiply each term of the equation by the LCD, 6.

$$\frac{x}{3} - \frac{x}{6} = \frac{2}{1}$$

$$6\left(\frac{x}{3}\right) - 6\left(\frac{x}{6}\right) = 6\left(\frac{2}{1}\right)$$

$$2x - x = 12$$

Example continued on next page.

Solve the resulting equation.

$$2x - x = 12$$
$$x = 12$$

Final answer: $x = 12$

The check is left to the student.

EXAMPLE 2:   Solve $\dfrac{y + 5}{4} - \dfrac{7 - y}{6} = \dfrac{7}{4}$.

SOLUTION: Multiply each term of the equation by the LCD, 12.

$$\frac{y + 5}{4} - \frac{7 - y}{6} = \frac{7}{4}$$

$$12\left(\frac{y + 5}{4}\right) - 12\left(\frac{7 - y}{6}\right) = 12\left(\frac{7}{4}\right)$$

$$3(y + 5) - 2(7 - y) = 3(7)$$

Solve the resulting equation.

$$3(y + 5) - 2(7 - y) = 3(7)$$
$$3y + 15 - 14 + 2y = 21$$
$$3y + 2y + 15 - 14 = 21$$
$$5y + 1 = 21$$
$$5y = 21 - 1$$
$$5y = 20$$
$$y = \frac{20}{5}$$
$$y = 4$$

Final answer: $y = 4$

The check is left to the student.

EXAMPLE 3:   Solve $\dfrac{3}{5} = \dfrac{x}{x + 2}$.

SOLUTION: Multiply by the LCD, $5(x + 2)$.

$$\frac{3}{5} = \frac{x}{x + 2}$$

$$(5)(x + 2)\left(\frac{3}{5}\right) = (5)(x + 2)\left(\frac{x}{x + 2}\right)$$

$$3(x + 2) = 5(x)$$

Solve the resulting equation.

$$3(x + 2) = 5(x)$$
$$3x + 6 = 5x$$
$$6 = 5x - 3x$$
$$6 = 2x$$
$$\frac{6}{2} = x$$
$$3 = x$$

Final answer: $x = 3$

The check is left to the student.

EXAMPLE 4: Solve $\dfrac{1}{y - 2} - \dfrac{5}{3y + 6} = \dfrac{2}{y^2 - 4}$.

SOLUTION: To identify the LCD, we factor the denominators.

$$\frac{1}{y - 2} - \frac{5}{3y + 6} = \frac{2}{y^2 - 4}$$

$$\frac{1}{(y - 2)} - \frac{5}{3(y + 2)} = \frac{2}{(y + 2)(y - 2)}$$

$$\text{LCD} = 3(y + 2)(y - 2)$$

Multiply by the LCD.

$$3(y + 2)(y - 2)\left(\frac{1}{y - 2}\right) - 3(y + 2)(y - 2)\left(\frac{5}{3(y + 2)}\right) = 3(y + 2)(y - 2)\left(\frac{2}{(y + 2)(y - 2)}\right)$$

$$3(y + 2)(1) - (y - 2)(5) = 3(2)$$

$$3(y + 2) - 5(y - 2) = 3(2)$$

Solve the resulting equation.

$$3(y + 2) - 5(y - 2) = 3(2)$$
$$3y + 6 - 5y + 10 = 6$$
$$3y - 5y + 6 + 10 = 6$$
$$-2y + 16 = 6$$
$$-2y = 6 - 16$$
$$-2y = -10$$
$$y = \frac{-10}{-2}$$
$$y = 5$$

Final answer: $y = 5$

The check is left to the student.

EXAMPLE 5: Solve the following problem.

One half a number is 5 more than $\frac{1}{3}$ the number. Find the number.

SOLUTION: We must first translate this problem into an algebraic equation.

Let x = the number.

One half a number is 5 more than $\frac{1}{3}$ the number

$$\frac{1}{2}x \qquad = \qquad 5 + \qquad \frac{1}{3}x$$

Solve the equation. (The LCD is 6.)

$$\frac{1}{2}x = 5 + \frac{1}{3}x$$

$$\frac{x}{2} = \frac{5}{1} + \frac{x}{3}$$

$$6\left(\frac{x}{2}\right) = 6\left(\frac{5}{1}\right) + 6\left(\frac{x}{3}\right)$$

$$3x = 30 + 2x$$
$$3x - 2x = 30$$
$$x = 30$$

Final answer: $x = 30$

The check is left to the student.

EXAMPLE 6:   Solve the following problem.

The denominator of a fraction is 12 more than its numerator. If 8 is added to the numerator (and the denominator is left unchanged), the resulting fraction is equal to $\frac{6}{7}$. Find the original fraction.

SOLUTION: We must first translate this problem into an algebraic equation.

Let $x =$ the numerator of the original fraction.
Then, $x + 12 =$ the denominator of the original fraction.

So, our original fraction is $\dfrac{x}{x + 12}$.

If 8 is added to the numerator . . . the resulting fraction is . . . $\dfrac{6}{7}$.

$$\dfrac{x + 8}{x + 12} \qquad = \dfrac{6}{7}$$

Solve this new fraction.

$$\frac{x + 8}{x + 12} = \frac{6}{7}$$

$$7(x + 12)\left(\frac{x + 8}{x + 12}\right) = 7(x + 12)\left(\frac{6}{7}\right)$$

$$7(x + 8) = 6(x + 12)$$

$$7x + 56 = 6x + 72$$

$$7x - 6x = 72 - 56$$

$$x = 16$$

Thus,

the original fraction is $\dfrac{x}{x + 12} = \dfrac{16}{28}$

Final answer: $\dfrac{16}{28}$

Do the following practice set. Check your answers with the answers in the right-hand margin.

PRACTICE SET 8-8

Solve the following equation. Check (see example 1 or 2).

1. $\dfrac{x}{5} = 2$

1. $x = 10$

2. $\dfrac{x}{9} = \dfrac{x}{6} - 4$

3. $\dfrac{2x + 3}{6} - \dfrac{x - 9}{4} = 5$

2. $x = 72$

3. $x = 27$

Solve the following equations. Check (see example 3 or 4).

4. $\dfrac{1}{x} + \dfrac{1}{3} = \dfrac{1}{2}$

5. $\dfrac{3}{4x} + \dfrac{1}{x} = \dfrac{7}{8}$

6. $\dfrac{4x}{x - 3} - 3 = \dfrac{x^2 + 15}{x^2 - 9}$

4. $x = 6$

5. $x = 2$

6. $x = -1$

Solve the following problems (see example 5 or 6).

7. One third a number is 2 less than $\frac{1}{2}$ the same number. Find the number.

8. The denominator of a fraction is 6 less than twice its numerator. The value of this fraction is equal to $\frac{3}{5}$. Find the fraction.

9. The numerator of a fraction is twice its denominator. If the denominator is increased by 4 (and the numerator is left unchanged), the value of the resulting fraction is $\frac{2}{3}$. Find the original fraction.

7. 12

8. $\frac{18}{30}$

9. $\frac{4}{2}$

EXERCISE 8-8

Solve the following equations. Check.

1. $\dfrac{x}{6} = 3$

2. $\dfrac{x}{9} = 3x - 5$

3. $\dfrac{x}{3} = 20$

4. $\dfrac{x}{6} + 23 = 7$

5. $\dfrac{4 + 3x}{8} = 15$

6. $\dfrac{8 + 3x}{4} = 20$

7. $\dfrac{3x - 12}{3} = 10$

8. $\dfrac{4x - 3}{3} = \dfrac{x + 3}{2}$

9. $\dfrac{2x - 3}{6} + \dfrac{x + 2}{12} = \dfrac{1}{4}$

10. $\dfrac{x + 2}{2} + 2 = -\dfrac{x + 3}{4}$

11. $\dfrac{3x + 1}{7} - 1 = x - 2$

12. $\dfrac{2x + 7}{9} - 4 = \dfrac{x - 7}{12}$

13. $\dfrac{2x + 1}{6} - \dfrac{2x - 15}{10} = 3$

14. $\dfrac{2x}{3} - \dfrac{1}{2} = \dfrac{2x + 5}{6}$

15. $\dfrac{3}{x} = \dfrac{12}{16}$

16. $\dfrac{y}{y + 2} = \dfrac{2}{3}$

17. $\dfrac{3}{x} + \dfrac{5}{3} = \dfrac{19}{3x}$

18. $\dfrac{4}{x} - \dfrac{5}{12} = \dfrac{1}{2} - \dfrac{3}{2x}$

19. $\dfrac{2}{3y} - \dfrac{9}{y} = 25$

20. $\dfrac{x+1}{x} = \dfrac{3}{2x} - 1$

21. $\dfrac{x}{x-6} - \dfrac{10}{x-6} = 3$

22. $\dfrac{8}{x-2} - \dfrac{2}{15} = \dfrac{1}{5}$

23. $\dfrac{23}{x^2-1} - \dfrac{x}{x+1} = \dfrac{5-x}{x-1}$

24. $\dfrac{3}{4x} + \dfrac{4}{4x-1} - \dfrac{2}{x} = 0$

25. The numerator of a certain fraction is 10 less than its denominator. When reduced, the fraction is equivalent to $\frac{3}{5}$. Find the fraction.

26. The sum of one half a number and one third the same number is 15. Find the number.

27. What number must be added to both the numerator and the denominator of the fraction $\frac{2}{7}$ so that the resulting fraction is equivalent to $\frac{1}{2}$?

28. Separate 100 into two parts such that one part is $\frac{3}{5}$ the other part.

29. One half the sum obtained when 6 is added to a certain number is equal to $\frac{1}{4}$ the difference that is obtained when 8 is subtracted from the number. Find the number.

30. What number must be added to both the numerator and denominator of the fraction $\frac{11}{23}$ to make the fraction equivalent to $\frac{7}{8}$?

8-9 SOLVING PROBLEMS THAT LEAD TO EQUATIONS INVOLVING FRACTIONS

In this section, we demonstrate how our work with algebraic fractions may be applied. We discuss one particular type of problem, namely, work problems.

WORK PROBLEMS

Work problems generally involve the amount of work a person(s)/machine(s) can do in a certain period of time. In solving work problems, we consider the following relationship.

$$\text{time of work} \times \text{rate of work} = \text{amount of work done}$$
$$t \quad \times \quad r \quad = \quad w$$

As an example, consider the following problem.

If Steve can paint a room in 90 minutes, then in 1 minute, he will complete $\frac{1}{90}$ of the job.

$$\text{We say Steve's rate of work is } \dfrac{1}{90}.$$

■ In 5 minutes, Steve will complete $\frac{5}{90}$ of the job.

$$t \cdot r = w$$
$$5 \cdot \left(\dfrac{1}{90}\right) = w$$
$$\dfrac{5}{90} = w$$

■ In x minutes, Steve will complete $x/90$ of the job.

$$t \cdot r = w$$
$$x \cdot \left(\dfrac{1}{90}\right) = w$$
$$\dfrac{x}{90} = w$$

■ In 90 minutes, Steve will complete the job.

$$t \cdot r = w$$

$$90 \cdot \frac{1}{90} = w$$

$$\frac{90}{90} = w$$

$$1 = w$$

NOTE:

When a job is completed, the amount of work done will equal 1.

If several people work together, the total amount of work done is the *sum* of the amount of work done by each person. For example:

Assume Don can paint the same room Steve is painting, in 60 minutes.

Then, Don's rate of work is $\dfrac{1}{60}$

However,

If Steve and Don work together, at the end of 1 minute, they will paint $\frac{1}{90} + \frac{1}{60}$ of the room (or $\frac{5}{180}$ of the room).

Steve	*Don*
$\dfrac{1}{90}$	$\dfrac{1}{60}$

$$\frac{1}{90} + \frac{1}{60}$$

$$= \frac{2}{180} + \frac{3}{180}$$

$$= \frac{5}{180}$$

In 36 minutes, Steve will paint $\frac{36}{90}$ of the room and Don will paint $\frac{36}{60}$ of the room.

Steve	*Don*
$t \cdot r = w$	$t \cdot r = w$
$36 \cdot \left(\dfrac{1}{90}\right) = w$	$36 \left(\dfrac{1}{60}\right) = w$
$\dfrac{36}{90} = w$	$\dfrac{36}{60} = w$

Together, they will finish $\frac{180}{180}$ of the room or the entire room. Again—note that when the whole job is completed, w is equal to 1.

Steve	*Don*
$\dfrac{36}{90}$	$\dfrac{36}{60}$

$$\frac{36}{90} + \frac{36}{60}$$

$$= \frac{72}{180} + \frac{108}{180}$$

$$= \frac{180}{180} = 1$$

We present three examples by using the problem-solving techniques that were developed in Chapter 5.

EXAMPLE 1: Working alone, Joe can insulate his attic in 4 hours. His son, Mike, determined he could do the job in 6 hours. How long would it take them if they worked together?

STEP 1: READ THE PROBLEM CAREFULLY.

STEPS 2/3: DETERMINE THE UNKNOWN QUANTITY AND SET x EQUAL TO IT.

There is one unknown in this problem.

The time it takes for Joe and Mike to insulate the attic.

> Let x = the number of hours required to complete the job.

STEP 4: SET UP A CHART SHOWING THE $t \cdot r = w$ RELATIONSHIP.

Person	Time	Rate of work	Amount of work done
Joe	x	$\dfrac{1}{4}$	$\dfrac{x}{4}$
Mike	x	$\dfrac{1}{6}$	$\dfrac{x}{6}$

STEP 5: DETERMINE THE RELATIONSHIP OF THE PROBLEM.

The relationship used in this problem is

$$\underset{\text{Joe does}}{\text{amount of work}} + \underset{\text{Mike does}}{\text{amount of work}} = \underset{\text{by both}}{\text{total work done}}$$

$$\frac{x}{4} \qquad + \qquad \frac{x}{6} \qquad = \qquad 1$$

STEP 6: SOLVE THE EQUATION.

$$\frac{x}{4} + \frac{x}{6} = 1$$

$$12\left(\frac{x}{4}\right) + 12\left(\frac{x}{6}\right) = 12(1)$$

$$3x + 2x = 12$$

$$5x = 12$$

$$x = \frac{12}{5} \qquad \text{or} \qquad x = 2\frac{2}{5} \text{ hours}$$

Example continued on next page.

STEP 7: CHECK.

■ In $2\frac{2}{5}$ hours, Joe will complete $\frac{6}{10}$ of the job.

$$\frac{x}{4} = \frac{\frac{12}{5}}{4} = \frac{6}{10}$$

■ In $2\frac{2}{5}$ hours, Mike will complete $\frac{4}{10}$ of the job.

$$\frac{x}{6} = \frac{\frac{12}{5}}{6} = \frac{4}{10}$$

■ In $2\frac{2}{5}$ hours, they will complete the entire job.

$$\frac{6}{10} + \frac{4}{10} = \frac{10}{10}$$
$$= 1$$

Final answer: $2\frac{2}{5}$ hours (or 2 hours, 24 minutes)

EXAMPLE 2: Howard has estimated that he can cut and split one cord of wood from a tree in 3 hours. It would take Glenn 5 hours to do the same job. Being a "take-charge" person, Howard began cutting but was able to work for only 1 hour before his back gave out. How long will it take Glenn to finish the job?

STEP 1: READ THE PROBLEM CAREFULLY.

STEPS 2/3: DETERMINE THE UNKNOWN QUANTITY AND SET x EQUAL TO IT.

There is one unknown quantity:

The length of time Glenn will require to finish cutting/splitting the wood.

Let x = the length of time Glenn will need to finish the job.

STEP 4: SET UP A CHART.

Person	t	r	w
Howard	1	$\frac{1}{3}$	$\frac{1}{3}$
Glenn	x	$\frac{1}{5}$	$\frac{x}{5}$

STEP 5: DETERMINE THE RELATIONSHIP OF THE PROBLEM.

The relationship found in this problem is

amount of work Howard did	+	amount of work Glenn will do	=	total work done by both
$\frac{1}{3}$	+	$\frac{x}{5}$	=	1

STEP 6: SOLVE THE EQUATION.

$$\frac{1}{3} + \frac{x}{5} = 1$$

$$15\left(\frac{1}{3}\right) + 15\left(\frac{x}{5}\right) = 15(1)$$

$$5 + 3x = 15$$

$$3x = 15 - 5$$

$$3x = 10$$

$$x = \frac{10}{3} \quad \left(\text{or } 3\frac{1}{3} \text{ hours}\right)$$

The check is left for the student.

Final answer: $3\frac{1}{3}$ hours (or 3 hours, 20 minutes)

EXAMPLE 3: A shower stall can be filled with water (when both hot and cold faucets are opened), in 8 minutes. When the drain is clogged (with hair and soap), it takes 10 minutes for the water to drain out. If you were taking a shower, how long do you have before the shower stall begins to overflow with water?

STEP 1: READ THE PROBLEM CAREFULLY.

STEPS 2/3: DETERMINE THE UNKNOWN QUANTITY AND SET x EQUAL TO IT.

There is one unknown quantity.

How long before the stall overflows?

Let x = The number of minutes it will take for the water to fill the shower stall.

STEP 4: SET UP A CHART.

Water	t	r	w
In (from shower head)	x	$\frac{1}{8}$	$\frac{x}{8}$
Out (drain)	x	$\frac{1}{10}$	$\frac{x}{10}$

Example continued on next page.

STEP 5: DETERMINE THE RELATIONSHIP OF THE PROBLEM.

The relationship in this problem is

$$\underset{\substack{\text{amount of water} \\ \text{filling up shower stall}}}{} - \underset{\substack{\text{amount of water} \\ \text{going out of drain}}}{} = \text{total amount of water in stall}$$

$$\frac{x}{8} \qquad - \qquad \frac{x}{10} \qquad = \qquad 1$$

STEP 6: SOLVE THE EQUATION.

$$\frac{x}{8} - \frac{x}{10} = 1$$

$$40\left(\frac{x}{8}\right) - 40\left(\frac{x}{10}\right) = 40(1)$$

$$5x - 4x = 40$$

$$x = 40 \quad \text{(it will take 40 minutes for the water to fill up the stall)}.$$

The check is left to the student.

Final answer: 40 minutes

EXERCISE 8-9

Solve the following problems.

1. It takes Milton 5 hours to cut and rake his lawn. It takes his wife, Nancy, and their three children 7 hours to do the same job. How long would it take if the whole family did the job together? (Round to nearest hour.)

2. Dave can load a certain amount of freight in 10 hours. It takes his brother Randy 12 hours to do the same job. How long would it take them to load the freight together? (Round to nearest tenth.)

3. Mary and her mother, Gladys, have decided to wallpaper a room together. Working alone, Mary would require 18 hours and Gladys, 12 hours. How long will it take them to complete the job together?

4. Shirley can shovel her driveway and walkway clear of snow in 2 hours. Her husband, Al, can do the same job in 30 minutes with a snow blower. How long would it take them to do the job together?

5. Joan can seal her driveway in 3 hours. It would take Bill half that long to do the same job. If Bill began the job and worked for 1 hour, how long will it take Joan to finish the job?

6. It takes Ricardo 20 minutes to fill one fuel tank of a truck when he uses pump No. 1 and 14 minutes when he uses pump No. 2. One day Ricardo is pressed for time and decides to use both pumps at the same time. How long will it take him to fill up the tank? (Round to nearest minute.)

7. Working alone, Pamela can type a report in 9 hours. It would take Joyan twice as long to type the same report. If Pamela works alone on the report for 3 hours and then quits, how long will it take Joyan to finish the report?

8. When Jim and Cynthia work together, they can clean their house in 3 hours. Working alone, however, Cynthia requires 4 hours. How long will Jim require?

9. Don and Greg own an automobile body shop. Together, they can "doll up" a car in 8 hours. If Don works 3 times faster than Greg, how long will it take Greg to doll up the car alone?

10. Steve and Danny are house painters. Steve is able to paint a house in 9 days, whereas it takes Danny 16 days to paint the same house. On Monday, they started working together on a house. Friday morning, before work, Steve and Danny have an argument and Danny walks off the job. How many days will it take Steve to finish the job alone?

11. Monica estimated she would need 2 hours to complete a math test if she did not use a calculator. However, if she used a calculator, she would only need 45 minutes. On the day of the test, Monica's instructor indicated she could use a calculator but she must complete the test in a 1-hour period. Halfway through the test, Monica's calculator "died." a. How long will it take Monica to complete the test without the calculator? b. Will she complete the test in the 1-hour period allowed?

12. An old clawfoot bathtub has separate hot and cold faucets. It takes the cold water faucet 10 minutes to fill the tub and the hot water faucet 14 minutes to fill the tub. If Margo wanted to draw a bath, how long would it take to fill the tub when both faucets are open? (Round to nearest minute.)

13. If the bathtub in problem 12 can be drained in 8 minutes, and Margo forgot to close the drain, how long would it take to fill the tub?

14. Bill, a mechanic, is changing the oil on a 10-wheel dump truck. It usually takes 5 minutes to fill the crankcase and 8 minutes to drain it. If after draining the crankcase, Bill forgets to replace the drain plug, how long will it take to fill the crankcase?

15. Kathy, LuAnn, and Marilyn, together, received a total of $635. Kathy's share was $50 more than $\frac{3}{5}$ of LuAnn's share. Marilyn's share was $\frac{1}{5}$ of LuAnn's share. Find the amount of money each woman received.

REVIEW EXERCISES

1. What is the LCD of $\dfrac{3}{4x^2}$, $\dfrac{-7}{6xy}$, $\dfrac{5}{8y^3}$?

In problems 2–4, find the missing numerator.

2. $\dfrac{3x}{x+1} = \dfrac{\rule{2cm}{0.4pt}}{x^2 - 2x - 3}$

3. $\dfrac{-2}{x-1} = \dfrac{\rule{1.5cm}{0.4pt}}{1-x}$

4. $\dfrac{x+1}{x-1} = \dfrac{\rule{1.5cm}{0.4pt}}{x^2 - 2x + 1}$

In problems 5–7, reduce the given fraction.

5. $\dfrac{2}{2x + 2}$

6. $\dfrac{(x + 1)^2}{x^2 + 3x + 2}$

7. $\dfrac{x^2 - 1}{x^2 - 4x + 3}$

In problems 8–16, perform the indicated operation.

8. $\dfrac{2x^3}{3y^2} \cdot \dfrac{9y}{4x}$

9. $\dfrac{2x + 2}{x^2 - x - 6} \cdot \dfrac{x^2 + x - 2}{4x + 4}$

10. $\dfrac{x^3}{2} \div 2x$

11. $\dfrac{x^2 - 4}{2x - 4} \div \dfrac{x^2 + 5x + 6}{x^2 + 3x}$

12. $\dfrac{2x}{x + 1} + \dfrac{2}{x + 1}$

13. $\dfrac{2x + 1}{x - 3} - \dfrac{x + 4}{x - 3}$

14. $\dfrac{3}{3x - 15} + \dfrac{4}{x^2 - 25}$

15. $\dfrac{2}{x} - \dfrac{2}{x - 1}$

16. $\dfrac{5}{x^2 + 3x + 2} - \dfrac{2}{x^2 - 4x + 3}$

In problems 17–18, simplify.

17. $2 + \dfrac{1}{x - 1}$

18. $\dfrac{3 + \dfrac{2}{x}}{2 - \dfrac{1}{x + 1}}$

In problems 19–22, solve and check.

19. $\dfrac{2x}{3} + \dfrac{1}{6} = \dfrac{x}{2}$

20. $\dfrac{x + 2}{x} = \dfrac{3}{2x} - 1$

21. $\dfrac{4}{2x + 4} = \dfrac{6}{x^2 - 4}$

22. One third of a number is two less than one half the same number plus one. Find the number.

CHAPTER 9

Square Roots and Radicals

9-1 INTRODUCTION TO RADICALS

A. Definitions and Symbols of Radicals

In section 7-4B, we stated that a perfect square is a number, expression, or term that can be expressed as the product of two equal factors. For example:

$$4 \text{ is a perfect square since } 4 = (2)(2).$$
$$25 \text{ is a perfect square since } 25 = (5)(5).$$
$$9x^2 \text{ is a perfect square since } 9x^2 = (3x)(3x).$$

One of the two equal factors of a perfect square is defined as the square root of the perfect square. That is,

$$b \text{ is called a square root of } a \text{ if } b^2 = a.$$

Thus,

$$2 \text{ is a square root of } 4 \text{ since } 2^2 = 4.$$
$$5 \text{ is a square root of } 25 \text{ since } 5^2 = 25.$$
$$3x \text{ is a square root of } 9x^2 \text{ since } (3x)^2 = 9x^2.$$

Notice also that:

$$-2 \text{ is a square root of } 4 \text{ since } (-2)^2 = 4.$$
$$-5 \text{ is a square root of } 25 \text{ since } (-5)^2 = 25.$$
$$-3x \text{ is a square root of } 9x^2 \text{ since } (-3x)^2 = 9x^2.$$

It appears then that each positive number has two square roots, one positive and one negative. This is indeed the case.

NOTE:

> The number 0 has just one square root, namely, 0 itself. Also, in our number system, negative numbers have no square roots. For example, we cannot find the square root of -4, since there is no real number b such that $b^2 = -4$. (b^2 will always be nonnegative.)

The positive square root of a number is commonly referred to as the principal square root and is denoted by the symbol "$\sqrt{}$" which is called a radical sign. Thus,

$$\sqrt{4} = +2$$
$$\sqrt{25} = +5$$
$$\sqrt{9x^2} = +3x$$

To indicate the negative square root, we place a negative sign in front of the radical symbol. Thus,

$$-\sqrt{4} = -2$$
$$-\sqrt{25} = -5$$
$$-\sqrt{9x^2} = -3x$$

In this text, we deal only with the principal root.

Typically, there are three parts of a radical expression:

The radical sign itself, the expression under the radical sign (called the radicand), and the index of the radical which indicates the root of the expression.

When dealing with square roots, the index is 2 and is commonly omitted.

$$\sqrt[2]{100} = 10$$
$$\sqrt{100} = 10$$

A cube root of a number has an index of 3. Further, b is a cube root of a if $b^3 = a$.

$$\sqrt[3]{8} = 2$$
since $2^3 = 8$

A fourth root of a number has an index of 4. Further, b is a fourth root of a if $b^4 = a$.

$$\sqrt[4]{16} = 2$$
since $2^4 = 16$

We deal only with square roots in this text.

Do the following practice set. Check your answers with the answers in the right-hand margin.

PRACTICE SET 9-1A

Evaluate each of the following.

1. $\sqrt{64}$ 2. $\sqrt{16x^2}$ 3. $\sqrt{(x+2)^2}$ 1. $+8$
4. $-\sqrt{49x^2}$ 5. $-\sqrt{(3y)^2}$ 6. $\sqrt{16x^2y^6}$ 2. $4x$

3. $(x+2)$
4. $-7x$
5. $-3y$
6. $4xy^3$

B. Computing Square Roots

Perfect squares (or simply, squares) are terms that have a square root. Thus, the perfect square numbers are

$$1, 4, 9, 16, 25, 36, 49, 64, 81, 100, \ldots$$

since each of these numbers have a square root.

Frequently in mathematics, it is necessary to evaluate the square root of a number that is not a perfect square (e.g., $\sqrt{30}$). In doing so, we are only able to give an approximate answer.

To approximate square roots of numbers that are not perfect squares, we may use a calculator or consult a table. (One such table may be found on the inside back cover of this text.) In the event that neither of these are available, we use a method referred to as the divide and average method.

THE DIVIDE AND AVERAGE METHOD FOR ESTIMATING SQUARE ROOTS

To find the square root of a number that is not a perfect square, and neither a calculator nor a square root table is available, we may utilize the following technique.

Find the square root of 19 to the nearest thousandth.

First, determine which two perfect square numbers the given number is between. (19 is between the perfect squares 16 and 25.)

$$16$$
$$19$$
$$25$$

Since 19 is between 16 and 25, then $\sqrt{19}$ must be between $\sqrt{16}$ and $\sqrt{25}$, or between 4 and 5. (Note: The symbol "$\approx$" means "is approximately equal to.")

$$\sqrt{16} = 4$$
$$\sqrt{19} \approx$$
$$\sqrt{25} = 5$$

Estimate $\sqrt{19}$ to one decimal place and divide 19 by this value. We estimate $\sqrt{19} \approx 4.4$.

$$\sqrt{16} = 4$$
$$\sqrt{19} \approx 4.4$$
$$\sqrt{25} = 5$$

(Since the divisor and quotient are not equal, the square root must be between the two.)

$$
\begin{array}{r}
4.318 \\
4.4\,\overline{)19.0.000} \\
176 \\
\overline{140} \\
132 \\
\overline{80} \\
44 \\
\overline{360} \\
352 \\
\overline{8}
\end{array}
$$

Now, find the average of the estimate and the quotient. This becomes our new estimate for $\sqrt{19}$.

$$\frac{4.4 + 4.318}{2}$$

$$= \frac{8.718}{2} = 4.359$$

This process may be continued to obtain as close an approximation as desired. Since we want our estimate to the nearest thousandth, we must apply this method once more.

Divide 19 by the new estimate.

$$
\begin{array}{r}
4.35879 \\
4.359\,\overline{)19.000.00000}
\end{array}
$$

Average the estimate and quotient.

$$\frac{4.359 + 4.35879}{2}$$

$$= \frac{8.71879}{2} = 4.359395$$

$$\approx 4.359$$

Thus, $\sqrt{19}$ approximately equals 4.359 to the nearest thousandth. Compare this answer to the table value.

NOTE:

> No matter what number we use as our first estimate, we will eventually arrive at the same square root for our answer.

Using the example as a guide, do the following practice set. Check your answers with the answers in the right-hand margin.

PRACTICE SET 9-1B

Evaluate each of the following to the place value given by using the divide/average method; compare your answers by using the table of square roots given on the inside back cover of the text.

1. The $\sqrt{15}$ to the nearest hundredth   1. 3.87
2. The $\sqrt{120}$ to the nearest thousandth   2. 10.954
3. The $-\sqrt{75}$ to the nearest tenth   3. −8.7
4. The $\sqrt{6}$ to the nearest tenth   4. 2.4
5. The $-\sqrt{28}$ to the nearest hundredth   5. −5.3
6. The $\sqrt{133}$ to the nearest thousandth   6. 11.5

EXERCISE 9-1

Perform the indicated operation.

1. Find the square root of $(2xy)^2$.   2. Find the square root of $(3x^2)^2$.
3. Find the square root of $(x+1)^2$.   4. Find the square root of $(2x-1)^2$.
5. $\sqrt{16y^2}$ 6. $-\sqrt{9x^2}$ 7. $\sqrt{4x^4}$ 8. $-\sqrt{x^6}$
9. $\sqrt{144x^6y^4}$ 10. $\sqrt{25x^2y^4z^6}$

Using the table of square roots, find the square roots of the following. Give the value of the answers to the nearest hundredth.

11. $\sqrt{21}$ 12. $\sqrt{7}$ 13. $\sqrt{140}$ 14. $\sqrt{2}$
15. $\sqrt{80}$ 16. $\sqrt{40}$ 17. $-\sqrt{101}$ 18. $-\sqrt{133}$
19. $\sqrt{121}$ 20. $\sqrt{67}$

Use the method of divide and average to find the square root of the following. Give the answer to the nearest tenth.

21. $\sqrt{6}$ 22. $\sqrt{10}$ 23. $\sqrt{11}$ 24. $\sqrt{15}$

25. $\sqrt{19}$ 26. $\sqrt{26}$ 27. $\sqrt{50}$ 28. $\sqrt{57}$

29. $\sqrt{110}$ 30. $\sqrt{130}$

9-2 SIMPLIFYING RADICALS

A. Simplifying Radicals When the Radicand Is an Integer

Consider the radical, $\sqrt{36}$. Since 36 is a perfect square, we know that $\sqrt{36} = 6$.

Now, consider the expression $(\sqrt{4})(\sqrt{9})$. Since both 4 and 9 are perfect squares, we can evaluate this expression with little difficulty.

$$(\sqrt{4})(\sqrt{9}) = (2)(3)$$
$$= 6$$

Since $\sqrt{36} = 6$ and since $(\sqrt{4})(\sqrt{6}) = 6$, then
$$\sqrt{36} = (\sqrt{4})(\sqrt{9}).$$

Moreover, since $36 = (4)(9)$, then

$$\sqrt{36} = \sqrt{(4)(9)}$$
$$= \sqrt{4}\sqrt{9}$$
$$= (2)(3)$$
$$= 6$$

This small example illustrates a very important property of radicals. That is, the square root of a product is equal to the square roots of the factors. In general:

$$\sqrt[N]{ab} = \sqrt[N]{a} \cdot \sqrt[N]{b}$$

provided a and b are positive numbers

Knowing this property, we can simplify radical expressions where the radicand is not a perfect square. To do this, we simply factor the radicand into two factors, one of which is the largest perfect square contained in the radicand. Then, we apply our property. For example, to simplify $\sqrt{50}$:

First, note that 50 can be factored into $(25)(2)$ and that 25 is the largest perfect square contained

$$\sqrt{50} = \sqrt{(25)(2)}$$

Now, we apply our property.

$$\sqrt{(25)(2)} = \sqrt{25} \cdot \sqrt{2}$$
$$= 5 \cdot \sqrt{2}$$

Thus,

$$\sqrt{50} \text{ is } 5 \cdot \sqrt{2}$$

We can also apply this property when the radicand contains variables. For example, to simplify $\sqrt{28x^3}$:

First, factor the radicand: 28 contains the perfect square 4; x^3 contains the perfect square x^2.

$$\sqrt{28x^3} = \sqrt{(4)(7)(x^2)(x)}$$

Then, apply our property. We first reorder the factors so that the perfect squares come first.

$$\sqrt{(4)(7)(x^2)(x)}$$
$$= \sqrt{(4)(x^2)(7)(x)}$$
$$= \sqrt{4}\sqrt{x^2}\sqrt{7}\sqrt{x}$$
$$= 2x\sqrt{7x}$$

Thus,

$$\sqrt{28x^3} \text{ simplifies to } 2x\sqrt{7x}$$

In summary, to simplify a radical when the radicand is an integer, we first find two factors of the radicand, one of which will be the largest perfect square contained in the radicand. Next, we form a separate square root for each factor and find the square root of any perfect squares. Finally, we multiply the factors outside the radical and combine with the radical.

PROCEDURE:

To simplify radicals when the radicand is an integer:

1. Find two factors of the radicand, one of which will be the largest perfect square contained in the radicand.

2. Form a separate square root for each factor.

3. Find the square root of each perfect square.

4. Multiply the factors outside the radical.

EXAMPLE 1: Simplify $\sqrt{27}$.

SOLUTION: Factor 27.

$$\sqrt{27} = \sqrt{(9)(3)}$$

Form two separate radicals and find the square root of the perfect square.

$$\sqrt{(9)(3)} = \sqrt{9}\sqrt{3}$$
$$= 3\sqrt{3}$$

Final answer: $3\sqrt{3}$

EXAMPLE 2: Simplify $\sqrt{12x^5}$.

SOLUTION: Factor the radicand.

$$\sqrt{12x^5} = \sqrt{(4)(3)(x^4)(x)}$$
$$= \sqrt{(4)(x^4)(3)(x)}$$

Form separate radicals and evaluate the perfect squares.

$$\sqrt{(4)(x^4)(3)(x)} = \sqrt{4}\sqrt{x^4}\sqrt{3}\sqrt{x}$$
$$= (2)(x^2)\sqrt{3}\sqrt{x}$$
$$= 2x^2\sqrt{3x}$$

Final answer: $2x^2\sqrt{3x}$

EXAMPLE 3: Simplify $2\sqrt{45}$.

SOLUTION:

$$2\sqrt{45} = 2\sqrt{(9)(5)}$$
$$= 2\sqrt{9}\sqrt{5}$$
$$= (2)(3)\sqrt{5}$$
$$= 6\sqrt{5}$$

Final answer: $6\sqrt{5}$

EXAMPLE 4: Simplify $-\sqrt{32x^2y^3}$.

SOLUTION:

$$-\sqrt{32x^2y^3} = (-1)\sqrt{(16)(2)(x^2)(y^2)(y)}$$
$$= (-1)\sqrt{(16)(x^2)(y^2)(2)(y)}$$
$$= (-1)\sqrt{16}\sqrt{x^2}\sqrt{y^2}\sqrt{2}\sqrt{y}$$
$$= (-1)(4)(x)(y)\sqrt{2}\sqrt{y}$$
$$= -4xy\sqrt{2y}$$

Final answer: $-4xy\sqrt{2y}$

Do the following practice set. Check your answers with the answers in the right-hand margin.

PRACTICE SET 9-2A

Simplify (see example 1 or 2).

1. $\sqrt{32}$

2. $\sqrt{9x^3y^5}$

3. $\sqrt{75x^4y^5}$

Simplify (see example 3 or 4).

4. $2\sqrt{18x^2}$

5. $-5\sqrt{24}$

6. $-2\sqrt{20x^4y^7}$

1. $4\sqrt{2}$
2. $3xy^2\sqrt{xy}$
3. $5x^2y^2\sqrt{3y}$

4. $6x\sqrt{2}$
5. $-10\sqrt{6}$
6. $-4x^2y^3\sqrt{5y}$

B.   Simplifying Radicals When the Radicand Is a Fraction

Consider the expression, $\sqrt{\frac{9}{16}}$. We know that $\sqrt{\frac{9}{16}} = \frac{3}{4}$ since $\left(\frac{3}{4}\right)^2 = \frac{9}{16}$. Notice also that

$$\frac{\sqrt{9}}{\sqrt{16}} = \frac{3}{4}$$

Thus,

$$\sqrt{\frac{9}{16}} = \frac{\sqrt{9}}{\sqrt{16}}$$

This example illustrates another property of radicals. That is, the square root of a fraction is equal to the square root of the numerator divided by the square root of the denominator. In general:

$$\sqrt[N]{\frac{a}{b}} = \frac{\sqrt[N]{a}}{\sqrt[N]{b}}$$

provided a and b are positive numbers

Using this property, we can simplify radicals when the radicand is a fraction. For example, to simplify $\sqrt{\frac{49}{9}}$:

Write the square root of the fraction as the square root of the numerator over the square root of the denominator.

$$\sqrt{\frac{49}{9}} = \frac{\sqrt{49}}{\sqrt{9}}$$

Now, simplify accordingly.

$$\frac{\sqrt{49}}{\sqrt{9}} = \frac{7}{3}$$

Thus,

$$\sqrt{\frac{49}{9}} = \frac{7}{3}$$

As a second example, consider $\sqrt{\frac{8}{25}}$. To simplify $\sqrt{\frac{8}{25}}$:

Write the square root of the numerator over the square root of the denominator and evaluate the perfect square.

$$\sqrt{\frac{8}{25}} = \frac{\sqrt{8}}{\sqrt{25}} = \frac{\sqrt{8}}{5}$$

Although $\sqrt{8}$ is not a perfect square, we can, nevertheless, simplify it.

$$\frac{\sqrt{8}}{5} = \frac{\sqrt{4}\sqrt{2}}{5} = \frac{2\sqrt{2}}{5}$$

Thus,

$$\sqrt{\frac{8}{25}} \text{ simplifies to } \frac{2\sqrt{2}}{5} \quad \text{or} \quad \frac{2}{5}\sqrt{2}$$

As a third example, consider the expression $\sqrt{\frac{3}{2}}$. Notice that in this example, the denominator (2) is not a perfect square.

Thus, when we simplify $\sqrt{\frac{3}{2}}$, our result will contain a square root (or radical) in the denominator of the fraction.

$$\sqrt{\frac{3}{2}} = \frac{\sqrt{3}}{\sqrt{2}}$$

Although this form, $\sqrt{3}/\sqrt{2}$, is quite acceptable, we can express the answer in another equivalent form that is more convenient in the event we wish to approximate the number with a decimal. This equivalent form can be obtained by employing a method called rationalizing the denominator. This method will be discussed in Section 9-5B, and the simplification of radicals that involve fractions where the denominator is not a perfect square is postponed until then.

So, in summary, to simplify radicals when the radicand is a fraction (and the denominator is a perfect square), we write the square root of the numerator over the square root of the denominator, and simplify.

PROCEDURE:

To simplify radicals when the radicand is a fraction (and the denominator is a perfect square):

1.   Write the square root of the numerator over the square root of the denominator.

2.   Simplify the square roots.

EXAMPLE 1: Simplify $\sqrt{\frac{1}{9}}$.

SOLUTION:

$$\sqrt{\frac{1}{9}} = \frac{\sqrt{1}}{\sqrt{9}} = \frac{1}{3}$$

Final answer: $\frac{1}{3}$

EXAMPLE 2: Simplify $\sqrt{\dfrac{27}{4}}$.

SOLUTION:

$$\sqrt{\frac{27}{4}} = \frac{\sqrt{27}}{\sqrt{4}} = \frac{\sqrt{9}\sqrt{3}}{2} = \frac{3\sqrt{3}}{2}$$

Final answer: $\dfrac{3\sqrt{3}}{2}$ or $\dfrac{3}{2}\sqrt{3}$

EXAMPLE 3: Simplify $6\sqrt{\dfrac{18}{25}}$.

SOLUTION:

$$6\sqrt{\frac{18}{25}} = \frac{6\sqrt{18}}{\sqrt{25}}$$

$$= \frac{6\sqrt{9}\sqrt{2}}{\sqrt{25}} = \frac{(6)(3)\sqrt{2}}{5} = \frac{18\sqrt{2}}{5}$$

Final answer: $\dfrac{18\sqrt{2}}{5}$ or $\dfrac{18}{5}\sqrt{2}$

EXAMPLE 4: Simplify $2x\sqrt{\dfrac{40x}{9y^2}}$.

SOLUTION:

$$2x\sqrt{\frac{40x}{9y^2}} = \frac{2x\sqrt{40x}}{\sqrt{9y^2}}$$

$$= \frac{2x\sqrt{40}\sqrt{x}}{\sqrt{9}\sqrt{y^2}}$$

$$= \frac{2x\sqrt{4}\sqrt{10}\sqrt{x}}{\sqrt{9}\sqrt{y^2}}$$

$$= \frac{(2x)(2)(\sqrt{10})(\sqrt{x})}{(3)(y)}$$

$$= \frac{4x\sqrt{10x}}{3y}$$

Final answer: $\dfrac{4x\sqrt{10x}}{3y}$

Do the following practice set. Check your answers with the answers in the right-hand margin.

PRACTICE SET 9-2B

Simplify (see example 1 or 2).

1. $\sqrt{\dfrac{1}{4}}$ 2. $\sqrt{\dfrac{4}{9}}$ 3. $\sqrt{\dfrac{27}{4}}$ 4. $\sqrt{\dfrac{75}{25}}$

Simplify (see example 3 or 4).

5. $3\sqrt{\dfrac{12}{16}}$ 6. $8\sqrt{\dfrac{2x^2}{9}}$ 6. $3x\sqrt{\dfrac{16x^2}{25}}$ 8. $5x\sqrt{\dfrac{27x^3}{4y^2}}$

1. $\frac{1}{2}$
2. $\frac{2}{3}$
3. $3\sqrt{3}/2$
4. $\sqrt{3}$
5. $3\sqrt{3}/2$
6. $8x\sqrt{2}/3$
7. $12x^2/5$
8. $15x^2\sqrt{3x}/2y$

EXERCISE 9-2

Simplify the following radicals.

1. $\sqrt{8}$ 2. $\sqrt{18}$ 3. $\sqrt{72}$ 4. $\sqrt{12}$

5. $\sqrt{20}$ 6. $\sqrt{45}$ 7. $\sqrt{32x^2}$ 8. $\sqrt{64x^5}$

9. $\sqrt{50x^5y^5}$ 10. $\sqrt{27x^2y^3}$ 11. $\sqrt{98x^2y^5z^6}$ 12. $\sqrt{49x^5y^7z^9}$

13. $3\sqrt{24}$ 14. $4\sqrt{48x^2}$ 15. $2\sqrt{125x^3}$ 16. $\dfrac{1}{2}\sqrt{32x^2}$

17. $\dfrac{2}{3}\sqrt{18x^3}$ 18. $\dfrac{1}{5}\sqrt{175x^3y^4}$ 19. $-\sqrt{36x^3}$ 20. $-\sqrt{54x^4}$

21. $-2\sqrt{20xy^2}$ 22. $-\dfrac{1}{2}\sqrt{20x^2y^3}$ 23. $-\sqrt{x^2yz^5}$ 24. $-2\sqrt{x^7y^8}$

25. $\sqrt{\dfrac{9}{16}}$ 26. $\sqrt{\dfrac{4}{9}}$ 27. $\sqrt{\dfrac{9}{25}}$ 28. $\sqrt{\dfrac{25}{36}}$

29. $\sqrt{\dfrac{1}{16}}$ 30. $2\sqrt{\dfrac{1}{16}}$ 31. $3\sqrt{\dfrac{16}{36}}$ 32. $\sqrt{\dfrac{2}{4}}$

33. $\sqrt{\dfrac{5}{4}}$ 34. $4\sqrt{\dfrac{50}{49}}$ 35. $5\sqrt{\dfrac{8}{49}}$ 36. $\dfrac{7}{9}\sqrt{\dfrac{7}{9}}$

37. $3\sqrt{\dfrac{4x^2}{16}}$ 38. $9\sqrt{\dfrac{16x^2}{100}}$ 39. $2x\sqrt{\dfrac{30}{81x^2}}$ 40. $7x\sqrt{\dfrac{40x^2}{100y^2}}$

41. $5xy\sqrt{\dfrac{60x^2}{25y^2}}$ 42. $\sqrt{\dfrac{x^2y^4}{4x^4}}$ 43. $\dfrac{2}{3}\sqrt{\dfrac{50x^2}{100}}$ 44. $\dfrac{9}{7}\sqrt{\dfrac{49x^2}{81y^2}}$

45. $\dfrac{3x}{4y} \sqrt{\dfrac{16y^2}{81x^2}}$

46. $\dfrac{6x}{5y} \sqrt{\dfrac{75y}{100x^4}}$

47. $x^2 \sqrt{\dfrac{30}{81x^2}}$

48. $\dfrac{2x}{5y} \sqrt{\dfrac{19}{64x^2y^2}}$

49. $\sqrt{\dfrac{x^5y^2z^3}{25}}$

50. $\sqrt{\dfrac{x^3y^5z^7}{100x^2y^4z^6}}$

9-3 ADDITION AND SUBTRACTION OF RADICALS

Radicals that have the same index and the same radicand are similar radicals. For example, $2\sqrt{3}$ and $7\sqrt{3}$ are similar radicals because

They are both square roots. $2 \boxed{\sqrt{3}}$ and $7 \boxed{\sqrt{3}}$

and

The radicands are the same. $2\sqrt{\boxed{3}}$ and $7\sqrt{\boxed{3}}$

Similar radicals are treated as like terms and may be combined by addition or subtraction in the same manner as like terms. For example, to combine $2\sqrt{3} + 7\sqrt{3}$:

Combine the coefficients by sign. $\boxed{2}\sqrt{3} + \boxed{7}\sqrt{3}$

$\boxed{2+7}$

9

and

Multiply this by the common radicand. $2\boxed{\sqrt{3}} + 7\boxed{\sqrt{3}}$

$9\sqrt{3}$

Thus,

$2\sqrt{3} + 7\sqrt{3}$ is $9\sqrt{3}$

When combining like terms such as $3x + x$, the coefficient of x is understood to be 1. Therefore, the sum of $3x + x$ equals $3x + 1x$ or $4x$.

This concept also applies to radicals. For example, in finding the sum of $3\sqrt{3} + \sqrt{3}$,

The coefficient of the first term is 3. $\boxed{3}\sqrt{3} + \sqrt{3}$

and

The coefficient of the second term is 1. $3\sqrt{3} + \boxed{}\sqrt{3}$

$(\sqrt{3} = 1\sqrt{3})$

Therefore, the sum is $4\sqrt{3}$ $\boxed{(3+1)}\sqrt{3}$

$4\sqrt{3}$

If our radicals are unlike, we are not permitted to combine the terms. For example, to add $2\sqrt{3}$ and $3\sqrt{2}$:

We cannot combine the two terms into a single term because they have different radicands. $2\sqrt{\boxed{3}} + 3\sqrt{\boxed{2}}$

The sum can only be indicated as $2\sqrt{3} + 3\sqrt{2}$.

Some radicals that may appear to be unlike can be transformed into like radicals by reducing or simplifying the radicals. For example, $\sqrt{18}$ and $\sqrt{2}$ are unlike radicals;

They have different radicands $\sqrt{\boxed{18}}, \ \sqrt{\boxed{2}}$

However:

By simplifying the radicals,
$$\sqrt{18} = 3\sqrt{2}$$
$$\sqrt{2} = \sqrt{2}$$

$\sqrt{18}$ and $\sqrt{2}$
$\sqrt{9 \cdot 2}$
$\sqrt{9} \cdot \sqrt{2}$
$3\sqrt{2}$

We now have similar radicals. $3\sqrt{2}$ and $\sqrt{2}$

When combining radicals, we must be sure the radicals have been reduced before we try to identify like radicals. For example, to combine $7\sqrt{2} - 3\sqrt{8}$:

First, simplify each radical.

$7\sqrt{2}$ is in the simplest form.

$3\sqrt{8}$ simplifies to $6\sqrt{2}$.

$7\sqrt{2} - 3\sqrt{8}$

$7\sqrt{2} - 3\sqrt{8}$
$3\sqrt{(4)(2)}$
$3\sqrt{4}\sqrt{2}$
$(3)(2)\sqrt{2}$
$6\sqrt{2}$

Now, combine like radicals according to their sign.

$7\sqrt{2} - 6\sqrt{2} = 1\sqrt{2}$
$= \sqrt{2}$

Thus,
$$7\sqrt{2} - 3\sqrt{8} = \sqrt{2}$$

In summary, to add or subtract radicals, we must first simplify each term that contains a radical. Then we combine all like terms according to their sign.

PROCEDURE:

To add or subtract radical expressions:

1. Simplify each term containing a radical.

2. Combine similar radicals by adding or subtracting coefficients, keeping the common radical.

EXAMPLE 1: Combine $2\sqrt{5} + 3\sqrt{5} - 9\sqrt{5}$.

SOLUTION: Since all the radicals are simplified and similar, we combine them according to their sign.

$$2\sqrt{5} + 3\sqrt{5} - 9\sqrt{5}$$
$$= 5\sqrt{5} - 9\sqrt{5}$$
$$= -4\sqrt{5}$$

Final answer: $-4\sqrt{5}$

EXAMPLE 2: Combine $5\sqrt{3} - 4\sqrt{2} + 2\sqrt{3} + \sqrt{2}$.

SOLUTION: All radicals are simplified as given. So, combine all like terms accordingly.

$$5\sqrt{3} - 4\sqrt{2} + 2\sqrt{3} + \sqrt{2}$$
$$= \underbrace{5\sqrt{3} + 2\sqrt{3}} - \underbrace{4\sqrt{2} + \sqrt{2}}$$
$$= \quad 7\sqrt{3} \quad - \quad 3\sqrt{2}$$

Final answer: $7\sqrt{3} - 3\sqrt{2}$

EXAMPLE 3: Combine $2\sqrt{50} - 3\sqrt{18}$.

SOLUTION: We first simplify the radicals.

$$2\sqrt{50} - 3\sqrt{18} = 2\sqrt{25}\sqrt{2} - 3\sqrt{9}\sqrt{2}$$
$$= (2)(5)\sqrt{2} - (3)(3)\sqrt{2}$$
$$= \quad 10\sqrt{2} \quad - \quad 9\sqrt{2}$$

Now, combine all like terms.

$$10\sqrt{2} - 9\sqrt{2} = 1\sqrt{2}$$

Final answer: $\sqrt{2}$

EXAMPLE 4: Combine $\sqrt{\dfrac{27}{4}} + 5\sqrt{12}$.

SOLUTION: Simplify the radicals.

$$\sqrt{\frac{27}{4}} + 5\sqrt{12} = \frac{\sqrt{27}}{\sqrt{4}} + 5\sqrt{4}\sqrt{3}$$

$$= \frac{\sqrt{9}\sqrt{3}}{\sqrt{4}} + 5\sqrt{4}\sqrt{3}$$

$$= \frac{3\sqrt{3}}{2} + (5)(2)\sqrt{3}$$

$$= \frac{3\sqrt{3}}{2} + 10\sqrt{3}$$

Combine like terms.

$$\frac{3\sqrt{3}}{2} + 10\sqrt{3} = \frac{3}{2}\sqrt{3} + 10\sqrt{3}$$

$$= \frac{3}{2}\sqrt{3} + \frac{20}{2}\sqrt{3}$$

$$= \frac{23}{2}\sqrt{3}$$

Final answer: $\dfrac{23}{2}\sqrt{3}$

Do the following practice set. Check your answers with the answers in the right-hand margin.

PRACTICE SET 9-3

Combine the following radicals (see example 1 or 2).

1. $3\sqrt{6} + \sqrt{6}$
2. $2\sqrt{3} - 5\sqrt{3} + \sqrt{3}$
3. $7\sqrt{7} + 2\sqrt{5} - 4\sqrt{7} - \sqrt{5}$

1. $4\sqrt{6}$
2. $-2\sqrt{3}$
3. $3\sqrt{7} + \sqrt{5}$

Combine the following radicals (see example 3 or 4).

4. $2\sqrt{18} - 3\sqrt{8}$

5. $\sqrt{\dfrac{3}{25}} + \sqrt{12}$

6. $5\sqrt{27} - 6\sqrt{\dfrac{5}{64}} + 2\sqrt{12}$

4. 0
5. $11\sqrt{3}/5$
6. $19\sqrt{3} - 3\sqrt{5}/4$

EXERCISE 9-3

Combine the following radicals.

1. $4\sqrt{3} + \sqrt{3}$
3. $\sqrt{2} + 2\sqrt{2}$
5. $4\sqrt{5} - \sqrt{5} - 5\sqrt{5}$
7. $3\sqrt{5} + \sqrt{5} + 4\sqrt{2} - \sqrt{2}$
9. $2\sqrt{3} + 2\sqrt{5} - 3\sqrt{5} + 3\sqrt{3}$

2. $2\sqrt{5} - 6\sqrt{5} - \sqrt{5}$
4. $2\sqrt{7} - 5\sqrt{7} + 3\sqrt{7}$
6. $\sqrt{2} + \sqrt{2} + \sqrt{2}$
8. $2\sqrt{6} - \sqrt{3} + 3\sqrt{6} + 4\sqrt{3}$
10. $\sqrt{5} + 2\sqrt{7} + \sqrt{5} - 2\sqrt{7}$

11. $\sqrt{8} + \sqrt{2}$

12. $\sqrt{3} - \sqrt{27}$

13. $3\sqrt{5} - \sqrt{20}$

14. $2\sqrt{24} - \sqrt{6}$

15. $\sqrt{32} - \sqrt{8} + \sqrt{2}$

16. $\sqrt{27} - \sqrt{48} + \sqrt{3}$

17. $5\sqrt{6} - 2\sqrt{24}$

18. $\sqrt{125} - 3\sqrt{5}$

19. $3\sqrt{8} - \sqrt{18}$

20. $2\sqrt{32} - 4\sqrt{8}$

21. $\dfrac{1}{2}\sqrt{8} + \sqrt{50}$

22. $2\sqrt{32} - 2\sqrt{50} + \sqrt{72}$

23. $2\sqrt{18} - \dfrac{1}{2}\sqrt{32}$

24. $3\sqrt{54} - 2\sqrt{24}$

25. $2\sqrt{12} - 3\sqrt{27}$

26. $\dfrac{1}{3}\sqrt{72} + \dfrac{1}{2}\sqrt{8}$

27. $-\sqrt{18} - \sqrt{72}$

28. $2\sqrt{50} - 3\sqrt{18}$

29. $\sqrt{45} + \dfrac{1}{2}\sqrt{80} - \sqrt{20}$

30. $-\sqrt{12} + \sqrt{48}$

31. $2\sqrt{96} - 3\sqrt{24}$

32. $\dfrac{1}{2}\sqrt{28} + \dfrac{1}{3}\sqrt{63}$

33. $4\sqrt{27} - 3\sqrt{48}$

34. $2\sqrt{125} - \sqrt{45}$

35. $\sqrt{99} - \sqrt{44}$

36. $3\sqrt{54} - 4\sqrt{96}$

37. $2\sqrt{3} - 2\sqrt{\dfrac{3}{4}}$

38. $3\sqrt{8} - 4\sqrt{\dfrac{2}{9}}$

39. $30\sqrt{\dfrac{1}{9}} - 2\sqrt{27}$

40. $\sqrt{45} - 10\sqrt{\dfrac{5}{4}}$

41. $-\sqrt{20} + 15\sqrt{\dfrac{5}{9}}$

42. $\sqrt{24} - 3\sqrt{\dfrac{9}{16}}$

43. $\sqrt{90} + 5\sqrt{\dfrac{40}{25}}$

44. $\sqrt{\dfrac{3}{4}} + 2\sqrt{\dfrac{28}{25}} - \sqrt{12}$

45. $4\sqrt{\dfrac{25}{4}} + 10\sqrt{100} - \sqrt{\dfrac{35}{16}}$

46. $3\sqrt{\dfrac{18}{9}} - 4\sqrt{\dfrac{6}{4}} + 5\sqrt{\dfrac{20}{49}}$

9-4   MULTIPLICATION OF RADICALS

We can use our first property of radicals, as presented in Section 9-2A, to multiply radicals. That is,

$$\text{If} \qquad \sqrt{ab} = \sqrt{a} \cdot \sqrt{b}$$
$$\text{Then} \qquad \sqrt{a} \cdot \sqrt{b} = \sqrt{ab}$$

Thus, the product of the square roots of two or more nonnegative numbers is equal to the square root of their product. For example, to find the product of $\sqrt{5} \cdot \sqrt{10}$:

First, write the radicands as factors of a product under the radical sign and multiply them.

$$\sqrt{5} \cdot \sqrt{10} = \sqrt{(5)(10)}$$
$$= \sqrt{50}$$

Now, simplify the radical.

$$\sqrt{50} = \sqrt{(25)(2)}$$
$$= \sqrt{25}\sqrt{2}$$
$$= 5\sqrt{2}$$

Thus,

$$\sqrt{5} \cdot \sqrt{10} = 5\sqrt{2}$$

If coefficients are involved, we simply multiply the coefficients together and then multiply the radicals together. For example, to multiply $2\sqrt{3} \cdot 3\sqrt{2}$:

Multiply the coefficients.

$$(2\sqrt{3})(3\sqrt{2}) = (2)(3)(\sqrt{3})(\sqrt{2})$$
$$= 6\sqrt{3}\sqrt{2}$$

Then, multiply the radicals.

$$= 6\sqrt{(3)(2)}$$
$$= 6\sqrt{6}$$

Thus,

$$(2\sqrt{3})(3\sqrt{2}) \text{ is } 6\sqrt{6}$$

PROCEDURE:

To multiply radical expressions:

1. Multiply the coefficients.
2. Multiply the radicands.
3. Simplify their product.

EXAMPLE 1:   Multiply $\sqrt{2} \cdot \sqrt{6}$.

SOLUTION: Multiply the radicands.

$$\sqrt{2} \cdot \sqrt{6} = \sqrt{(2)(6)}$$
$$= \sqrt{12}$$

Simplify the radical.

$$\sqrt{12} = \sqrt{(4)(3)}$$
$$= \sqrt{4}\sqrt{3}$$
$$= 2\sqrt{3}$$

Final answer: $2\sqrt{3}$

EXAMPLE 2:   Multiply $(2\sqrt{3})(4)$.

SOLUTION: We only need to multiply coefficients.

$$(2\sqrt{3})(4) = (2)(4)\sqrt{3}$$
$$= 8\sqrt{3}$$

Final answer: $8\sqrt{3}$

EXAMPLE 3: Multiply $(5\sqrt{8})(7\sqrt{3})$.

SOLUTION: Multiply coefficients.

$$(5\sqrt{8})(7\sqrt{3}) = (5)(7)\sqrt{8}\sqrt{3}$$
$$= 35\sqrt{8}\sqrt{3}$$

Multiply radicands.

$$35\sqrt{8}\sqrt{3} = 35\sqrt{(8)(3)}$$
$$= 35\sqrt{24}$$

Simplify the radical.

$$35\sqrt{24} = 35\sqrt{(4)(6)}$$
$$= 35\sqrt{4}\sqrt{6}$$
$$= (35)(2)\sqrt{6}$$
$$= 70\sqrt{6}$$

Final answer: $70\sqrt{6}$

EXAMPLE 4: Multiply $6\sqrt{\dfrac{1}{3}} \cdot 4\sqrt{12}$.

SOLUTION: Multiply coefficients.

$$\left(6\sqrt{\frac{1}{3}}\right) \cdot (4\sqrt{12}) = (6)(4)\sqrt{\frac{1}{3}}\sqrt{12}$$
$$= 24\sqrt{\frac{1}{3}}\sqrt{12}$$

Multiply radicands.

$$24\sqrt{\frac{1}{3}}\sqrt{12} = 24\sqrt{\left(\frac{1}{3}\right)(12)}$$
$$= 24\sqrt{4}$$

Simplify.

$$24\sqrt{4} = (24)(2)$$
$$= 48$$

Final answer: 48

EXAMPLE 5: Multiply $(2\sqrt{3x^2y})(\sqrt{6y})$.

SOLUTION:

$$
\begin{aligned}
(2\sqrt{3x^2y})(\sqrt{6y}) &= 2\sqrt{3x^2y}\sqrt{6y} \\
&= 2\sqrt{(3x^2y)(6y)} \\
&= 2\sqrt{(3)(6)(x^2)(y)(y)} \\
&= 2\sqrt{18x^2y^2} \\
&= 2\sqrt{18}\sqrt{x^2}\sqrt{y^2} \\
&= 2\sqrt{(9)(2)}\sqrt{x^2}\sqrt{y^2} \\
&= 2\sqrt{9}\sqrt{2}\sqrt{x^2}\sqrt{y^2} \\
&= 2\sqrt{9}\sqrt{x^2}\sqrt{y^2}\sqrt{2} \\
&= (2)(3)(x)(y)\sqrt{2} \\
&= 6xy\sqrt{2}
\end{aligned}
$$

Final answer: $6xy\sqrt{2}$

EXAMPLE 6: Multiply $(3\sqrt{6})^2$.

SOLUTION:

$$
\begin{aligned}
(3\sqrt{6})^2 &= (3\sqrt{6})(3\sqrt{6}) \\
&= (3)(3)\sqrt{6}\sqrt{6} \\
&= 9\sqrt{36} \\
&= (9)(6) \\
&= 54
\end{aligned}
$$

Final answer: 54

EXAMPLE 7: Multiply $2\sqrt{3}(2\sqrt{5} - 3\sqrt{20} + \sqrt{45})$.

SOLUTION: First, use the distributive property.

$$
2\sqrt{3}(2\sqrt{5} - 3\sqrt{20} + \sqrt{45})
$$
$$
= (2\sqrt{3})(2\sqrt{5}) + (2\sqrt{3})(-3\sqrt{20}) + (2\sqrt{3})(+\sqrt{45})
$$
$$
= \quad 4\sqrt{15} \qquad\qquad -6\sqrt{60} \qquad\qquad +2\sqrt{135}
$$

Example continued on next page.

Now, clean things up.

$$4\sqrt{15} - 6\sqrt{60} + 2\sqrt{135}$$
$$= 4\sqrt{15} - 6\sqrt{(4)(15)} + 2\sqrt{(9)(15)}$$
$$= 4\sqrt{15} - 6\sqrt{4}\sqrt{15} + 2\sqrt{9}\sqrt{15}$$
$$= 4\sqrt{15} - (6)(2)\sqrt{15} + (2)(3)\sqrt{15}$$
$$= 4\sqrt{15} - 12\sqrt{15} + 6\sqrt{15}$$
$$= \quad\quad -8\sqrt{15} \quad\quad + \quad 6\sqrt{15}$$
$$= \quad\quad\quad\quad -2\sqrt{15}$$

Final answer: $-2\sqrt{15}$

Do the following practice set. Check your answers with the answers in the right-hand margin.

PRACTICE SET 9-4

Multiply the following radicals, and express the product in its simplest form (see example 1, 2, or 3).

1. $2\sqrt{6} \cdot \sqrt{3}$

2. $2\sqrt{5} \cdot 3\sqrt{10}$

3. $(2\sqrt{8})(-2\sqrt{2})$

1. $6\sqrt{2}$

2. $30\sqrt{2}$

3. -16

Multiply the following radicals and express the product in its simplest form (see example 4, 5, 6, or 7).

4. $\sqrt{27} \cdot \sqrt{\dfrac{1}{3}}$

5. $3\sqrt{4x} \cdot \sqrt{3xy^2}$

6. $(3\sqrt{10})^2$

7. $2\sqrt{2}(\sqrt{6} + 2\sqrt{24})$

4. 3

5. $6xy\sqrt{3}$

6. 90

7. $20\sqrt{3}$

EXERCISE 9-4

Multiply the following radicals and express the product in its simplest form.

1. $\sqrt{3} \cdot \sqrt{3}$

2. $\sqrt{21} \cdot \sqrt{21}$

3. $\sqrt{x} \cdot \sqrt{x}$

4. $\sqrt{x+1} \cdot \sqrt{x+1}$

5. $\sqrt{8} \cdot \sqrt{2}$

6. $\sqrt{12} \cdot \sqrt{3}$

7. $\sqrt{4} \cdot \sqrt{9}$

8. $\sqrt{4} \cdot \sqrt{25}$

9. $3 \cdot 2\sqrt{5}$

10. $7 \cdot 2\sqrt{10}$

11. $(-2)(6\sqrt{3})$

12. $(-2)(-2\sqrt{5})$

13. $\sqrt{3} \cdot 8\sqrt{5}$

14. $2\sqrt{3} \cdot \sqrt{2}$

15. $2\sqrt{2} \cdot 3\sqrt{8}$

16. $2\sqrt{10} \cdot 6\sqrt{10}$

17. $6\sqrt{2} \cdot 3\sqrt{5}$

18. $4\sqrt{7} \cdot 3\sqrt{3}$

19. $(-\sqrt{3})(2\sqrt{2})$

20. $(-2\sqrt{5})(3\sqrt{2})$

21. $\sqrt{6} \cdot \sqrt{3}$

22. $\sqrt{4} \cdot \sqrt{3}$

23. $2\sqrt{2} \cdot \sqrt{24}$

24. $\sqrt{15} \cdot 2\sqrt{5}$

25. $3\sqrt{2} \cdot 3\sqrt{10}$

26. $2\sqrt{40} \cdot \dfrac{1}{2}\sqrt{2}$

27. $(-\sqrt{6})(2\sqrt{12})$

28. $(-\sqrt{10})(-\sqrt{5})$

29. $\sqrt{8} \cdot 2\sqrt{10}$

30. $\sqrt{15} \cdot 7\sqrt{3}$

31. $\dfrac{1}{2}\sqrt{3} \cdot \sqrt{8}$

32. $3 \cdot 2\sqrt{18}$

33. $5\sqrt{2} \cdot 3\sqrt{14}$

34. $9\sqrt{3} \cdot \sqrt{21}$

35. $\sqrt{30} \cdot \dfrac{\sqrt{3}}{3}$

36. $2\sqrt{22} \cdot 2\sqrt{2}$

37. $\sqrt{12} \cdot \sqrt{\dfrac{1}{3}}$

38. $\sqrt{\dfrac{2}{3}} \cdot \sqrt{24}$

39. $\sqrt{\dfrac{2}{3}} \cdot \sqrt{27}$

40. $\sqrt{16} \cdot \sqrt{\dfrac{3}{4}}$

41. $\sqrt{\dfrac{3}{2}} \cdot \sqrt{\dfrac{1}{2}}$

42. $\sqrt{\dfrac{2}{5}} \cdot \sqrt{\dfrac{4}{5}}$

43. $\sqrt{\dfrac{3}{5}} \cdot \sqrt{\dfrac{5}{3}}$

44. $\sqrt{\dfrac{4}{3}} \cdot \sqrt{\dfrac{3}{2}}$

45. $\sqrt{\dfrac{7}{9}} \cdot \sqrt{\dfrac{3}{4}}$

46. $\sqrt{\dfrac{3}{5}} \cdot \sqrt{\dfrac{18}{5}}$

47. $2\sqrt{x} \cdot 3\sqrt{x}$

48. $2\sqrt{2x} \cdot \sqrt{6x}$

49. $\sqrt{2x} \cdot \sqrt{10xy}$

50. $2\sqrt{6x^2y} \cdot \sqrt{3xy^2}$

51. $3\sqrt{x^3y} \cdot 2\sqrt{49x^3y^3}$

52. $2\sqrt{4x^5} \cdot 5\sqrt{3y^3}$

53. $(\sqrt{3})^2$

54. $(2\sqrt{5})^2$

55. $(-3\sqrt{2})^2$

56. $(2\sqrt{x})^2$

57. $(2\sqrt{2x})^2$

58. $(3\sqrt{2x})^2$

59. $2\sqrt{3}(\sqrt{2} - \sqrt{5})$

60. $\sqrt{2}(\sqrt{3} + \sqrt{5})$

61. $3\sqrt{2}(\sqrt{2} + 2\sqrt{8})$

62. $\sqrt{3}(2\sqrt{3} + \sqrt{27})$

63. $2\sqrt{5}(2\sqrt{15} + \sqrt{10} - 2\sqrt{60})$

64. $2\sqrt{8}(\sqrt{3} - \sqrt{2} + \sqrt{32})$

9-5   DIVISION OF RADICALS

A.   Radicands that Divide Evenly

We can use our second property of radicals, as presented in Section 9-2B, to divide radicals. That is,

$$\text{If} \qquad \sqrt[N]{\dfrac{a}{b}} = \dfrac{\sqrt[N]{a}}{\sqrt[N]{b}}$$

$$\text{Then} \qquad \dfrac{\sqrt[N]{a}}{\sqrt[N]{b}} = \sqrt[N]{\dfrac{a}{b}}$$

Thus, to divide the square root of a number by the square root of another number, we simply find the square root of the quotient of the numbers. For example, to divide $\sqrt{8}$ by $\sqrt{2}$:

First, write the radicands as a fraction (or quotient) under the radical sign and divide them.

$$\sqrt{8} \div \sqrt{2} = \frac{\sqrt{8}}{\sqrt{2}}$$

$$= \sqrt{\frac{8}{2}}$$

Now, we simplify.

$$= \sqrt{4}$$

$$= 2$$

Thus,

$$\sqrt{8} \div \sqrt{2} \text{ is } 2$$

If coefficients are involved, we simply divide the coefficients and then the radicals. Thus, we must perform two operations.

For example, to divide $6\sqrt{10}$ by $2\sqrt{5}$:

First, divide the coefficients.
Then, divide the radicals.

$$6\sqrt{10} \div 2\sqrt{5} = \frac{6\sqrt{10}}{2\sqrt{5}}$$

$$= \left(\frac{6}{2}\right)\left(\sqrt{\frac{10}{5}}\right)$$

$$= 3\sqrt{2}$$

Thus,

$$6\sqrt{10} \div 2\sqrt{5} \text{ is } 3\sqrt{2}$$

PROCEDURE:

> To divide radical expressions where the radicand of the denominator divides evenly into the radicand of the numerator:
>
> 1. Divide the coefficients.
>
> 2. Divide the radicands.
>
> 3. Simplify.

EXAMPLE 1: Divide $\sqrt{20}$ by $\sqrt{5}$.

SOLUTION:

$$\sqrt{20} \div \sqrt{5} = \frac{\sqrt{20}}{\sqrt{5}}$$

$$= \sqrt{4}$$

$$= 2$$

Final answer: 2

EXAMPLE 2:   Divide $15\sqrt{14}$ by $3\sqrt{2}$.

SOLUTION:

$$15\sqrt{14} \div 3\sqrt{2} = \frac{15\sqrt{14}}{3\sqrt{2}}$$

$$= \left(\frac{15}{3}\right)\left(\sqrt{\frac{14}{2}}\right)$$

$$= 5\sqrt{7}$$

Final answer: $5\sqrt{7}$

EXAMPLE 3:   Divide $9\sqrt{10}$ by $27\sqrt{5}$.

SOLUTION:

$$9\sqrt{10} \div 27\sqrt{5} = \frac{9\sqrt{10}}{27\sqrt{5}}$$

$$= \left(\frac{9}{27}\right)\left(\sqrt{\frac{10}{5}}\right)$$

$$= \left(\frac{1}{3}\right)(\sqrt{2})$$

$$= \frac{\sqrt{2}}{3}$$

Final answer: $\dfrac{\sqrt{2}}{3}$

EXAMPLE 4:   Divide $6\sqrt{27x^5}$ by $2\sqrt{3x^3}$.

SOLUTION:

$$6\sqrt{27x^5} \div 2\sqrt{3x^3} = \frac{6\sqrt{27x^5}}{2\sqrt{3x^3}}$$

$$= \left(\frac{6}{2}\right)\sqrt{\frac{27x^5}{3x^3}}$$

$$= 3\sqrt{9x^2}$$

$$= (3)(3x)$$

$$= 9x$$

Final answer: $9x$

Do the following practice set. Check your answers with the answers in the right-hand margin.

PRACTICE SET 9-5A

Divide and simplify (see example 1, 2, 3, or 4).

1. $\dfrac{4\sqrt{18}}{2\sqrt{2}}$

2. $\dfrac{3\sqrt{24}}{\sqrt{2}}$

3. $\dfrac{2\sqrt{54}}{6\sqrt{3}}$

4. $\dfrac{\sqrt{12xy^2}}{\sqrt{3xy}}$

1. 6
2. $6\sqrt{3}$
3. $\sqrt{2}$
4. $2\sqrt{y}$

B. Rationalizing the Denominator

Let us now consider the expression $\sqrt{3} \div \sqrt{2}$. As we stated earlier, it is perfectly acceptable to write the answer to this division as $\sqrt{3}/\sqrt{2}$. However, another equivalent form is more convenient if we want to approximate this value. This equivalent form can be obtained by multiplying both the numerator and denominator by $\sqrt{2}$.

$$\frac{\sqrt{3}}{\sqrt{2}} = \frac{\sqrt{3}}{\sqrt{2}} \cdot \frac{\sqrt{2}}{\sqrt{2}}$$

$$= \frac{\sqrt{(3)(2)}}{\sqrt{(2)(2)}} = \frac{\sqrt{6}}{\sqrt{4}} = \frac{\sqrt{6}}{2}$$

Thus,

$$\frac{\sqrt{3}}{\sqrt{2}} = \frac{\sqrt{6}}{2}$$

This procedure is called rationalizing the denominator. Note that $\sqrt{6}/2$ can be approximated more easily than $\sqrt{3}/\sqrt{2}$. (We only have one square root to deal with and do not have to divide by a "decimal.")

$$\frac{\sqrt{6}}{2} \approx \frac{2.449}{2} = 1.224$$

$$\frac{\sqrt{3}}{\sqrt{2}} \approx \frac{1.732}{1.414} = 1.224$$

The process of rationalizing the denominator, in effect, makes the radicand of the denominator a perfect square. Here is another example. Rationalize $1/\sqrt{3}$.

To make the radicand of the denominator a perfect square, $\dfrac{1}{\sqrt{3}}$

Multiply both the numerator and denominator by $\sqrt{3}$.

$$= \frac{1}{\sqrt{3}} \cdot \frac{\sqrt{3}}{\sqrt{3}}$$

$$= \frac{\sqrt{3}}{\sqrt{9}} = \frac{\sqrt{3}}{3}$$

Thus,

$$\frac{1}{\sqrt{3}} = \frac{\sqrt{3}}{3}$$

We can now use this technique to completely simplify radical expressions. (We do not consider a radical expression to be simplified when any radical appears in the denominator.) For example, to simplify $2\sqrt{\frac{75}{10}}$:

First, express the radical as the square root of the numerator over the square root of the denominator.

$$2\sqrt{\frac{75}{10}} = 2\frac{\sqrt{75}}{\sqrt{10}}$$

$$= \frac{2\sqrt{75}}{\sqrt{10}}$$

Simplify the radical.

$$\frac{2\sqrt{75}}{\sqrt{10}} = \frac{2\sqrt{25}\sqrt{3}}{\sqrt{10}}$$

$$= \frac{(2)(5)\sqrt{3}}{\sqrt{10}}$$

$$= \frac{10\sqrt{3}}{\sqrt{10}}$$

Since a radical is still in the denominator, we must rationalize the denominator.

$$\frac{10\sqrt{3}}{\sqrt{10}} = \frac{10\sqrt{3}}{\sqrt{10}} \cdot \frac{\sqrt{10}}{\sqrt{10}}$$

$$= \frac{10\sqrt{30}}{\sqrt{100}}$$

$$= \frac{10\sqrt{30}}{10}$$

$$= \sqrt{30}$$

Thus,

$$2\sqrt{\frac{75}{10}} = \sqrt{30}$$

PROCEDURE:

To rationalize the denominator of a radical expression:

1. Multiply the numerator and the denominator by the smallest radical that will make the denominator's radicand a perfect square.
2. Simplify the radicals.
3. Reduce the fraction.

EXAMPLE 1: Rationalize the denominator of $\dfrac{\sqrt{5}}{\sqrt{2}}$.

SOLUTION: To rationalize this denominator, we must multiply both the numerator and denominator by $\sqrt{2}$.

$$\frac{\sqrt{5}}{\sqrt{2}} = \frac{\sqrt{5}}{\sqrt{2}} \cdot \frac{\sqrt{2}}{\sqrt{2}}$$

$$= \frac{\sqrt{10}}{\sqrt{4}}$$

$$= \frac{\sqrt{10}}{2}$$

Final answer: $\dfrac{\sqrt{5}}{\sqrt{2}} = \dfrac{\sqrt{10}}{2}$

EXAMPLE 2: Rationalize the denominator of $\dfrac{4}{\sqrt{12}}$.

SOLUTION: Before we hastily multiply this fraction by $\sqrt{12}$, notice that we can first "clean up" $\sqrt{12}$.

$$\frac{4}{\sqrt{12}} = \frac{4}{\sqrt{4}\sqrt{3}}$$

$$= \frac{4}{2\sqrt{3}}$$

$$= \frac{2}{\sqrt{3}}$$

Now, we multiply the fraction by $\sqrt{3}$ to rationalize the denominator.

$$\frac{2}{\sqrt{3}} = \frac{2}{\sqrt{3}} \cdot \frac{\sqrt{3}}{\sqrt{3}}$$

$$= \frac{2\sqrt{3}}{\sqrt{9}}$$

$$= \frac{2\sqrt{3}}{3}$$

Final answer: $\dfrac{4}{\sqrt{12}} = \dfrac{2\sqrt{3}}{3}$

NOTE:

In example 2, we could have immediately multiplied the fraction by $\sqrt{12}$.

$$\frac{4}{\sqrt{12}} \cdot \frac{\sqrt{12}}{\sqrt{12}} = \frac{4\sqrt{12}}{\sqrt{144}}$$

However, when doing this, frequently we end up with very large numbers under the radical, which make our simplification process a bit more involved. Thus, the rule of thumb is to simplify all radicals before rationalizing a denominator.

EXAMPLE 3: Simplify $\sqrt{\dfrac{5}{8}}$.

SOLUTION: First, "clean up" the radical.

$$\sqrt{\frac{5}{8}} = \frac{\sqrt{5}}{\sqrt{8}}$$

$$= \frac{\sqrt{5}}{\sqrt{4}\sqrt{2}}$$

$$= \frac{\sqrt{5}}{2\sqrt{2}}$$

Since we still have a radical in the denominator, we rationalize the denominator.

$$\frac{\sqrt{5}}{2\sqrt{2}} = \frac{\sqrt{5}}{2\sqrt{2}} \cdot \frac{\sqrt{2}}{\sqrt{2}}$$

$$= \frac{\sqrt{10}}{2\sqrt{4}}$$

$$= \frac{\sqrt{10}}{(2)(2)}$$

$$= \frac{\sqrt{10}}{4}$$

Final answer: $\dfrac{\sqrt{10}}{4}$ or $\dfrac{1}{4}\sqrt{10}$

EXAMPLE 4: Simplify $\sqrt{\dfrac{7}{14}}$.

SOLUTION: Notice here that the fraction $\frac{7}{14}$ reduces to $\frac{1}{2}$.

$$\sqrt{\frac{7}{14}} = \sqrt{\frac{1}{2}}$$

So, instead of simplifying $\sqrt{\frac{7}{14}}$, we simplify $\sqrt{\frac{1}{2}}$.

$$\sqrt{\frac{1}{2}} = \frac{\sqrt{1}}{\sqrt{2}} = \frac{1}{\sqrt{2}}$$

$$(\text{rationalize}) = \frac{1}{\sqrt{2}} \cdot \frac{\sqrt{2}}{\sqrt{2}}$$

$$= \frac{\sqrt{2}}{\sqrt{4}} = \frac{\sqrt{2}}{2}$$

Final answer: $\dfrac{\sqrt{2}}{2}$

Do the following practice set. Check your answers with the answers in the right-hand margin.

PRACTICE SET 9-5B

Rationalize the denominators and simplify the resulting fraction (see example 1 or 2).

1. $\dfrac{1}{\sqrt{3}}$

2. $\dfrac{1}{\sqrt{5}}$

3. $\dfrac{2\sqrt{5}}{\sqrt{6}}$

4. $\dfrac{\sqrt{3}}{\sqrt{8}}$

5. $\dfrac{2\sqrt{16}}{\sqrt{18}}$

6. $\dfrac{5\sqrt{3}}{\sqrt{54}}$

Simplify the following expressions completely (see example 3 or 4).

7. $\sqrt{\dfrac{2}{6}}$

8. $\sqrt{\dfrac{3}{5}}$

9. $\dfrac{1}{6}\sqrt{\dfrac{30}{5}}$

10. $2\sqrt{\dfrac{8}{5}}$

11. $\sqrt{\dfrac{49}{50}}$

12. $4\sqrt{\dfrac{9}{8}}$

1. $\sqrt{3}/3$
2. $\sqrt{5}/5$
3. $\sqrt{30}/3$
4. $\sqrt{6}/4$
5. $4\sqrt{2}/3$
6. $5\sqrt{2}/6$

7. $\sqrt{3}/3$
8. $\sqrt{15}/5$
9. $\sqrt{6}/6$
10. $4\sqrt{10}/5$
11. $7\sqrt{2}/10$
12. $3\sqrt{2}$

EXERCISE 9-5

Divide the following radicals and simplify.

1. $\dfrac{\sqrt{27}}{\sqrt{3}}$

2. $\dfrac{4\sqrt{32}}{\sqrt{2}}$

3. $\dfrac{2\sqrt{125}}{5\sqrt{5}}$

4. $\dfrac{\sqrt{54}}{6\sqrt{6}}$

5. $\dfrac{2\sqrt{175}}{4\sqrt{7}}$

6. $\dfrac{4\sqrt{147}}{2\sqrt{3}}$

7. $\dfrac{2\sqrt{3}}{\sqrt{3}}$

8. $\dfrac{6\sqrt{2}}{2\sqrt{2}}$

9. $\dfrac{\sqrt{24}}{\sqrt{2}}$

10. $\dfrac{3\sqrt{6}}{\sqrt{3}}$

11. $\dfrac{\sqrt{24}}{2\sqrt{2}}$

12. $\dfrac{3\sqrt{24}}{2\sqrt{3}}$

13. $\dfrac{5\sqrt{40}}{2\sqrt{5}}$

14. $\dfrac{3\sqrt{36}}{\sqrt{2}}$

15. $\dfrac{\sqrt{5x^3}}{\sqrt{x}}$

16. $\dfrac{\sqrt{27x}}{\sqrt{3x}}$

17. $\dfrac{\sqrt{20x^2}}{\sqrt{5x^2}}$

18. $\dfrac{\sqrt{45x^3y}}{\sqrt{15xy}}$

19. $\dfrac{\sqrt{8xy^3}}{\sqrt{2y}}$

20. $\dfrac{\sqrt{36x^3y^3}}{\sqrt{4x^2y}}$

Rationalize the denominator and simplify the resulting fraction.

21. $\dfrac{\sqrt{1}}{\sqrt{3}}$

22. $\dfrac{\sqrt{3}}{\sqrt{6}}$

23. $\dfrac{\sqrt{2}}{\sqrt{3}}$

24. $\dfrac{3\sqrt{7}}{\sqrt{3}}$

25. $\dfrac{8\sqrt{3}}{8}$

26. $\dfrac{\sqrt{4}}{\sqrt{5}}$

27. $\dfrac{8\sqrt{2}}{2\sqrt{12}}$

28. $\dfrac{\sqrt{5}}{2\sqrt{2}}$

29. $\dfrac{\sqrt{2}}{\sqrt{8}}$

30. $\dfrac{\sqrt{6}}{\sqrt{12}}$

31. $\dfrac{1}{\sqrt{2}}$

32. $\dfrac{1}{\sqrt{3}}$

33. $\dfrac{1}{\sqrt{5}}$

34. $\dfrac{4}{\sqrt{2}}$

35. $\dfrac{5}{\sqrt{2}}$

36. $\dfrac{9}{\sqrt{3}}$

37. $\dfrac{9}{3\sqrt{3}}$

38. $\dfrac{12}{2\sqrt{3}}$

39. $\dfrac{4}{3\sqrt{2}}$

40. $\dfrac{4}{3\sqrt{8}}$

41. $\dfrac{7}{4\sqrt{7}}$

42. $\dfrac{14}{7\sqrt{2}}$

43. $\dfrac{5}{2\sqrt{15}}$

44. $\dfrac{20}{3\sqrt{20}}$

45. $\dfrac{20}{2\sqrt{5}}$

46. $\dfrac{4}{2\sqrt{6}}$

Simplify each of the following completely.

47. $2\sqrt{3} - 2\sqrt{\dfrac{3}{4}}$

48. $3\sqrt{8} - 4\sqrt{\dfrac{1}{2}}$

49. $30\sqrt{\dfrac{1}{3}} - 2\sqrt{27}$

50. $\sqrt{45} - 10\sqrt{\dfrac{4}{5}}$

51. $-\sqrt{20} + 15\sqrt{\dfrac{1}{5}}$

52. $\sqrt{24} - 3\sqrt{\dfrac{2}{3}}$

53. $\sqrt{90} + 5\sqrt{\dfrac{2}{5}}$

54. $\sqrt{\dfrac{3}{4}} + \sqrt{\dfrac{1}{3}}$

55. $4\sqrt{\dfrac{5}{2}} + 10\sqrt{\dfrac{1}{10}}$

56. $3\sqrt{\dfrac{2}{3}} - 4\sqrt{\dfrac{3}{2}}$

9-6 RADICAL EQUATIONS

Radical equations are equations in which the unknown (variable) is part of the radicand. For example, the following are radical equations:

Since x is found in the radicand
$$\begin{cases} \sqrt{x} = 4 \\ \sqrt{x+1} = 6 \\ 3\sqrt{2x+1} = 14 \end{cases}$$

To solve radical equations for the unknown, we transform our radical equation to a linear equation that we already know how to solve.

Earlier in this chapter, if we had a radical in the denominator of a fraction, we removed the radical by making the radicand a perfect square. In most cases, this was done by multiplying the radical by itself or squaring the radical. For example:

- The $\sqrt{2}$ multiplied by $\sqrt{2} = (\sqrt{2})^2$ or 2
- The $\sqrt{5}$ multiplied by $\sqrt{5} = (\sqrt{5})^2$ or 5
- The $\sqrt{x}$ multiplied by $\sqrt{x} = (\sqrt{x})^2$ or x
- The $\sqrt{x+1}$ multiplied by $\sqrt{x+1} = (\sqrt{x+1})^2 = x+1$

In summary, if we square a square root (radical), the answer is the radicand.

In our previous work with equations, we learned that our equality could be maintained by adding, subtracting, multiplying, or dividing each side by the same quantity. We are now expanding our operations to include squaring each side of an equation.

For example, we can solve the radical equation, $\sqrt{x} = 5$, by

Removing the radical as a result of squaring each side of the equation

$$\sqrt{x} = 5$$
$$(\sqrt{x})^2 = (5)^2$$
$$x = 25$$

Thus,

the solution of $\sqrt{x} = 5$ is $x = 25$.

The solution for radical equations, as well as for all other equations, should be checked.

Check:

$\sqrt{x}$	5
$\sqrt{25}$	5
5	

As a second example, consider the equation, $3\sqrt{x} - 6 = 9$. Before quickly squaring both sides, notice that if we square the left-hand side, we have to square the binomial $3\sqrt{x} - 6$.

$$3\sqrt{x} - 6 = 9$$
$$(3\sqrt{x} - 6)^2 = (9)^2$$

In doing this, we still do not eliminate the radical.

$$\begin{aligned}
(3\sqrt{x} - 6)^2 &= (3\sqrt{x} - 6)(3\sqrt{x} - 6) \\
&= (3\sqrt{x})(3\sqrt{x}) - (6)(3\sqrt{x}) - (6)(3\sqrt{x}) \quad (-6)(-6) \\
&= \quad 9\sqrt{x^2} \quad - \quad 18\sqrt{x} \quad - \quad 18\sqrt{x} \quad\quad + 36 \\
&= 9x - 36\sqrt{x} + 36
\end{aligned}$$

Thus, we have to square the equation again.

Notice further that we do *not* simply square each term of the equation; we must multiply each side of the equation as a quantity.

WRONG	RIGHT
$3\sqrt{x} - 6 = 9$	$3\sqrt{x} - 6 = 9$
$(3\sqrt{x})^2 - (6)^2 = (9)^2$	$(3\sqrt{x} - 6)^2 = (9)^2$

If the radical of an equation is to be removed by squaring, it will be extremely beneficial to get the radical term by itself on one side of the equation, before squaring.

$$3\sqrt{x} - 6 = 9$$
$$3\sqrt{x} = 9 + 6$$
$$3\sqrt{x} = 15$$

Now, square each side, and solve the equation.

$$3\sqrt{x} = 15$$
$$(3\sqrt{x})^2 = (15)^2$$
$$9x = 225$$
$$x = \frac{225}{9}$$
$$x = 25$$

Always check the results.

Check:

$3\sqrt{x} - 6$	9
$3\sqrt{25} - 6$	9
$3(5) - 6$	
$15 - 6$	
9	

Thus,

the solution of $3\sqrt{x} - 6 = 9$ is $x = 25$

Sometimes, when squaring both sides of an equation, the result we obtain is not a solution of the original equation. Such results are called extraneous solutions. For example, to solve $\sqrt{2x - 4} = -4$ for x:

Square both sides.

$$\sqrt{2x - 4} = -4$$
$$(\sqrt{2x - 4})^2 = (-4)^2$$
$$2x - 4 = 16$$

And solve the equation for x.

$$2x - 4 = 16$$
$$2x = 20$$
$$x = 10$$

However, when checking $x = 10$ into our original equation, we get a false statement.

$\sqrt{2x - 4}$	-4
$\sqrt{2(10) - 4}$	-4
$\sqrt{20 - 4}$	
$\sqrt{16}$	
4	
	$4 = -4$ (false)

Thus,

there is no solution.

Consequently, the check is an important part of solving radical equations.

PROCEDURE:

> To solve an equation containing radicals (square roots):
>
> 1. Isolate the term containing the radical (square root), and combine like terms.
> 2. Square each side of the equation.
> 3. Solve the resulting linear equation.
> 4. Check the solution in the original equation.

EXAMPLE 1: Solve $3\sqrt{x} = 6$.

SOLUTION: Since the radical term is already by itself, we simply square both sides and solve the equation.

$$3\sqrt{x} = 6$$
$$(3\sqrt{x})^2 = (6)^2$$
$$9x = 36$$
$$x = \frac{36}{9}$$
$$x = 4$$

Check:

$3\sqrt{x}$	6
$3\sqrt{4}$	6
$3(2)$	
6	

Final answer: $x = 4$

EXAMPLE 2: Solve $2\sqrt{x-1} = 4$.

SOLUTION: Since the radical term is already by itself, square both sides and solve the equation.

$$2\sqrt{x-1} = 4$$
$$(2\sqrt{x-1})^2 = (4)^2$$
$$4(x-1) = 16$$
$$4x - 4 = 16$$
$$4x = 16 + 4$$
$$4x = 20$$
$$x = \frac{20}{4}$$
$$x = 5$$

Check:

$$\begin{array}{c|c} 2\sqrt{x-1} & 4 \\ \hline 2\sqrt{(5)-1} & 4 \\ 2\sqrt{5-1} & \\ 2\sqrt{4} & \\ 2\,(2) & \\ \hline 4 & \end{array}$$

Final answer: $x = 5$

EXAMPLE 3:   Solve $2\sqrt{3x+4} + 2 = -8$.

SOLUTION: First, isolate the radical term.

$$2\sqrt{3x+4} + 2 = -8$$
$$2\sqrt{3x+4} = -8 - 2$$
$$2\sqrt{3x+4} = -10$$

Square both sides and solve the equation.

$$2\sqrt{3x+4} = -10$$
$$(2\sqrt{3x+4})^2 = (-10)^2$$
$$4(3x+4) = 100$$
$$12x + 16 = 100$$
$$12x = 100 - 16$$
$$12x = 84$$

$$x = \frac{84}{12}$$

$$x = 7$$

Check:

$$\begin{array}{c|c} 2\sqrt{3x+4} + 2 & -8 \\ \hline 2\sqrt{3(7)+4} + 2 & -8 \\ 2\sqrt{21+4} + 2 & \\ 2\sqrt{25} + 2 & \\ 2\,(5) + 2 & \\ 10 + 2 & \\ \hline 12 & \end{array}$$

Final answer: Since $12 \neq -8$, there is no solution

EXAMPLE 4:   Solve $3\sqrt{x} - 12 = -\sqrt{x}$.

SOLUTION: Isolate the radical term.

$$3\sqrt{x} - 12 = -\sqrt{x}$$
$$3\sqrt{x} + \sqrt{x} = 12$$
$$4\sqrt{x} = 12$$

Example continued on next page.

Square both sides and solve the equation.

$$4\sqrt{x} = 12$$
$$(4\sqrt{x})^2 = (12)^2$$
$$16x = 144$$
$$x = \frac{144}{16}$$
$$x = 9$$

Check:

$3\sqrt{x} - 12$	$-\sqrt{x}$
$3\sqrt{9} - 12$	$-\sqrt{9}$
$3(3) - 12$	-3
$9 - 12$	
-3	

Final answer: $x = 9$

EXAMPLE 5:   Solve the following problem.

Four times the square root of 5 times a number is 20. Find the number.

SOLUTION: First, translate this problem into an algebraic equation.

Four times   the square root of 5 times a number   is 20.

$$4 \cdot \qquad\qquad \sqrt{5x} \qquad\qquad = 20$$

or

$$4\sqrt{5x} = 20$$

Now, solve the equation.

$$4\sqrt{5x} = 20$$
$$(4\sqrt{5x})^2 = (20)^2$$
$$16(5x) = 400$$
$$80x = 400$$
$$x = \frac{400}{80}$$
$$x = 5$$

Check:

$4\sqrt{5x}$	20
$4\sqrt{(5)(5)}$	20
$4\sqrt{25}$	
$4(5)$	
20	

Final answer: $x = 5$

Do the following practice set. Check your answers with the answers in the right-hand margin.

PRACTICE SET 9-6

Solve and check the following radical equations (see example 1, 2, 3 or 4).

1. $\sqrt{2x} = 6$

2. $\sqrt{3x + 1} + 1 = 5$

3. $\sqrt{x} = 4 - \sqrt{x}$

4. $\sqrt{x - 1} = -3$

1. $x = 18$

2. $x = 5$

3. $x = 4$

4. No solution

Solve the following problems (see example 5).

5. One less than 3 times the square root of a number equals 1. Find the number.

6. The square root of 1 more than 4 times a number is 3. Find the number.

7. Three more than the square root of twice a number is -1. Find the number.

5. Number is $\frac{4}{9}$

6. Number is 2

7. No solution

EXERCISE 9-6

Solve and check the following radical equations.

1. $\sqrt{x} = 7$

2. $\sqrt{x + 1} = 2$

3. $\sqrt{x + 1} = 1$

4. $\sqrt{3x} = 6$

5. $\sqrt{2x - 3} = 3$

6. $\sqrt{2x} = -4$

7. $\sqrt{2x - 1} = -7$

8. $\sqrt{3a + 6} = 6$

9. $\sqrt{3x + 3} = 15$

10. $\sqrt{2x - 4} = 4$

11. $\sqrt{5x + 1} = -4$

12. $\sqrt{2x - 1} - 2 = 1$

13. $2\sqrt{3x} = 6$

14. $2\sqrt{x - 1} = 6$

15. $4\sqrt{2x + 3} = 12$

16. $2\sqrt{3x - 2} = 4$

17. $2\sqrt{2x - 6} + 2 = 6$

18. $4\sqrt{y + 1} = 8$

19. $3\sqrt{5x + 1} - 2 = 10$

20. $3\sqrt{3x - 3} + 1 = 10$

21. $\sqrt{x + 1} + 1 = -\sqrt{x + 1} + 5$

22. $2\sqrt{2x - 1} - 4 = -\sqrt{2x - 1} + 5$

23. The square root of 4 less than a number is 9. Find the number.

24. Three less than twice the square root of a number is 5. Find the number.

25. Twice the square root of 6 times a number less 6 is 12. Find the number.

26. The difference of the square root of 5 times a number less 1, and 3, is equal to 0.

27. The difference of 4 times the square root of 5 more than 5 times a number, and 5, is equal to 15.

REVIEW EXERCISES

Using the table of square roots, find the square roots of the following.

1. The $\sqrt{29}$ to the nearest tenth

2. The $\sqrt{130}$ to the nearest hundredth

Use the method of divide and average to find the square root of the following.

3. The $\sqrt{23}$ to the nearest tenth

4. The $\sqrt{73}$ to the nearest tenth

Simplify the following radicals.

5. $\sqrt{196}$

6. $\sqrt{54x^2}$

7. $\sqrt{112x^5y^7}$

8. $\sqrt{\dfrac{3}{7}}$

9. $3\sqrt{\dfrac{3}{6}}$

10. $2\sqrt{\dfrac{8}{5}}$

Combine the following radicals.

11. $3\sqrt{5} + \sqrt{5} - 2\sqrt{5}$

12. $4\sqrt{2} + 3\sqrt{18}$

13. $4\sqrt{27} - \sqrt{75}$

14. $\sqrt{8} - 2\sqrt{\dfrac{1}{2}}$

15. $\sqrt{24} - 9\sqrt{\dfrac{2}{3}}$

Multiply the following radicals.

16. $2\sqrt{3} \cdot 3\sqrt{6}$

17. $3\sqrt{2} \cdot 7\sqrt{3} \cdot \sqrt{8}$

18. $\sqrt{\dfrac{2}{3}} \cdot \sqrt{\dfrac{1}{4}}$

19. $(2\sqrt{5})^2$

Divide the following radicals.

20. $\dfrac{\sqrt{20}}{\sqrt{5}}$

21. $\dfrac{\sqrt{2}}{\sqrt{8}}$

22. $\dfrac{4\sqrt{3}}{\sqrt{2}}$

23. $\dfrac{3}{\sqrt{3}}$

Solve the following radical equations.

24. $\sqrt{x-1} = 5$

25. $\sqrt{x+2} + 4 = 2$

26. Six less than the square root of a number is 2. Find the number.

CHAPTER 10

Linear Equations and Inequalities in Two Variables

10-1 THE RECTANGULAR COORDINATE SYSTEM

In Section 1-1, we introduced the number line. We learned that whenever a number represents a point on the number line, the number is commonly referred to as the **coordinate** of that point. For example, on the following number line:

■ The coordinate of point A is 0.

■ The coordinate of point B is $\frac{5}{2}$.

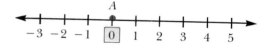

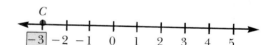

■ The coordinate of point C is -3.

We now draw *two* number lines, one horizontal and one vertical, and let them cross at their respective zero points. These two lines are referred to as the *axes of a rectangular coordinate system.*

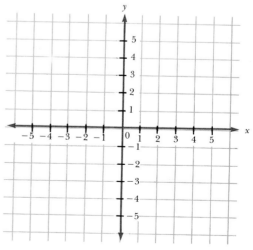

In the rectangular coordinate system, the horizontal number line represents the horizontal axis and is called the *x* axis, the vertical number line represents the vertical axis and is called the *y* axis, and the point where the *x* axis and *y* axis cross is called the origin.

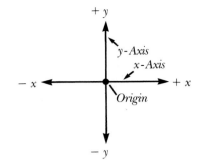

The *x* and *y* axes divide this *coordinate plane* into four regions. These regions are called *quadrants* and are numbered in a counterclockwise order, using Roman numerals.

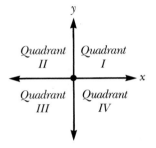

Since we now have two lines, the *coordinates of a point* is represented by a pair of numbers. For example:

The coordinates of point *A*, may be represented by the pair of numbers (4, 6).

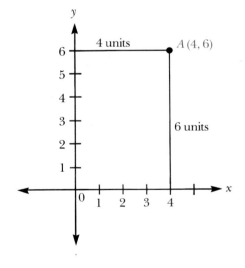

The coordinates of a point that is given by a pair of numbers is called an *ordered pair*. Each ordered pair is given in the form (x, y), where *x* and *y* are enclosed within parentheses.

The first number of an ordered pair (x, y) represents movement along the *x* axis. That is, we move left, right, or not at all.

■   If the *first* number is *positive*, we move right.

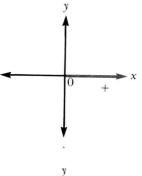

■   If the *first* number is *negative*, we move left.

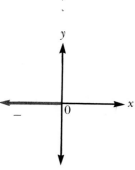

■   If the *first* number is zero, then we do not move at all.

The second number of an ordered pair, (x, y), represents movement along the y axis. That is, we move up, down, or not at all.

■   If the *second* number is *positive*, we move up.

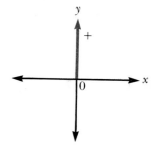

■   If the *second* number is *negative*, we move down.

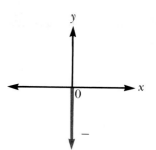

■   If the *second* number is zero, we do not move at all.

Thus, to locate the point with coordinates $(3, 5)$:

First, start at the origin and move 3 units to the right. The first number is positive (3 , 5).

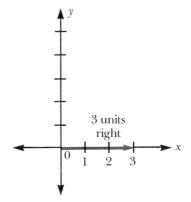

Now, from that spot, move 5 units up. The second number is positive (3, 5).

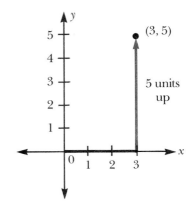

The point $(3, 5)$ is in quadrant I.

To locate the point with coordinates $(-4, 2)$:

First, start at the origin and move 4 units to the left. The first number is negative (−4 , 2).

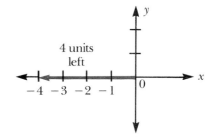

Now, move 2 units up. The second number is positive $(-4, 2)$.

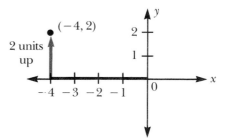

The point $(-4, 2)$ is in quadrant II.

Notice that in both the preceding examples, to locate a point, given its coordinates, we always begin at the origin and move first either right or left (depending on if the first number is positive or negative). From that spot, we move up or down (depending on if the second number is positive or negative). Thus, the *order* in which the numbers are written is important, hence the name, ordered pair.

PROCEDURE:

To graph the point with coordinates (x, y):

1. Begin at the origin and move "x" units along the x axis.
    a. If x is positive, move right.
    b. If x is negative, move left.
    c. If x is zero, make no horizontal movement.

2. From the location arrived at in (1), move "y" units along the y axis.
    a. If y is positive, move up.
    b. If y is negative, move down.
    c. If y is zero, make no vertical movement.
    Plot the point at this location.

EXAMPLE 1:   Graph the point $(1, -3)$.

SOLUTION: Beginning at the origin, we move 1 unit to the right.

Now, we move 3 units down.

Thus,

$(1, -3)$ is in the fourth quadrant

EXAMPLE 2: Graph the point $(-4, -6)$.

SOLUTION: Beginning at the origin, move 4 units to the left.

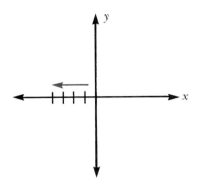

Now, move 6 units down.

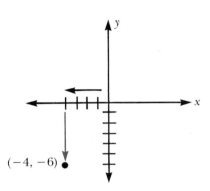

Thus,

$(-4, -6)$ is in the third quadrant.

When graphing points in the x/y plane, if the x value of the ordered pair is zero, then the graph of the point will be on the y axis. And if the y value of the ordered pair is zero, then the graph of the point will be on the x axis. Consider the two examples that follow.

EXAMPLE 3: Graph the point $(0, 3)$.

SOLUTION: Since the first number is zero, we move neither left nor right; we simply move 3 units up from the origin.

Thus,

$(0, 3)$ is on the y axis.

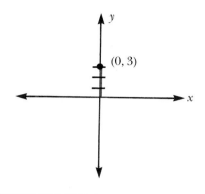

EXAMPLE 4: Graph the point $(-4, 0)$.

SOLUTION: Since the second number is zero, we move neither up nor down; we simply move 4 units left from the origin.
Thus,

$(-4, 0)$ is on the x axis.

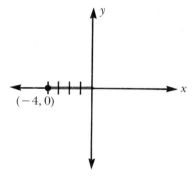

We can also determine the coordinates of a point, given its graph.

EXAMPLE 5: What are the coordinates of point A in the following figure?

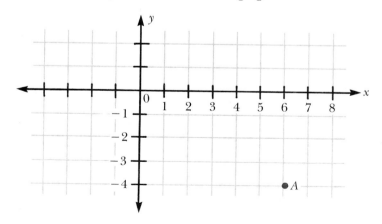

Example continued on next page.

SOLUTION: Since we always move either left or right first, the only way to get to point A is to move 6 units to the right.
Thus,

our x value is 6.

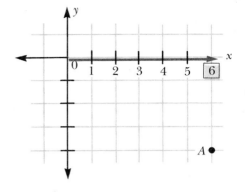

Now, from this spot, we must move 4 units down.
Thus,

our y value is -4.

Final answer: $(6, -4)$

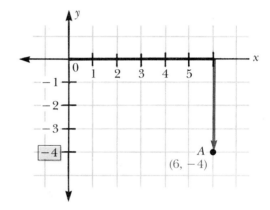

EXAMPLE 6: What are the coordinates of point B in the following figure?

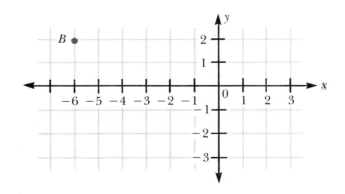

SOLUTION: To get to point B (starting from the origin), first, move 6 units to the left. Thus,

our x value is -6.

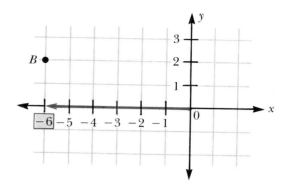

Now, from this spot, move 2 units up. Thus,

our y value is 2.

Final answer: $(-6, 2)$

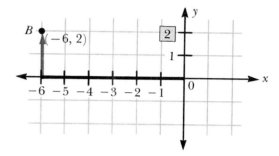

Do the following practice set. Check your answers with the answers in the right-hand margin.

PRACTICE SET 10-1

Graph each of the following points and state either the quadrant in which the point is located or the axis on which the point is located (see example 1, 2, 3, or 4).

1. $(5, 3)$

2. $(-3, 2)$

3. $(0, -5)$

4. $(-3, -2)$

5. $(2, -1)$

6. $(4, 0)$

1.

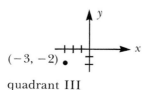

quadrant I

2.

quadrant II

3.

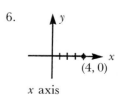

y axis

4.

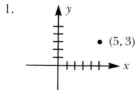

quadrant III

5.

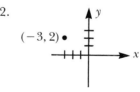

quadrant IV

6.

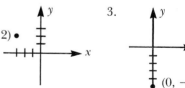

x axis

Using the following graph, state the coordinates for each of the following points (see example 5).

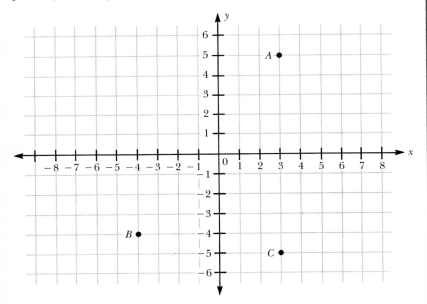

7. *A*	8. *B*	9. *C*	7. $(3, 5)$
			8. $(-4, -4)$
			9. $(3, -5)$

EXERCISE 10-1

Label the following points on the given set of axes.

1. $A(5, -1)$
2. $B(-2, 3)$
3. $C(-3, 2)$
4. $D(1, 3)$
5. $E(0, -4)$
6. $F(-1, 4)$
7. $G(-3, -2)$
8. $H(-2, -2)$
9. $I(-3, 0)$
10. $J(-5, 4)$

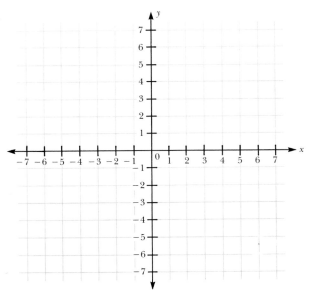

Plot each set of ordered pairs on the axes given.

11. (0, 0)
    (2, 3)
    (−2, −4)
    (4, 3)
    (−1, −1)

12. (−3, −2)
    (3, −1)
    (1, 2)
    (2, 4)
    (0, −4)

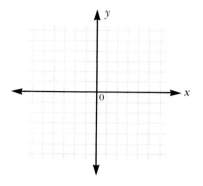

Using the following graph, name the coordinates of each of the following points.

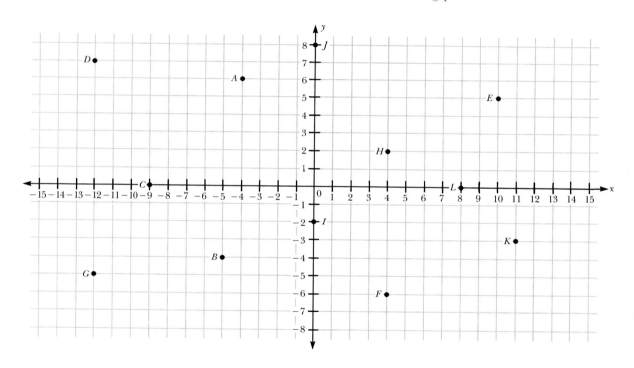

13. A   14. B   15. C   16. D   17. E   18. F
19. G   20. H   21. I   22. J   23. K   24. L

25. If a point is on the y axis, what is the value of the x coordinate?

26. What is the y value for every point that lies on the x axis?

27. Assume x to be negative and y to be negative. Name the quadrant in which the following points lie.
    a. (x, −y)   b. (−x, −y)

On a sheet of graph paper, draw a pair of coordinate axes. Plot each set of the following points on your paper, in the order given, from left to right. After you plot a point, connect it to the previously plotted point by a straight line. If all points are plotted (and connected) correctly, a picture will appear.

28. $(-6, 1)$ $(-3, 3)$ $(-1, 4)$ $(-6, 6)$ $(-8, 9)$ $(-9, 11)$ $(-10, 15)$ $(-7, 13)$ $(-4, 10)$ $(0, 6)$ $(1, 4)$ $(2, 7)$
$(3, 11)$ $(6, 14)$ $(7, 10)$ $(7, 6)$ $(4, 4)$ $(5, 1)$ $(5, -3)$ $(4, -5)$ $(0, -7)$ $(0, -5)$ $(-3, -7)$ $(-2, -5)$
$(-6, -3)$ $(-7, -1)$

Lift pencil and start fresh.

$(-4, 0)$ $(-4, 1)$ $(-2, 1)$ $(-2, 0)$

Lift pencil and start fresh.

$(-4, -9)$ $(-4, -7)$ $(-1, -10)$ $(-1, -7)$ $(-4, -9)$

Lift pencil and start fresh.

$(0, -9)$ $(0, -8)$ $(4, -7)$ $(4, -8)$ $(0, -9)$

29. $(1, 4)$ $(1, 1)$ $(2, 0)$ $(4, -5)$ $(6, -7)$ $(4, -6)$ $(3, -7)$ $(2, -7)$ $(2, -8)$ $(3, -8)$ $(4, -9)$ $(-1, -9)$
$(0, -8)$ $(1, -8)$ $(1, -7)$ $(1, -8)$ $(-2, -8)$ $(-1, -7)$ $(0, -7)$ $(0, -6)$ $(-2, -4)$ $(-2, -2)$
$(-1, 0)$ $(-1, 3)$ $(-2, 4)$ $(-3, 3)$ $(-9, 4)$ $(-10, 3)$ $(-11, 4)$ $(-10, 5)$ $(-9, 4)$ $(-10, 5)$ $(-10, 7)$
$(-9, 8)$ $(-3, 9)$ $(-2, 10)$ $(1, 10)$ $(3, 9)$ $(4, 7)$ $(4, 4)$ $(3, 3)$ $(2, 3)$ $(1, 4)$ $(0, 6)$ $(0, 8)$

Lift pencil and start fresh.

$(-4, 8)$ $(-3, 7)$

10-2 FINDING ORDERED PAIRS THAT SATISFY EQUATIONS IN TWO VARIABLES

Equations that involve two variables, such as x and y, are called equations in two variables. For example:

$$y = 3x - 8 \qquad x + y = 16 \qquad \text{and} \qquad 3x - 4y - 20 = 0$$

are all considered equations in two variables.

There are many ordered pairs of numbers that will make an equation in two variables true. Such ordered pairs are said to satisfy the equation and are called solutions of the equation. For example:

The ordered pair $(2, 5)$ satisfies the equation $y = 2x + 1$, since we get a true statement when we substitute 2 for x and 5 for y in the equation.

$$y = 2x + 1$$
$$5 \overset{?}{=} 2(2) + 1$$
$$5 \overset{?}{=} 4 + 1$$
$$5 = 5 \text{ (true)}$$

NOTE:

If we reverse the numbers in the ordered pair $(2, 5)$, we get a different ordered pair $(5, 2)$. Notice that $(5, 2)$ does not satisfy the equation $y = 2x + 1$.

$$y = 2x + 1$$
$$2 \overset{?}{=} 2(5) + 1$$
$$2 \overset{?}{=} 10 + 1$$
$$2 = 11 \text{ (false)}$$

Thus, it is imperative that we maintain the proper order of the numbers of an ordered pair.

In our work with equations in two variables, it is necessary to find ordered pairs of numbers that satisfy a given equation. One way to do this is to substitute any value for x into the equation and then solve the equation for y. For example:

In the equation $3x + y = 10$, by letting $x = 1$, we can find a corresponding value for y by simply solving the equation for y.

$$3x + y = 10$$
$$3(1) + y = 10$$
$$3 + y = 10$$
$$y = 7$$

Thus, the ordered pair $(1, 7)$ satisfies the equation $3x + y = 10$ and can be checked by substituting back into the given equation.

$3x + y$	10
$3(1) + (7)$	10
$3 + 7$	10
10	10 (true)

Notice that *we* determined what we would like x to be. No one gave us a value to try. By arbitrarily selecting any value for x and then finding its corresponding y value, we are able to generate an infinite number of ordered pairs that satisfy a given equation. (Note: We may also select a value for y, then solve for x.)

PROCEDURE:

To find an ordered pair that satisfies an equation in two variables:

1. Select a value for x.
2. Substitute this value for x into the given equation.
3. Solve the equation for y.
4. Check to see that the ordered pair (x, y) satisfies the given equation.

EXAMPLE 1: Find an ordered pair that satisfies the equation $y = 2x + 2$. (Let $x = 3$.)

Example continued on next page.

SOLUTION: Substituting 3 for x and then solving for y, we get a corresponding y value of 8.

$$y = 2x + 2$$
$$y = 2(3) + 2$$
$$y = 6 + 2$$
$$y = 8$$

The ordered pair $(3, 8)$ can now be checked to see if it indeed satisfies the equation.

y	$2x + 2$
8	$2(3) + 2$
8	$6 + 2$
8	8 (true)

Final answer: $(3, 8)$

EXAMPLE 2:   Find two ordered pairs that satisfy the equation $3x - y + 4 = 0$. (Let $x = 0$ and $x = 4$.)

SOLUTION:

Substituting 0 for x	Substituting 4 for x
$3x - y + 4 = 0$	$3x - y + 4 = 0$
$3(0) - y + 4 = 0$	$3(4) - y + 4 = 0$
$0 - y + 4 = 0$	$12 - y + 4 = 0$
$-y + 4 = 0$	$16 - y = 0$
$4 = y$	$16 = y$
Thus, $(0, 4)$ is one ordered pair.	Thus, $(4, 16)$ is a second ordered pair.

The check is left to the student.

Final answer: $(0, 4)$ and $(4, 16)$

Do the following practice set. Check your answers with the answers in the right-hand margin.

PRACTICE SET 10-2

State whether the given ordered pair satisfies the given equation.

1. $y = 3x - 6$, $(2, 0)$ 1.   Yes

2. $2x + y = 16$, $(4, 8)$ 2.   Yes

3. $4x - y + 8 = 0$, $(-1, 12)$ 3.   No

Find an ordered pair(s) that satisfies the given equation. Use the given value for x (see example 1 or 2).

4. $y = 2x - 1$, (Let $x = 0$)	4. $(0, -1)$
5. $3x = y + 4$, (Let $x = 2$)	5. $(2, 2)$
6. $5x + y = 9$, (Let $x = -2$ and $x = 1$)	6. $(-2, 19)$ $(1, 4)$

EXERCISE 10-2

State whether the given ordered pair satisfies the given equation. Answer yes or no.

1. $y = 2x$, $(3, 6)$ 2. $y = -5x$, $(10, 2)$

3. $y = 4x - 3$, $(6, 27)$ 4. $y = 3x + 1$, $(4, 13)$

5. $2x + 3y = 12$, $(5, 1)$ 6. $5x - 8y = 17$, $(3, -1)$

7. $3x - 2y = 0$, $(4, 6)$ 8. $3y = 2x - 4$, $(3, 1)$

9. $3x = y - 4$, $(-7, -2)$ 10. $x - 4y = 16$, $(0, -4)$

Given the x value, find an ordered pair of numbers that will satisfy each of the following equations.

11. $y = x + 4$, $(x = 3)$ 12. $y = x - 7$, $(x = 2)$

13. $y = 2x - 6$, $(x = 4)$ 14. $y = 3x + 5$, $(x = -3)$

15. $2y = x - 5$, $(x = 1)$ 16. $5y = 2x + 1$, $(x = 7)$

17. $x - y = 0$, $(x = -8)$ 18. $x + 3y = 0$, $(x = 0)$

19. $2x + 3y = 8$, $(x = 7)$ 20. $3x - 2y = 5$, $(x = 3)$

21. $-2x - 3y = 5$, $(x = -1)$ 22. $-x + 2y = -3$, $(x = 5)$

23. $y = 3$, $(x = 5)$ 24. $y = -4$, $(x = 12)$

Find two ordered pairs for each of the following equations by arbitrarily choosing *any* x value.

25. $y = 2x + 1$ 26. $y = 3x + 5$ 27. $x + 2y = 6$

28. $3x - y = 8$ 29. $4x - 3y = 0$ 30. $y = x$

10-3 GRAPHING STRAIGHT LINES

A. Graphing the Equation's Solution

Given the equation $y = -3x + 10$, if we let x take on the values $-1, 0, 1, 2, 3$, and 4, we generate six corresponding y values and, ultimately, six ordered pairs. For example:

If $x = -1$,	If $x = 0$,	If $x = 1$,
then $y = -3(-1) + 10$	then $y = -3(0) + 10$	then $y = -3(1) + 10$
$y = 3 + 10$	$y = 0 + 10$	$y = -3 + 10$
$y = 13$	$y = 10$	$y = 7$

If $x = 2$,	If $x = 3$,	If $x = 4$,
then $y = -3(2) + 10$	then $y = -3(3) + 10$	then $y = -3(4) + 10$
$y = -6 + 10$	$y = -9 + 10$	$y = -12 + 10$
$y = 4$	$y = 1$	$y = -2$

Thus,

$(-1, 13)$, $(0, 10)$, $(1, 7)$, $(2, 4)$, $(3, 1)$, and $(4, -2)$ all satisfy the equation $y = -3x + 10$.

If we plot these points on the rectangular coordinate axes, notice that they all appear to line up on a straight line.

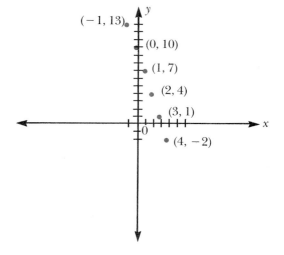

This is true! All ordered pairs that satisfy the equation $y = -3x + 10$ lie on this line and all ordered pairs that do not satisfy the equation $y = -3x + 10$ do not lie on this line. Thus,

this line is called the *graph* of the equation $y = -3x + 10$

In general, the graph of any equation that is of the form

$$Ax + By = C$$

where A and B both are not zero, is a straight line and is called a **linear equation.**

To graph a straight line, then, we first choose values for x and find the corresponding y value for each chosen x value. Next, we plot the resulting ordered pairs, and, finally, we draw a straight line through the points.

NOTE:

A straight line may be determined by plotting only two points that satisfy the equation. However, it is good practice to choose a third point to serve as a check; if the third point also lies on the line, then we are reasonably assured that the graph is correct.

PROCEDURE:

> To graph a straight line:
>
> 1.   Select a minimum of three values for x.
>
> 2.   Determine the corresponding y value for each x value selected.
>
> 3.   Plot each resulting ordered pair on the same set of axes.
>
> 4.   Draw the straight line through the points.

EXAMPLE 1:   Graph the equation $y = 2x - 1$.

SOLUTION: In choosing our x values, we use $x = 0$, 1, and 2.

If $x = 0$	If $x = 1$	If $x = 2$
then $y = 2(0) - 1$	then $y = 2(1) - 1$	then $y = 2(2) - 1$
$y = 0 - 1$	$y = 2 - 1$	$y = 4 - 1$
$y = -1$	$y = 1$	$y = 3$
Thus,	Thus,	Thus,
$(0, -1)$	$(1, 1)$	$(2, 3)$

Plotting these points and drawing a straight line through them will result in the graph of the equation $y = 2x - 1$.

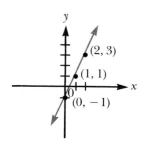

EXAMPLE 2:   Graph the equation $2x + y = 4$.

Example continued on next page.

SOLUTION: If we choose values for x to be $x = 2$, $x = 4$, and $x = -2$, we get the ordered pairs $(2, 0)$, $(4, -4)$, and $(-2, 8)$. Check these ordered pairs. Plotting and drawing the line through these points will give us the correct graph.

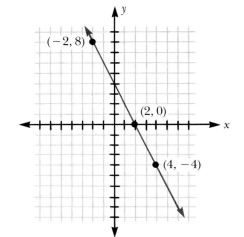

EXAMPLE 3: Graph the equation $3x + 4y = 12$.

SOLUTION: Choosing $x = 0$, -4, and 4, we get the ordered pairs $(0, 3)$, $(-4, 6)$, and $(4, 0)$. The graph looks as follows.

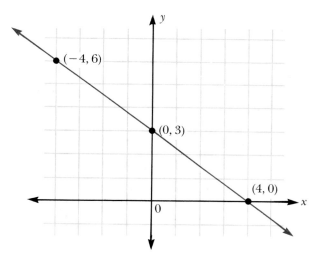

Do the following practice set. Check your answers with the answers in the right-hand margin. (Note: Although the ordered pairs used to graph the lines in the answers may be different from yours, the lines should still be the same.)

PRACTICE SET 10-3A

Graph each of the following equations (see example 1, 2, or 3).

1. $y = 2x$

2. $y = x + 4$

3. $y = 3 - x$

4. $3x + y = 5$

5. $x - y = 3$

6. $y - 2x = -1$

7. $x - 2y = 0$

8. $3x - 2y = 6$

9. $2x + 3y = 6$

1.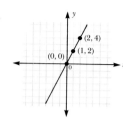

2.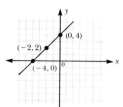

3.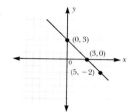

4.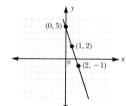

5.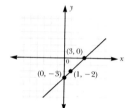

6.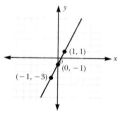

7.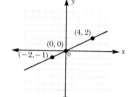

8.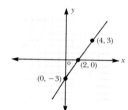

9.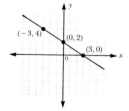

B.   Graphing a Linear Equation by Its Intercepts

The point where a line crosses the x axis is called the **x intercept** and the point where a line crosses the y axis is called the **y intercept.** For example:

The graph of the line $y = x + 4$ crosses the x axis at the point $(-4, 0)$ and the line crosses the y axis at the point $(0, 4)$.

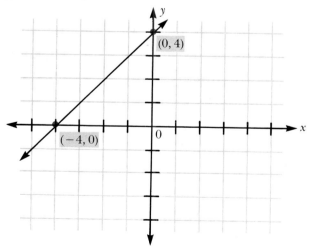

Thus, for the equation, $y = x + 4$, the x intercept is $(-4, 0)$ and the y intercept is $(0, 4)$.

Notice for the x intercept, the y value is zero. $(-4, \boxed{0})$

And for the y intercept, the x value is zero. $(\boxed{0}, 4)$

When graphing the equation of a straight line, we often find it easier to locate the intercepts of an equation and plot these points than to choose values for x and solve for the corresponding values of y (as was done in part A).

To find the intercepts of an equation, we simply substitute zero into the equation on two different occasions: once for x and once for y. For example, in the equation, $2x + 3y = 6$, to find the x intercept:

Let $y = 0$ and solve for x. Thus, $(3, 0)$ is the x intercept.

$$2x + 3y = 6$$
$$2x + 3(0) = 6$$
$$2x + 0 = 6$$
$$2x = 6$$
$$x = 3$$

To find the y intercept:

Let $x = 0$ and solve for y. Thus, $(0, 2)$ is the y intercept.

$$2x + 3y = 6$$
$$2(0) + 3y = 6$$
$$0 + 3y = 6$$
$$3y = 6$$
$$y = 2$$

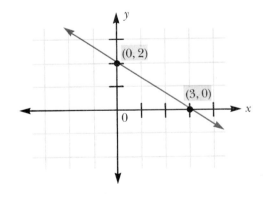

Plotting these two points and drawing a straight line through them produces the graph of $2x + 3y = 6$.

NOTE:

Although two points are enough to determine a straight line, a third point should be used to serve as a check.

PROCEDURE:

To graph a linear equation by its intercepts:

1. Find the x intercept by setting y equal to zero in the equation and solving for x.
2. Find the y intercept by setting x equal to zero in the equation and solving for y.
3. Plot the points on the same set of axes and draw the straight line through them.

EXAMPLE 1:   Graph $2x - y = 8$, using the intercept method.

SOLUTION:

If $\quad\quad x = 0$,	If $\quad\quad y = 0$
then $2(0) - y = 8$	then $2x - 0 = 8$
$0 - y = 8$	$2x = 8$
$-y = 8$	$x = 4$
$y = -8$	

Thus,

the intercepts are $(0, -8)$ and $(4, 0)$

We get the following line.

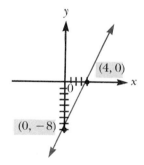

EXAMPLE 2:   Graph $3x + 4y = -12$, using the intercept method.

SOLUTION:

If $\quad\quad x = 0$	If $\quad\quad y = 0$
then $3(0) + 4y = -12$	then $3x + 4(0) = -12$
$0 + 4y = -12$	$3x + 0 = -12$
$4y = -12$	$3x = -12$
$y = -3$	$x = -4$

Example continued on next page.

Thus,

the intercepts are $(0, -3)$ and $(-4, 0)$

Plotting, we get

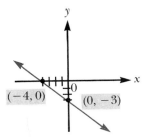

Do the following practice set. Check your answers with the answers in the right-hand margin.

PRACTICE SET 10-3B

Graph, using the intercept method (see example 1 or 2).

1. $y = 2x + 4$

2. $x - y = 5$

3. $4x + 3y = 19$

1.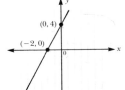

2.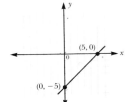

3.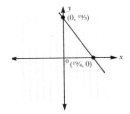

C.   Graphing Special Equations (Equations with Only One Variable)

In part A, we learned that, in general, a linear equation is of the form $Ax + By = C$.

If we let $A = 0$ or $B = 0$ (but not both), we get two special types of straight lines. The equations of these lines will have only one variable and the graphs of such equations are either vertical or horizontal lines.

HORIZONTAL LINES (LINES PARALLEL TO THE *x* AXIS)

Consider the equation $y = 4$. Although this equation is missing the x variable, we may still write $y = 4$ with an x variable as $y = 0 \cdot x + 4$. Notice that no matter what value for x we substitute into the equation, y will always be 4.

If $x = -1$	If $x = 0$	If $x = 18$	
then $y = 0(-1) + 4$	then $y = 0(0) + 4$	then $y = 0(18) + 4$	and so forth
$y = 0 + 4$	$y = 0 + 4$	$y = 0 + 4$	
$y = 4$	$y = 4$	$y = 4$	

Thus,
(−1, 4)

Thus,
(0, 4)

Thus,
(18, 4)

Since y always equals 4 for any given x value, the graph of $y = 4$ is a horizontal line that crosses the y axis at $(0, 4)$.

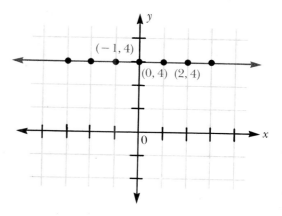

In general, the graph of $y = a$ is a horizontal line that crosses the y axis at $(0, a)$.

VERTICAL LINES (LINES PARALLEL TO THE y AXIS)

Consider the equation $x = 3$. Although this equation is missing the y variable, we may still write $x = 3$ with a y variable as $x = 0 \cdot y + 3$. Notice that no matter what value for y we substitute into the equation, x will always be 3. (Be careful in the ordering of x and y.)

If $y = -1$
then $x = 0(-1) + 3$
$x = 0 + 3$
$x = 3$

Thus,
(3, −1)

If $y = 0$
then $x = 0(0) + 3$
$x = 0 + 3$
$x = 3$

Thus,
(3, 0)

If $y = 12$
then $x = 0(12) + 3$
$x = 0 + 3$
$x = 3$

Thus,
(3, 12)

and so forth

Since x will always be equal to 3, for any given y value, the graph of $x = 3$ will be a vertical line that crosses the x axis at $(3, 0)$.

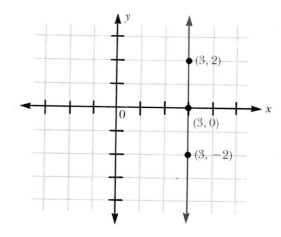

In general, the graph of $x = a$ is a vertical line that crosses the y axis at $(a, 0)$.

Do the following practice set. Check your answers with the answers in the right-hand margin.

PRACTICE SET 10-3C

Graph the following equations.

1. $x = -2$

2. $y + 3 = 0$

3. $x - 4 = 0$

1.

2.

3.

EXERCISE 10-3

Graph each of the following equations by arbitrarily selecting a minimum of three points.

1. $y = 3x$
2. $y = -2x$
3. $y = x + 4$
4. $y = 2x + 1$

5. $y = -2x - 1$
6. $y = 5x - 3$
7. $3y = x - 9$
8. $2y = 6x + 4$

9. $y + x = 3$
10. $y - x = -3$
11. $x + 2y = 12$
12. $x - 3y = 0$

13. $2x + 4y = 8$
14. $2x + 3y = -6$
15. $3y - 2x = 12$
16. $4y - 6x = 24$

17. $y = \frac{1}{2}x$
18. $y = \frac{2}{3}x$
19. $y = -\frac{3}{2}x + 1$
20. $y = -\frac{1}{4}x - 2$

Graph each of the following equations by the intercept method.

21. $x + y = 3$
22. $x - y = 5$
23. $y = 2x + 4$
24. $y = -3x - 9$

25. $x = 3y - 6$
26. $x = 2y + 8$
27. $3x + 2y = 12$
28. $2x - 3y = 6$

29. $5x - 3y = 15$

30. $4x + 5y = 10$

Graph each of the following special equations.

31. $x = 3$
32. $y = -4$
33. $y = 7$
34. $x = -5$
35. $y = 0$

36. $x = 0$
37. $y = \frac{3}{5}$
38. $x = \frac{1}{2}$
39. $2x - 1 = 0$
40. $4y = -3$

10-4   THE SLOPE OF A LINE

A.   Finding the Slope of a Line

The slope of a line may be thought of as the measure of the *steepness* of the line. For example, if we think of a line as a hill, then we can see that:

The steepness of this line . . .

Appears to be greater than the steepness of this line . . . .

We can express the slope (or steepness) of a line between two points as a ratio of rise (or height), divided by run.

Rise represents the change in vertical distance.

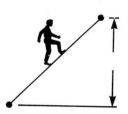

Run represents the change in horizontal distance.

For example:

■ The slope of the line at the right may be expressed as the ratio $\frac{1}{1}$. Thus,

the slope is equal to 1

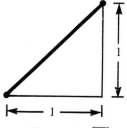

$$\text{Slope} = \frac{1}{1} = \boxed{1}$$

■ The slope of this next line may be expressed as the ratio $\frac{2}{1}$. Thus,

the slope of this line is equal to 2

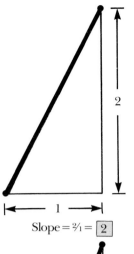

Slope $= \frac{2}{1} = \boxed{2}$

Notice that if we maintain a constant length, then the greater the height is, the steeper the line is and, consequently, the greater the slope is.

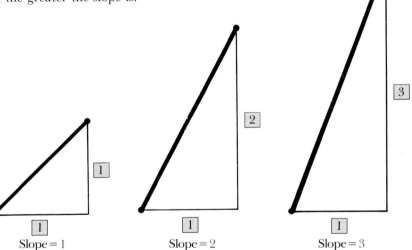

Slope $= 1$ Slope $= 2$ Slope $= 3$

This idea of slope (i.e., slope is equal to the ratio of rise to the run) may be extended to that of lines that are graphed in the coordinate system. For example, consider the line, $3x - 4y = -1$.

The graph of this line looks like the graph shown here.

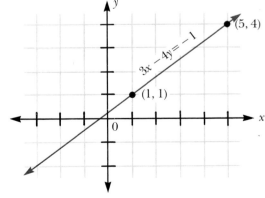

To find the slope of this line, we again find the ratio of rise to the run.

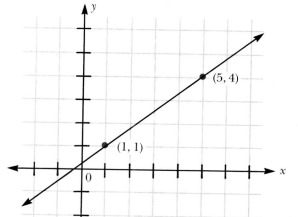

First, we select any two points on the line. [We choose (1, 1) and (5, 4).]

■ The rise or height, is considered the vertical distance between these two points. (This distance represents the change in the y values and is equal to 3.)

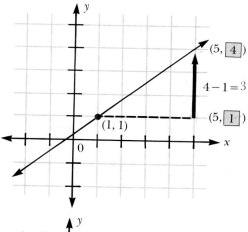

■ The run is considered the horizontal distance between these two points. (This distance represents the change in the x values and is equal to 4.)

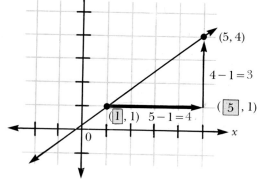

The slope is then the vertical change (3) divided by the horizontal change (4).

$$\text{Slope} = \frac{\text{change in } y}{\text{change in } x} = \frac{3}{4}$$

Thus,

the slope of the line with the equation $3x - 4y = -1$, is equal to $\dfrac{3}{4}$

With this illustration in mind, we define the slope in general terms as follows.

Let the first point be represented by (x_1, y_1), and let the second point be represented by (x_2, y_2).

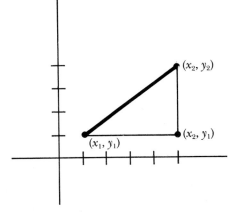

From the diagram, we can see that the change in y is given by $y_2 - y_1$.

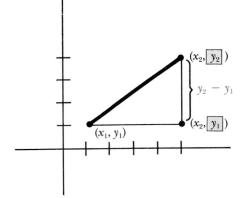

And the change in x is given by $x_2 - x_1$.

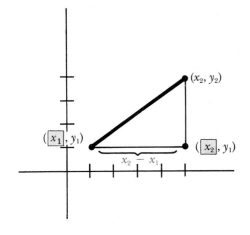

Consequently, we may find the slope of a line by the following formula.

$$\text{Slope} = \frac{\text{the change in } y}{\text{the change in } x} = \frac{y_2 - y_1}{x_2 - x_1}$$

If we let m represent the slope, we have

$$m = \frac{y_2 - y_1}{x_2 - x_1}$$

To find the slope of a line, then, we first find the difference between the y values of any two points (called the change in y and represented as $y_2 - y_1$). Next, we find the difference between the x values of the same two points (called the change in x and represented as $x_2 - x_1$). And finally, we divide the change in y by the change in x to determine the slope [or $m = (y_2 - y_1)/(x_2 - x_1)$].

PROCEDURE:

To find the slope of a line:

1. Select a first point and a second point.

2. Subtract the y value of the first point from the y value of the second point $(y_2 - y_1)$.

3. Subtract the x value of the first point from the x value of the second point $(x_2 - x_1)$.

4. Divide the result from (2) by the result from (3) $[(y_2 - y_1)/(x_2 - x_1)]$. This is slope, m.

___

EXAMPLE 1:   Find the slope of the line that passes through the points $(-3, 1)$ and $(6, 4)$.

SOLUTION: Let $(-3, 1)$ be the first point and $(6, 4)$ be the second point.

First Point	Second Point
$(-3, \quad 1)$	$(6, \quad 4)$
$\downarrow \quad \downarrow$	$\downarrow \quad \downarrow$
$x_1 \quad y_1$	$x_2 \quad y_2$

Find the change in y. Find the change in x.

$$y_2 - y_1 = 4 - 1 \qquad\qquad x_2 - x_1 = 6 - (-3)$$
$$= 3 \qquad\qquad\qquad\qquad = 6 + 3$$
$$= 9$$

Example continued on next page.

Divide the change in y by the change in x to determine slope, m.

$$m = \frac{y_2 - y_1}{x_2 - x_1}$$

$$m = \frac{3}{9}$$

$$m = \frac{1}{3}$$

Final answer: $m = \dfrac{1}{3}$

NOTE:

The order of choosing points (x_1, y_1) and (x_2, y_2) is not important. However, the way in which they are calculated is important. Using example 1, we find that

If $(x_1, y_1) = (-3, 1)$ and $(x_2, y_2) = (6, 4)$,

the slope is $\dfrac{1}{3}$.

$$\frac{1 - 4}{-3 - 6} = \frac{-3}{-9}$$

$$= \frac{1}{3}$$

If $(x_1, y_1) = (6, 4)$ and $(x_2, y_2) = (-3, 1)$,

the slope is still $\dfrac{1}{3}$.

$$\frac{4 - 1}{6 - (-3)} = \frac{3}{9}$$

$$= \frac{1}{3}$$

EXAMPLE 2: Find the slope of the line that passes through the points $(-4, 4)$ and $(2, 1)$.

SOLUTION: Let $(-4, 4)$ be the first point and $(2, 1)$ be the second point.

First Point   Second Point

$(-4, \quad 4)$ $(2, \quad 1)$

$\quad x_1 \quad y_1$ $\quad x_2 \quad y_2$

Using the equation, $m = (y_2 - y_1)/(x_2 - x_1)$, we can find the slope directly.

$$m = \frac{y_2 - y_1}{x_2 - x_1}$$

$$= \frac{1 - 4}{2 - (-4)}$$

$$= \frac{-3}{2 + 4}$$

$$= \frac{-3}{6}$$

$$= \frac{-1}{2}$$

Final answer: $m = -\dfrac{1}{2}$

EXAMPLE 3: Find the slope of the line that passes through the points $(-3, -3)$ and $(-5, -6)$.

SOLUTION: Identify (x_1, y_1) and (x_2, y_2).

First Point   Second Point

$(-3, \quad -3)$ $(-5, \quad -6)$

$\quad \downarrow \qquad \downarrow$ $\quad \downarrow \qquad \downarrow$

$\quad x_1 \qquad y_1$ $\quad x_2 \qquad y_2$

Find slope by the slope equation.

$$m = \frac{y_2 - y_1}{x_2 - x_1}$$

$$= \frac{-6 - (-3)}{-5 - (-3)}$$

$$= \frac{-6 + 3}{-5 + 3}$$

$$= \frac{-3}{-2}$$

$$= \frac{3}{2}$$

Final answer: $m = \dfrac{3}{2}$

EXAMPLE 4: Find the slope of the line that passes through the points $(-2, -1)$ and $(3, -1)$.

SOLUTION: Identify (x_1, y_1) and (x_2, y_2).

First Point

$(-2, \quad -1)$
$\downarrow \qquad \downarrow$
$x_1 \qquad y_1$

Second Point

$(3, \quad -1)$
$\downarrow \qquad \downarrow$
$x_2 \qquad y_2$

Find the slope by the slope equation.

$$m = \frac{y_2 - y_1}{x_2 - x_1}$$

$$= \frac{-1 - (-1)}{3 - (-2)}$$

$$= \frac{-1 + 1}{3 + 2}$$

$$= \frac{0}{5}$$

$$= 0$$

Final answer: $m = 0$

EXAMPLE 5: Find the slope of the line that passes through the points $(5, 3)$ and $(5, -6)$.

SOLUTION: Identify (x_1, y_1) and (x_2, y_2).

First Point

$(5, \quad 3)$
$\downarrow \quad \downarrow$
$x_1 \quad y_1$

Second Point

$(5, \quad -6)$
$\downarrow \qquad \downarrow$
$x_2 \qquad y_2$

Find the slope by the slope equation.

$$m = \frac{y_2 - y_1}{x_2 - x_1}$$

$$= \frac{-6 - 3}{5 - 5}$$

$$= \frac{-9}{0}$$

Division by zero is undefined.

Final answer: Undefined

NOTE:

Examples 4 and 5 are representative of slopes of horizontal and vertical lines, respectively. Specifically:

■ In example 4, since the y values of the two points are equal, the line that passes through them is horizontal. Consequently, horizontal lines have zero slopes.

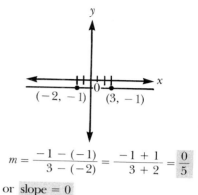

$$m = \frac{-1 - (-1)}{3 - (-2)} = \frac{-1 + 1}{3 + 2} = \frac{0}{5}$$

or slope = 0

■ In example 5, since the x values of the two points are equal, the line that passes through them is vertical. Consequently, vertical lines have no, or undefined, slopes.

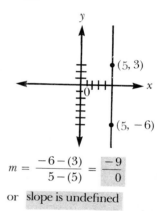

$$m = \frac{-6 - (3)}{5 - (5)} = \frac{-9}{0}$$

or slope is undefined

Do the following practice set. Check your answers with the answers in the right-hand margin.

PRACTICE SET 10-4A

Find the slope of the line that passes through each of the following pairs of points (see example 1, 2, 3, or 4).

1. $(8, 1)$ and $(3, 6)$

2. $(4, -3)$ and $(-2, 6)$

3. $(-6, -3)$ and $(2, -5)$

4. $(3, 6)$ and $(3, 0)$

5. $(0, 0)$ and $(-6, -3)$

6. $(2, -2)$ and $(3, -2)$

1. $m = 5/-5$ or -1
2. $m = 9/-6$ or $-\frac{3}{2}$
3. $m = -2/8$ or $-\frac{1}{4}$
4. $m = $ undefined
5. $m = -3/-6$ or $\frac{1}{2}$
6. $m = 0$

B.   The Slope and *y*-Intercept of a Line

Consider the equation $y = 2x + 4$.

Graphing this equation by
its intercepts yields the
graph as shown. Notice
that the slope of this line
is equal to 2, and further,
the *y* intercept is $(0, 4)$.

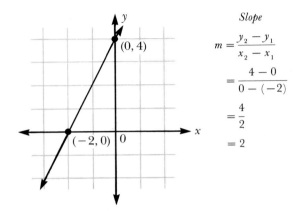

Slope

$$m = \frac{y_2 - y_1}{x_2 - x_1}$$

$$= \frac{4 - 0}{0 - (-2)}$$

$$= \frac{4}{2}$$

$$= 2$$

Relating these observations back to the equation itself,
we find that the coefficient of the *x* term is the slope,
and the constant term is the 4 coordinate of the *y*
intercept.

$$y = \boxed{2}\,x + \boxed{4}$$
$$\quad\quad\;\; | \quad\quad |$$
$$\quad\quad \text{slope} \quad y \text{ intercept}$$

This is true! In fact, any equation that is expressed in the form $y = mx + b$ has its slope
equal to the coefficient of *x* (namely, *m*), and its *y* intercept is the constant $(0, b)$. $y = mx + b$
is called the slope/intercept form of a line.

EXAMPLE 1:   Identify the slope and *y* intercept of $y = 2x - 16$.

SOLUTION: Since the equation is in the form $y = 2x - 16$, the slope is 2 and the *y* intercept is -16.

$$y = \boxed{2}\,x - \boxed{16}$$
$$\quad\quad\;\; | \quad\quad\quad |$$
$$\quad\quad \text{slope} \quad y \text{ intercept}$$

Final answer: $m = 2$; $b = -16$

EXAMPLE 2:   Identify the slope and *y* intercept of $3y + 4x = 16$.

SOLUTION: We must first put this in the form $y = mx + b$.

$$3y + 4x = 6$$
$$3y = -4x + 6$$
$$y = \frac{-4x + 6}{3}$$
$$y = \frac{-4x}{3} + \frac{6}{3}$$
$$y = \frac{-4x}{3} + 2$$

Now, the slope is $\dfrac{-4}{3}$ and the y intercept is 2.

$$y = \boxed{\dfrac{-4}{3}}x + \boxed{2}$$

slope y intercept

Final answer: $m = \dfrac{-4}{3}; \quad b = 2$

Do the following practice set. Check your answers with the answers in the right-hand margin.

PRACTICE SET 10-4B

Find the slope and y intercept for each of the following lines (see example 1 or 2).

1. $y = -2x + 3$

2. $y = \dfrac{5x + 8}{16}$

3. $2x - y = 8$

4. $3x + 2y = -10$

1. $m = -2, \quad b = 3$
2. $m = \frac{5}{16}, \quad b = \frac{1}{2}$
3. $m = 2, \quad b = -8$
4. $m = -\frac{3}{2}, \quad b = -5$

C. Graphing a Line by Its Slope

As we have previously learned, the slope of a line is given as the change in y divided by the change in x. In effect, the change in y constitutes movement along the y axis, and the change in x constitutes movement along the x axis. For example, a slope of $\frac{3}{4}$ means for every 3 units we move along the y axis, we would move 4 units along the x axis.

The significance of this is that if we are given a point through which a line passes and the slope of the line, we can ultimately graph the line by moving "slope units" along the x

and y axes from the given point—to get additional points—and then simply draw the line through these points. As an example, we graph the line that passes through the point $(5, 4)$ and has a slope of $\frac{3}{4}$.

A slope of $\frac{3}{4}$ means the change in y is 3 units and the change in x is 4 units.

$\dfrac{3}{4}$ —change in y
—change in x

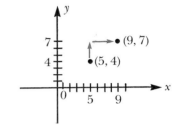

Thus, from the point $(5, 4)$ we move 3 units along y (go "up" since 3 is positive), and 4 units along x (go "right" since 4 is positive). We now arrive at the point $(9, 7)$.

Additional points can be obtained in a similar manner, resulting in the graph of the line.

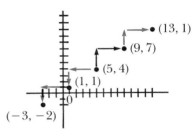

NOTE:

Since $\frac{3}{4}$ is positive, we can also obtain points by the equivalent slope $-3/-4$ (which means we move "down" on y and to the "left" on x).

PROCEDURE:

To graph a line given only a point it passes through and its slope:

1. Plot the given point.

2. From this point, move as many units as indicated by the numerator of the slope, along the y axis.
    a. Move up if the numerator is positive.
    b. Move down if the numerator is negative.

3. From the location arrived at from (2), move as many units as indicated by the denominator of the slope, along the x axis.
    a. Move right if the denominator is positive.
    b. Move left if the denominator is negative.
    Plot a point at this final location.

4. Draw the straight line through the given point and the point arrived at via the slope. Additional points may be obtained in a similar manner.

EXAMPLE 1:   Graph the line that passes through the point $(1, -4)$ and whose slope is $\frac{2}{3}$.

SOLUTION: First, plot the point $(1, -4)$.

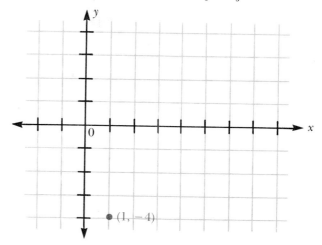

Now, move slope units from this point to graph the line.

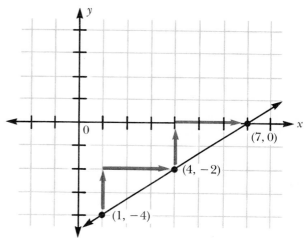

EXAMPLE 2:   Graph the equation $x + 2y = -6$ by using its slope.

SOLUTION: We can find the slope and y intercept of this equation by writing it in the form $y = mx + b$.

$$x + 2y = -6$$
$$2y = -x - 6$$
$$y = \frac{-x - 6}{2}$$
$$y = \frac{-x}{2} - \frac{6}{2}$$
$$y = \frac{-1}{2} x - 3$$

Example continued on next page.

Now, with a y intercept of -3 and a slope of $-\frac{1}{2}$, we can graph the line.

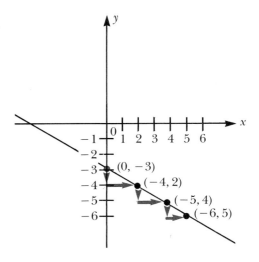

Do the following practice set. Check your answers with the answers that follow the problems.

PRACTICE SET 10-4C

Graph the line that passes through the given point and has the given slope (see example 1).

1. $(2, 3)$, $m = \dfrac{2}{1}$

2. $(3, -2)$, $m = \dfrac{-1}{1}$

3. $(0, -4)$, $m = \dfrac{1}{2}$

Graph the following equations by using their slope and y intercept (see example 2).

4. $y = 2x - 3$

5. $x - 2y = 4$

6. $2x + 3y = 6$

1.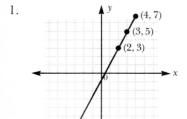

2.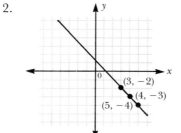

3.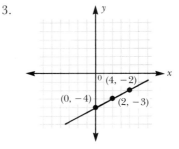

4.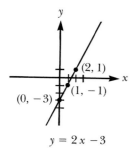

$$y = 2x - 3$$

5.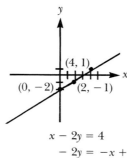

$$x - 2y = 4$$
$$-2y = -x + 4$$
$$y = -\tfrac{1}{2} - \tfrac{1}{2}$$
$$y = -\tfrac{1}{2}x - 2$$

6.

$$2x + 3y = 6$$
$$3y = -2x + 6$$
$$y = -\tfrac{2}{3}x + \tfrac{6}{3}$$
$$y = -\tfrac{2}{3}x + 2$$

EXERCISE 10-4

Find the slope of the line that passes through each of the following pairs of points.

1. $(0, 0)$ and $(4, 2)$
2. $(3, 6)$ and $(0, 0)$
3. $(3, 2)$ and $(0, 5)$
4. $(1, 4)$ and $(3, 4)$
5. $(9, -3)$ and $(4, 1)$
6. $(-1, -2)$ and $(3, 4)$
7. $(5, 3)$ and $(5, -5)$
8. $(-4, 2)$ and $(2, -8)$
9. $(2, 3)$ and $(-4, -1)$
10. $(3, -6)$ and $(-2, -4)$

Graph the line with the given slope and the given point.

11. $(0, 0)$,   slope $= 1$
12. $(2, 1)$,   slope $= 3$
13. $(1, -2)$,   slope $= \dfrac{1}{2}$
14. $(2, -3)$,   slope $= \dfrac{1}{3}$
15. $(-3, 1)$,   slope $= \dfrac{2}{5}$
16. $(-2, 0)$,   slope $= \dfrac{3}{4}$
17. $(0, 6)$,   slope $= \dfrac{-3}{2}$
18. $(-2, -3)$,   slope $= -\dfrac{1}{3}$
19. $(2, -4)$,   slope $= 0$
20. $(4, -1)$,   slope is undefined

Find the slope and y intercept for each of the following equations. Graph the equations by using the slope and y intercept.

21. $y = 2x + 1$
22. $y = x - 4$
23. $y = 4x - 3$
24. $y = 2x$
25. $y = x$
26. $y = -x + 3$
27. $y = \dfrac{1}{2}x + 6$
28. $y = 0$
29. $y = -\dfrac{2}{3}x + 4$
30. $y = \dfrac{3}{5}x - 6$
31. $x - y = 3$
32. $x + y = -3$

33. $2x + 3y = 8$ 34. $x - 4y = 0$ 35. $3x - 4y = 16$ 36. $5x + 2y = -7$

37. $3y = -2x + 4$ 38. $-2y = 3x - 8$ 39. $\dfrac{2}{3}x - \dfrac{1}{3}y = 2$ 40. $3y = 5(x - 1)$

10-5   WRITING EQUATIONS OF STRAIGHT LINES

In previous sections of this chapter, we have been given the equation of a line and our goal was to graph that equation. In this section, we do the opposite. That is, given the graph of a line or specific properties of a line (e.g., slope), our aim is to determine the equation of that line.

A.   Writing the Equation of a Line Knowing Its Slope and y-Intercept

A linear equation in the form $y = mx + b$ has a slope of m and a y intercept of $(0, b)$. So, if we know (or are given) the slope of a line and its y intercept, we simply replace m with the slope of the line and b with its y intercept.

As an example,

consider the line whose slope is -4 and has a y intercept of 2. The equation of this line (which satisfies these properties) is $y = -4x + 2$.

$$y = mx + b$$
$$y = -4x + 2$$

Do the following practice set. Check your answers with the answers in the right-hand margin.

PRACTICE SET 10-5A

Write the equation of the line with the given slope m and which crosses the y axis at the given point.

1. $m = 3$, $(0, 2)$

2. $m = \dfrac{1}{2}$, $(0, -3)$

3. $m = \dfrac{-3}{5}$, $\left(0, \dfrac{2}{3}\right)$

1. $y = 3x + 2$

2. $y = \frac{1}{2}x - 3$

3. $y = -\frac{3}{5}x + \frac{2}{3}$

B.   Writing the Equation of a Line Knowing Its Slope and Any Point It Passes Through

In part A, we learned that we can use the equation $y = mx + b$ to find the equation of a line given the line's slope and its y intercept. If, however, we are given a point other than the y intercept, then the slope/intercept form of a line cannot be so easily applied. In such cases we use another form—called the point/slope form—to help us find the equation of a line. This form is given as

$$y - y_1 = m(x - x_1)$$

where $m =$ slope and x_1 and y_1 represent the x and y values of any point on the line.

For example, to write the equation of the line that has a slope of -2 and passes through the point $(2, 4)$, we simply:

Substitute -2 for m, 2 for x_1, and 4 for y_1 into the point/slope form.

$$y - y_1 = m(x - x_1)$$
$$\downarrow \quad \downarrow \quad \downarrow$$
$$y - 4 = -2(x - 2)$$

Next, solve the equation for y, to get into the form $y = mx + b$.

$$y - 4 = -2(x - 2)$$
$$y - 4 = -2x + 4$$
$$y = -2x + 8$$

NOTE:

We want to express the equation in the form $y = mx + b$ because this form lends itself to being graphed as we learned in Section 10-4C.

PROCEDURE:

To write the equation of a line knowing only its slope and a point it passes through:

1. Use the point/slope form equation for a line

$$y - y_1 = m(x - x_1)$$

and substitute the given slope for m, the x value of the point for x_1, and the y value of the point for y_1.

2. Solve the equation for y, arranging it in the form, $y = mx + b$.

EXAMPLE 1: Write the equation of the line that has a slope of -3 and passes through the point $(-5, 1)$.

SOLUTION: Replace -3 for m, -5 for x_1, and 1 for y_1.

$$y - y_1 = m(x - x_1)$$
$$y - 1 = -3(x - (-5))$$
$$y - 1 = -3(x + 5)$$

Solve for y.

$$y - 1 = -3(x + 5)$$
$$y - 1 = -3x - 15$$
$$y = -3x - 14$$

Final answer: $y = -3x - 14$

EXAMPLE 2: Write the equation of the line that passes through the point $(0, 6)$ and has a slope of $\frac{1}{3}$.

SOLUTION: Replace $\frac{1}{3}$ for m, 0 for x, and 6 for y.

$$y - y_1 = m(x - x_1)$$

$$y - 6 = \frac{1}{3}(x - 0)$$

Solve for y.

$$y - 6 = \frac{1}{3}(x - 0)$$

$$y - 6 = \frac{1}{3}x - 0$$

$$y = \frac{1}{3}x + 6$$

Final answer: $y = \dfrac{1}{3}x + 6$

In example 2, since the point given, $(0, 6)$, is really the y intercept, we also could have used the slope/intercept form $(y = mx + b)$ to find the equation satisfying the given properties. The significance here is that the point/slope form, $y - y_1 = m(x - x_1)$, is the more universally applied form in which we can find the equation of a line. Here is one more example.

EXAMPLE 3: Write the equation of the line that passes through the two points $(1, 4)$ and $(3, 8)$.

SOLUTION: Since the slope is not given, we must first find it.

First point   Second point

$(1, \quad 4)$ $(3, \quad 8)$

$x_1 \quad y_1$ $x_2 \quad y_2$

$$m = \frac{y_2 - y_1}{x_2 - x_1}$$

$$m = \frac{8 - 4}{3 - 1}$$

$$m = \frac{4}{2}$$

$$m = 2$$

Now, we substitute into $y - y_1 = m(x - x_1)$ accordingly and solve for y.

$$y - y_1 = m(x - x_1)$$
$$\downarrow \quad \downarrow \quad \downarrow$$
$$y - 4 = 2(x - 1)$$
$$y - 4 = 2x - 2$$
$$y = 2x + 2$$

Final answer: $y = 2x + 2$

Do the following practice set. Check your answers with the answers in the right-hand margin.

PRACTICE SET 10-5B

Write the equation of the line (in the form $y = mx + b$) that has a given slope m and which passes through the given point (see examples 1 and 2).

1. $m = 2,$ $(2, 3)$

2. $m = -5,$ $(1, -4)$

3. $m = \dfrac{1}{3},$ $(-3, 5)$

Write the equation of the line (in the form $y = mx + b$) that passes through each of the given pairs of points (see example 3).

4. $(3, 6)$ and $(2, 4)$

5. $(1, 2)$ and $(2, 0)$

6. $(-2, 1)$ and $(-3, -5)$

1. $y = 2x - 1$

2. $y = -5x + 1$

3. $y = \frac{1}{3}x + 6$

4. $y = 2x$

5. $y = -2x + 4$

6. $y = 6x + 13$

EXERCISE 10-5

Write the equation of the line (in the form, $y = mx + b$) that has the given slope and passes through the given point.

1. $m = 2,$ $(2, 3)$

2. $m = -1,$ $(0, 0)$

3. $m = -4,$ $(-1, -3)$

4. $m = 3,$ $(-3, 2)$

5. $m = \dfrac{1}{2},$ $(3, 0)$

6. $m = \dfrac{3}{-4},$ $(0, -2)$

7. $m = \dfrac{-5}{4},$ $(-2, 4)$

8. $m = -\dfrac{3}{5},$ $(2, 3)$

9.   No slope, $(3, 2)$

10. $m = 0,$ $(2, 4)$

Write the equation of the line (in the form, $y = mx + b$) that passes through the given points.

11. $(2, 4)$ and $(5, 1)$

12. $(2, -1)$ and $(3, 8)$

13. $(0, 0)$ and $(2, -3)$

14. $(-2, 4)$ and $(-1, 3)$

15. $(-2, 0)$ and $(4, 6)$

16. $(0, -1)$ and $(3, 10)$

17. $(-2, -3)$ and $(2, -1)$

18. $(-3, 5)$ and $(-1, -8)$

19. $(-5, -2)$ and $(-3, -1)$

20. $(-2, 0)$ and $(0, -2)$

In problems 61–65, use the fact that parallel lines have equal slopes to write the equation of the line with the given description.

21. The line parallel to the line, $y = 2x + 3$, that has a y intercept of 1.

22. The line parallel to the line, $y = -2x + 1$, that has a y intercept of -3.

23. The line parallel to the line, $x - y = 2$, that has an x intercept of 2.

24. The line parallel to the line, $2x + 3y = 3$, and passes through the point $(0, 0)$.

25. The line parallel to the line, $3y - 2x = 2$, and passes through the point $(-3, -1)$.

10-6   SOLVING LINEAR INEQUALITIES IN TWO VARIABLES

A.   Introduction

When we graph the line of a linear equation in two variables, the line divides the plane into three regions—the *line itself*, the *region above the line*, and the *region below the line*.

Using the equation, $y = 2x + 4$, we illustrate this.

1.   THE LINE ITSELF

The line itself is the *boundary* of the two regions and contains all points that satisfy the equation $y = 2x + 4$. For example, the coordinates of points A $(-3, -2)$, B $(1, 6)$, and C $(3, 10)$ all satisfy the equation $y = 2x + 4$ (see chart).

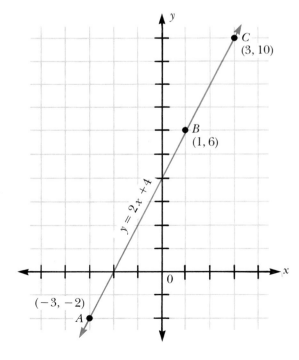

	x	$2x + 4$	y
A	-3	$2(-3) + 4$ $-6 + 4$ -2	-2
B	1	$2(1) + 4$ $2 + 4$ 6	6
C	3	$2(3) + 4$ $6 + 4$ 10	10

2.   THE REGION ABOVE THE LINE

The region *above* the line, $y = 2x + 4$, contains all the points whose y values are *greater* than "twice the x values plus four." That is, $y > 2x + 4$. For example:

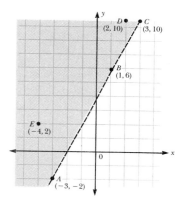

At the point D (2, 10), the y value, 10, is greater than twice the x value plus four (see preceding graph).

$y > 2x + 4$
$10 > 2(2) + 4$
$10 > 4 + 4$
$10 > 8$

The point E (−4, 2) also satisfies $y > 2x + 4$ because $2 > 2(−4) + 4$ is true (see preceding graph).

$y > 2x + 4$
$2 > 2(−4) + 4$
$2 > −8 + 4$
$2 > −4$

Both D and E are *above* the line.

3.   THE REGION BELOW THE LINE

The region below the line $y = 2x + 4$ contains all the points whose y values are *smaller* than twice the x values plus four. That is, $y < 2x + 4$. For example:

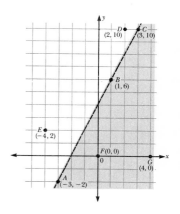

At the point F (0, 0), the y value, 0 is *less* than twice the x value plus 4 (see preceding graph).

$y < 2x + 4$
$0 < 2(0) + 4$
$0 < 0 + 4$
$0 < 4$

The point G (4, 0) also satisfies $y < 2x + 4$ because $0 < 2(4) + 4$ is true (see preceding graph).

$y < 2x + 4$
$0 < 2(4) + 4$
$0 < 8 + 4$
$0 < 12$

Both F (0, 0) and G (4, 0) are *below* the line.

NOTE:

When dealing with an inequality, such as $y > 2x + 4$, we graph the boundary line, $y = 2x + 4$, using a *dashed* line. This shows that the line is not part of the graph.

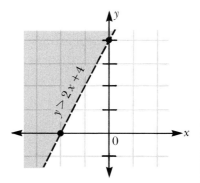

If the inequality were $y \geq 2x + 4$, then we would graph the boundary line, using a *solid* line to show that the line is part of the graph.

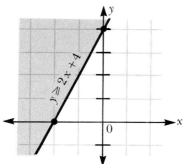

Also,

When dealing with an inequality, such as $y < 2x + 4$, we graph the boundary line, $y = 2x + 4$, using a *dashed* line. This shows that the line is not part of the graph.

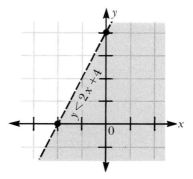

If the inequality were $y \leq 2x + 4$, then we would graph the boundary line, using a *solid* line to show that the line is part of the graph.

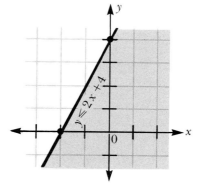

In summary, a linear equation in two variables, such as $y = 2x + 4$, divides the plane into three sets of points.

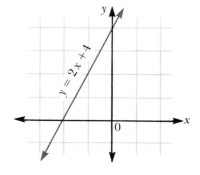

■ The set of points on the line (i.e., all points that satisfy the equation $y = 2x + 4$).

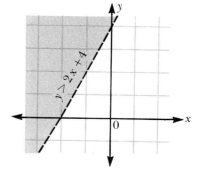

■ The set of points above the line (i.e., all points that satisfy the inequality $y > 2x + 4$).

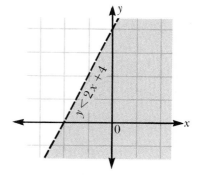

■ The set of points below the line (i.e., all points that satisfy the inequality $y < 2x + 4$).

With this in mind, we are now able to graph inequalities in two variables.

B. Graphing Inequalities in Two Variables

To graph an inequality in two variables, we basically do two things: Graph the boundary line and then determine the correct region that will satisfy the given inequality. For example,

To graph the inequality, $y > x + 5$:

First, graph the boundary line, $y = x + 5$; we graph this by the intercept method. (Note that since $y \neq x + 5$, we use a dashed line.)

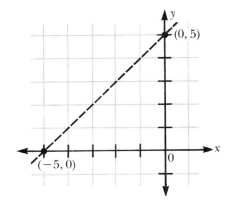

Now, simply pick a test point either above or below the boundary line; we pick $(0, 0)$ which is below the line.

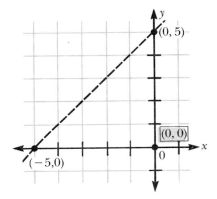

Substitute the x and y values of the test point into the inequality and solve.

$y > x + 5$
$0 > 0 + 5$
$0 > 5$ (false)

Since the answer is false, the region *below* the boundary line does *not* satisfy the inequality. Thus, the region *above* the inequality is our solution. We show this by shading that region.

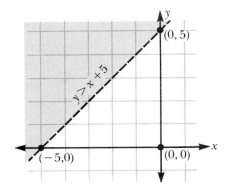

PROCEDURE:

> To graph inequalities in two variables:
>
> 1. Treat the inequality as an equation and graph this equation. This line is the boundary line.
>   a. Use a dashed line if the inequality is < or >.
>   b. Use a solid line if the inequality is ≤ or ≥.
>
> 2. Choose a test point either above or below the boundary line. You may use $(0, 0)$ provided it is not part of the boundary line.
>
> 3. Substitute the x and y values of the test point into the inequality and solve the inequality.
>   a. If the solution is true, shade the region that contains the test point.
>   b. If the solution is false, shade the region that does not contain the test point.

EXAMPLE 1:   Graph $y > 3x$.

SOLUTION: Graph the boundary line, $y = 3x$. Notice that we use a dashed line.

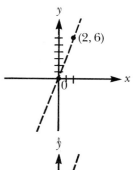

Choose a test point either above or below the boundary line. (We choose $(2, 0)$ which is below the boundary line.)

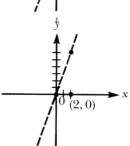

Substitute the x and y values of the test point into the inequality and solve.

$$y > 3x$$
$$(0) > 3(2)$$
$$0 > 6 \text{ (false)}$$

Example continued on next page.

Since the result is false, shade the region that does not contain the test point.

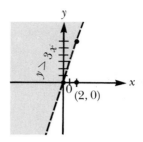

EXAMPLE 2:   Graph $x + y \leq 3$.

SOLUTION: Graph the boundary line, $x + y = 3$. (We use the intercept method.) The line is solid.

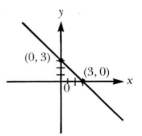

Choose a test point. (We choose $(0, 0)$ which is below the boundary line.)

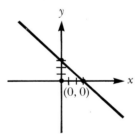

Substitute the x and y values of the test point into the inequality and solve.

$$x + y < 3$$
$$(0) + (0) < 3$$
$$0 < 3 \text{ (true)}$$

Since the result is true, shade the region that contains the test point.

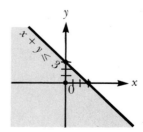

EXAMPLE 3:   Graph $3x - 4y < 0$.

SOLUTION: The boundary line, $3x - 4y = 0$, is dashed and the test point $(-3, 0)$ yields a true solution.

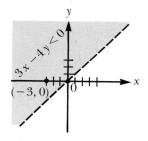

$$3x - 4y < 0$$
$$3(-3) - 4(0) < 0$$
$$-9 - 0 < 0$$
$$-9 < 0 \text{ (true)}$$

EXAMPLE 4:   Graph $2x + 3y \geq -6$.

SOLUTION: The boundary line, $2x + 3y = -6$, is solid and the test point $(0, 0)$ yields a true solution.

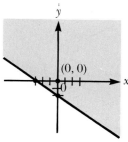

$$2x + 3y \geq -6$$
$$2(0) + 3(0) \geq -6$$
$$0 + 0 \geq -6$$
$$0 \geq -6 \text{ (true)}$$

Do the following practice set. Check your answers with the answers that follow the problems.

PRACTICE SET 10-6B

Graph the following inequalities (see example 1, 2, 3, or 4).

1. $y \leq -2x$

2. $x - 2y > 0$

3. $2x + 3y \geq -6$

1.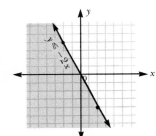

2.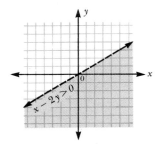

3.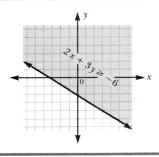

EXERCISE 10-6

Graph the following inequalities in the rectangular coordinate system.

1. $y > 2x$
2. $y < x + 2$
3. $y \leq 2x - 3$
4. $y \geq 3$

5. $x < 4$
6. $x + y > 3$
7. $x - 2y \leq 4$
8. $y - 2x \geq 4$

9. $2x + y \geq 0$
10. $2x + 3y \leq 6$
11. $x + 3y - 4 > 0$
12. $2x + 3y \leq 0$

13. $5 \geq 2x + 3y$
14. $6 < x - 3y$
15. $x \leq 2y$
16. $y > 8 + x$

17. $x - y < 2$
18. $x + 2y \leq 6$
19. $\frac{1}{2}x - y > 4$
20. $(x + y) + 6 < 0$

REVIEW EXERCISES

1. Plot each of the following ordered pairs on the axes given. Label each point.

$A(3, 6)$
$B(-3, 2)$
$C(5, -1)$
$D(-3, -3)$
$E(2, 0)$
$F(0, -4)$

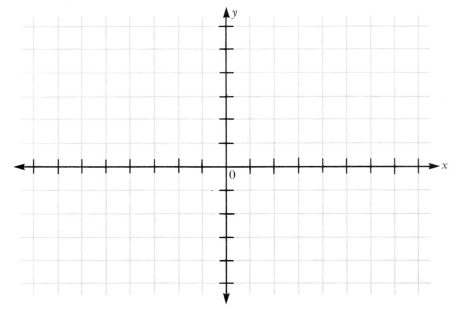

In problems 2–15, graph the given equation in the x/y plane.

2. $x = 2$ 3. $y = -3$ 4. $y + 4 = 0$ 5. $x + 6 = 0$

6. $y = -3x + 4$ 7. $y = 2x - 6$ 8. $y = 4x$ 9. $y = -3x$

10. $y = 5x + 3$ 11. $y = 5x + 6$ 12. $3y - 2x = -6$ 13. $2y + x = 5$

14. $-y + 3x + 4 = 0$ 15. $3y + 5x - 6 = 0$

In problems 16–25:
a. Graph the line that satisfies the stated conditions.
b. Write the equation of the line.

16. Slope of zero and passes through the point $(2, 4)$

17. Passes through the point $(-3, -3)$ and has no slope

18. Slope of 2/5 and passes through the point $(-2, 4)$

19. Slope of 1/3 and passes through the point $(5, -6)$

20. Slope of -1 and passes through the origin

21. Slope of $-7/3$ and passes through the point $(2, 0)$

22. Passes through the points $(-2, 4)$ and $(2, 0)$

23. Passes through the points $(4, 5)$ and $(-6, -3)$

24. Passes through the points $(-3, 0)$ and $(3, 0)$

25. Passes through the points $(4, -6)$ and $(4, 3)$

In problems 26–35, graph the given inequality.

26. $x < 8$ 27. $y \geq -2$ 28. $x + 5 < 0$ 29. $y - 6 > 2$

30. $x + y > 4$ 31. $y \leq 3x$ 32. $2x - 5y \geq 10$ 33. $4x - 3y > 12$

34. $3x + 5y \leq 15$ 35. $2x - 8 < 16 + y$

CHAPTER 11

Solving Systems of Equations in Two Variables

11-1 INTRODUCTION

In Chapter 4, we learned how to solve equations in one variable. For example, to solve the equation $2x + 4 = 16$ for x:

Subtract 4 from both sides.

$$2x + 4 = 16$$
$$\underline{-4 \quad\quad -4}$$
$$2x = 12$$

Then, divide both sides by 2.

$$2x = 12$$
$$\frac{\overset{1}{2x}}{\underset{1}{2}} = \frac{\overset{6}{12}}{\underset{1}{2}}$$
$$x = 6$$

Thus,

$$x = 6$$

Further, $x = 6$ is referred to as the *solution* to the equation, $2x + 4 = 16$. Thus, when 6 is substituted for x, the two sides of the equation are equal. (This can be accomplished by a check.)

	$2x + 4$	16
Check:		
Substitute 6 for x	$2(6) + 4$	16
and add.	$12 + 4$	
	16	

Therefore,

$$x = 6 \ \textit{satisfies} \ \text{the equation,} \ 2x + 4 = 16$$

We now learn how to solve *two* linear equations in *two variables* (where each equation contains the same two variables). Whenever we have two (or more) equations that are to be considered together, we refer to them as a system of simultaneous equations (or just a system of equations). The solution for a system of two equations in two variables is an ordered pair, an infinite number of ordered pairs, or no solution. If the solution is an ordered pair, or an infinite

420

number of ordered pairs, then the solution(s) must satisfy both equations. For example:

$$\left.\begin{array}{l} y = x - 3 \\ y = 2x - 5 \end{array}\right\} \text{ is a system of equations in two variables.}$$

Generating a list of ordered pairs that satisfy $y = x - 3$, we have:

x	$x - 3$	y
0	$0 - 3$ -3	-3
1	$1 - 3$ -2	-2
* 2	$2 - 3$ -1	-1
3	$3 - 3$ 0	0
4	$4 - 3$ 1	1
5	$5 - 3$ 2	2

Generating a list of ordered pairs that satisfy $y = 2x - 5$, we have:

x	$2x - 5$	y
0	$2(0) - 5$ $0 - 5$ -5	-5
1	$2(1) - 5$ $2 - 5$ -3	-3
* 2	$2(2) - 5$ $4 - 5$ -1	-1
3	$2(3) - 5$ $6 - 5$ 1	1
4	$2(4) - 5$ $8 - 5$ 3	3

Notice that both lists contain the ordered pair $(2, -1)$. Since $(2, -1)$ satisfies both equations, we can say the ordered pair $(2, -1)$ is a solution to the system of equations $y = x - 3$ and $y = 2x - 5$.

Thus,

$$(2, -1) \text{ satisfies the system } \begin{cases} y = x - 3 \\ y = 2x - 5 \end{cases}$$

In this chapter, we learn two methods to solve such systems of equations: a graphical method and an algebraic method.

EXERCISE 11-1

Determine whether the given ordered pair is a solution of the given system of equations by checking if the ordered pair satisfies both equations.

1. $\begin{cases} x + y = 6 \\ y = x - 3 \end{cases}$ $(2, 1)$

2. $\begin{cases} y = 2x + 1 \\ y = x - 1 \end{cases}$ $(-2, -3)$

3. $\begin{cases} x = y \\ 3x = 3y \end{cases}$ $(0, 2)$

4. $\begin{cases} 2x + y = 0 \\ x - 2y = 0 \end{cases}$ $(-2, 1)$

5. $\begin{cases} y = 2x + 4 \\ 3x + y = -6 \end{cases}$ $(-2, 0)$

6. $\begin{cases} 3y + 2x = -1 \\ 2y = 3x - 18 \end{cases}$ $(4, -3)$

7. $\begin{cases} 3x + 2y = 0 \\ 2x + 2y = 8 \end{cases}$ $(1, 3)$

8. $\begin{cases} x = 3y + 13 \\ 3x + 2y = 16 \end{cases}$ $(-2, -5)$

9. $\begin{cases} -5y + 2x = -16 \\ 2x - 5y = -7 \end{cases}$ $(3, 4)$

10. $\begin{cases} 3x - 2y - 13 = 0 \\ 4y + x = -19 \end{cases}$ $(1, -5)$

11-2 SOLUTION BY GRAPHING

Each equation in the system of equations that we will be studying is a linear equation in two variables. Thus, the graph of each equation is a straight line.

If we graph each equation on the same set of axes, our solution would be the point at which the two lines cross. Consider the following example.

To solve the system $\begin{cases} y = x + 1 \\ y = 2x - 4 \end{cases}$ graphically, graph each equation on the same set of axes.

Graph $y = x + 1$.
(We use the intercept method.)

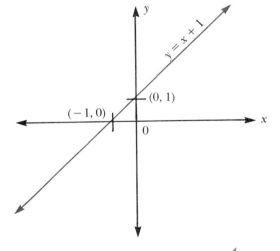

Graph $y = 2x - 4$.
(We use the intercept method.)

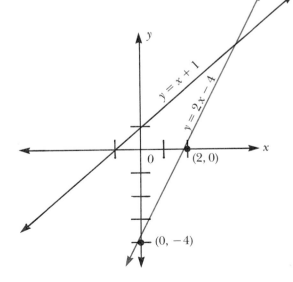

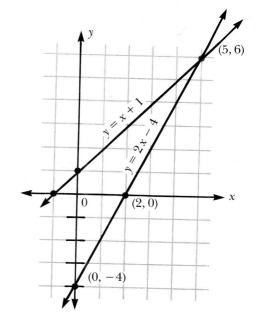

The solution is the point where
both lines cross.
(5, 6)

The solution (5, 6) may be confirmed (or checked) by substituting it into the two equations of the system.

Check: Substitute the ordered pair (5, 6) into each equation.
(Note: $x = 5$, $y = 6$)

y	$x + 1$	y	$2x - 4$
6	(5) + 1	6	2(5) − 4
	5 + 1		10 − 4
	6		6

Thus,

$$(5, 6) \text{ satisfies both equations}$$

In summary, we may solve a system of two linear equations graphically by graphing each equation on the same set of axes. The solution is the ordered pair at the point of intersection of the two graphs. Moreover, the solution may be confirmed by substituting the x and y values of the ordered pair into the two equations of the system.

PROCEDURE:

To solve a system of two linear equations in two variables graphically:

1. Graph each equation separately on the same set of axes.

2. Locate the solution, the ordered pair found at the point of intersection of the two graphs.

EXAMPLE 1: Solve $\begin{cases} y = x - 1 \\ y = -x + 5 \end{cases}$ graphically.

SOLUTION: Graph each equation on the same set of axes.

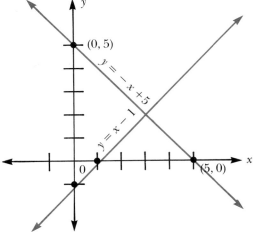

The ordered pair $(3, 2)$ is the point of intersection.

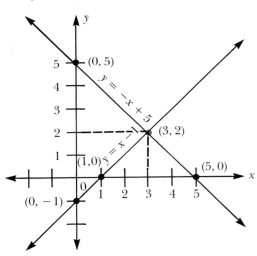

Check:

y	$x - 1$
2	$3 - 1$
	2

y	$-x + 5$
2	$-(3) + 5$
	$-3 + 5$
	2

Final answer: $(3, 2)$

EXAMPLE 2: Solve $\begin{cases} 2x + y = 4 \\ 2x + 3y = 12 \end{cases}$ graphically.

SOLUTION: Graph each equation on the same set of axes.

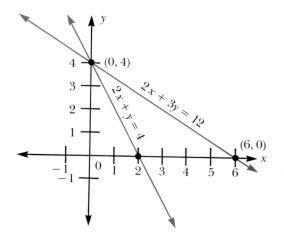

The ordered pair $(0, 4)$ is the point of intersection.

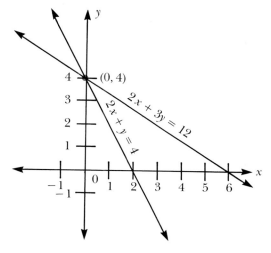

Check:

$2x + y$	4
$2(0) + 4$	4
$0 + 4$	
4	

$x + 3y$	12
$(0) + 3(4)$	12
$0 + 12$	
12	

Final answer: $(0, 4)$

NOTE:

In examples 1 and 2, the graphs of each system of linear equations intersected in exactly one point. The equations of such systems are commonly referred to as being *consistent* or *independent*.

EXAMPLE 3: Solve $\begin{cases} -2x + y = 4 \\ 6x - 3y = 12 \end{cases}$ graphically.

SOLUTION: Graphing each equation on the same set of axes, notice that there is no point of intersection; the lines are parallel.

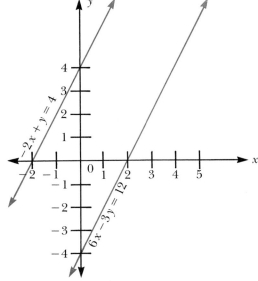

Since the graphs of the equations are parallel, there is no point of intersection and hence no solution. The equations of a system that yield no solution are considered to be *inconsistent*.

EXAMPLE 4: Solve $\begin{cases} x + y = 2 \\ 2x + 2y = 4 \end{cases}$ graphically.

SOLUTION: Graphing each equation on the same set of axes, notice that both graphs are equal. That is, one is "on top" of the other.

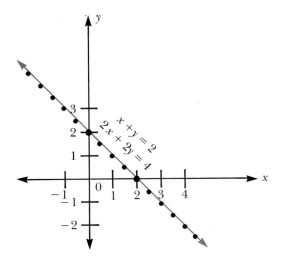

Since the graph of each equation yields the same line, any solution of one equation is also a solution of the other. Thus, there are an infinite number of solutions. The equations of such a system are said to be *dependent*.

As shown in the previous examples, there are three kinds of systems of equations in two variables.

1. CONSISTENT EQUATIONS

The graphs of each equation intersect at exactly one point—thus, there is only one common solution (examples 1 and 2).

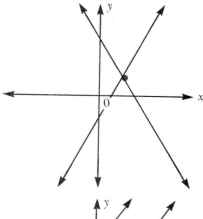

2. INCONSISTENT EQUATIONS

The graphs of each equation do not intersect (i.e., the lines are parallel)—thus, there is no common solution (example 3).

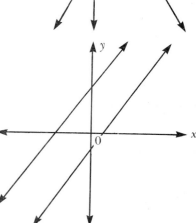

3. DEPENDENT EQUATIONS

The graphs of each equation are identical (i.e., the lines are the same)—thus, there are an infinite number of common solutions (example 4).

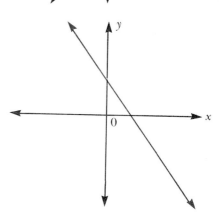

Do the following practice set. Check your answers with the answers that follow the problems.

PRACTICE SET 11-2

Solve graphically (see example 1 or 2).

1. $\begin{cases} y = x \\ y = 2x + 2 \end{cases}$

2. $\begin{cases} y = -2x - 1 \\ 3x - 2y = -5 \end{cases}$

3. $\begin{cases} 2x - y = 3 \\ x + 4y = 6 \end{cases}$

Solve graphically (see example 3).

4. $\begin{cases} 2x + 3y = -1 \\ 3y = -2x + 6 \end{cases}$

5. $\begin{cases} y = 3x - 3 \\ y = 3x - 6 \end{cases}$

6. $\begin{cases} x - 4y = 4 \\ 3x - 12y = -3 \end{cases}$

Solve graphically (see example 4).

7. $\begin{cases} y = 3x \\ 3y = 9x \end{cases}$

8. $\begin{cases} 2y + x = 2 \\ 4y + 2x = 4 \end{cases}$

9. $\begin{cases} 4x - 12y = 24 \\ x - 3y = 6 \end{cases}$

1.
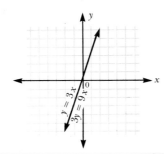

2.
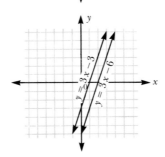

3.
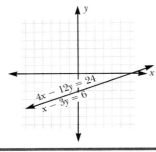

4.
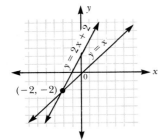

5.
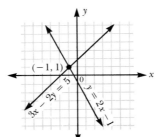

6.
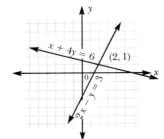

7.
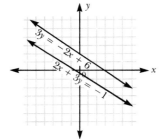

8.
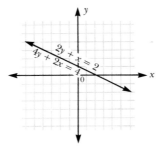

9.
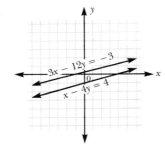

EXERCISE 11-2

Solve the following systems of equations graphically. State whether the system is consistent, inconsistent, or dependent.

1. $\begin{cases} y = 2x \\ y = -2x + 4 \end{cases}$
2. $\begin{cases} y = -3x - 2 \\ y = 2x + 8 \end{cases}$
3. $\begin{cases} 2x - y = 1 \\ 2x - y = -4 \end{cases}$
4. $\begin{cases} x + y = 1 \\ 2x - y = -1 \end{cases}$

5. $\begin{cases} x - 4y = -10 \\ 2x + 6y = 22 \end{cases}$
6. $\begin{cases} y = -x + 3 \\ 2y = 2x + 6 \end{cases}$
7. $\begin{cases} y = -x + 3 \\ 3x - 4y = 2 \end{cases}$
8. $\begin{cases} 2x + y = -2 \\ 2x - 3y = 6 \end{cases}$

9. $\begin{cases} x + y = 2 \\ 5x + 5y = 10 \end{cases}$
10. $\begin{cases} 4x + 2y = -8 \\ 2x + 4y = -10 \end{cases}$
11. $\begin{cases} y - 3x = 2 \\ 6x - y = 4 \end{cases}$
12. $\begin{cases} 2x - y + 3 = 0 \\ y = -x + 3 \end{cases}$

13. $\begin{cases} 4x + 2y = 2 \\ x = 1 \end{cases}$
14. $\begin{cases} y = 3 \\ x + y = 6 \end{cases}$
15. $\begin{cases} x = 1 \\ x = 5 \end{cases}$
16. $\begin{cases} x = -1 \\ y = 2 \end{cases}$

17. $\begin{cases} x = 0 \\ y = 0 \end{cases}$
18. $\begin{cases} 3x + 2y = 5 \\ 4y = -6x + 10 \end{cases}$
19. $\begin{cases} x = 3y - 14 \\ 3x + 2y = -9 \end{cases}$
20. $\begin{cases} 4y + x = 11 \\ 12y = x + 5 \end{cases}$

11-3   SOLUTION BY ELIMINATION

In Section 11-2, we learned how to solve a system of equations in two variables graphically. However, this method will not always allow us to determine accurately the correct solution. For example, if we solve the following system graphically:

$$\begin{cases} 3x + 2y = 11 \\ 6x - 2y = 10 \end{cases}$$

Notice that it is difficult to determine *accurately* the coordinates of the point of intersection.

$$\begin{bmatrix} x \approx 2.3 \\ y \approx 2.1 \end{bmatrix}$$

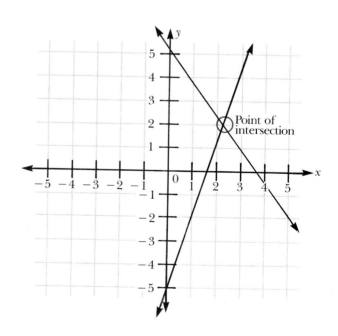

Thus, we must learn algebraic methods for solving systems of equations in two variables. As you will see, algebraic solutions usually take less time to do and give us more accurate results than a graphical method. One such algebraic method is called elimination.

Part I

As an example of the elimination method, consider the following system of equations.

$$\begin{cases} 3x + 2y = 11 \\ 6x - 2y = 10 \end{cases}$$

Notice that the coefficient of the y term in the first equation $(+2)$ and the coefficient of the y term in the second equation (-2) are opposite in value.

$$3x + 2\,y = 11$$
$$6x - 2\,y = 10$$

Thus, if we add these two equations together, we can eliminate the y terms $(2y + (-2y) = 0)$. Our result is $9x = 21$.

$$3x + 2y = 11$$
$$6x - 2y = 10$$
$$\overline{9x + 0y = 21}$$
$$9x = 21$$

Once one of the variables has been eliminated, we can then solve for the remaining variable. Our x value, then, is $x = \frac{7}{3}$.

$$9x = 21$$
$$x = \frac{21}{9}$$
$$x = \frac{7}{3}$$

By substituting $x = \frac{7}{3}$ into either equation of the given system, we can determine the corresponding y value. (We use the first equation.)

$$3x + 2y = 11$$
$$3\left(\frac{7}{3}\right) + 2y = 11$$
$$7 + 2y = 11$$
$$2y = 11 - 7$$
$$2y = 4$$
$$y = \frac{4}{2}$$
$$y = 2$$

The solution to the system $\begin{cases} 3x + 2y = 11 \\ 6x - 2y = 10 \end{cases}$ is the ordered pair $(\frac{7}{3}, 2)$. This should be checked in both equations.

Check:

$3x + 2y$	11
$3(\frac{7}{3}) + 2(2)$	11
$7 + 4$	
11	

$6x - 2y$	10
$6(\frac{7}{3}) - 2(2)$	10
$14 - 4$	
10	

So,

$(\frac{7}{3}, 2)$ is indeed the correct solution

EXAMPLE 1:   Solve $\begin{cases} x + y = 4 \\ 2x - y = 5 \end{cases}$ by elimination.

SOLUTION: Since the coefficients of the y terms are opposite in value, add the equations to eliminate the y terms.

$$
\begin{array}{r}
x + y = 4 \\
2x - y = 5 \\
\hline
3x + 0y = 9 \\
3x = 9
\end{array}
$$

Solve for x.

$$3x = 9$$

$$x = \frac{9}{3}$$

$$x = 3$$

Substitute $x = 3$ into either one of the equations given in the system and solve for y. (We use the first equation.)

$$x + y = 4$$
$$(3) + y = 4$$
$$y = 4 - 3$$
$$y = 1$$

Final answer: $(3, 1)$

The check is left for the student.

EXAMPLE 2:   Solve $\begin{cases} -5x + 2y = 4 \\ 5x - 3y = 4 \end{cases}$ by elimination.

SOLUTION: Since the coefficients of the x terms are opposite in value, we can eliminate the x terms and solve for y.

$$
\begin{array}{r}
-5x + 2y = 4 \\
5x - 3y = 4 \\
\hline
0x - y = 8 \\
-y = 8 \\
y = -8
\end{array}
$$

Example continued on next page.

We can now substitute $y = -8$ into either one of the given equations and solve for x. (We use the second equation.)

$$5x - 3y = 4$$
$$5x - 3(-8) = 4$$
$$5x + 24 = 4$$
$$5x = 4 - 24$$
$$5x = -20$$
$$x = \frac{-20}{5}$$
$$x = -4$$

Final answer: $(-4, -8)$

The check is left for the student.

NOTE:

> In examples 1 and 2, since we have only one common solution for each system, the equations of the systems are consistent.

Do the following practice set. Check your answers with the answers in the right-hand margin.

PRACTICE SET 11-3 (PART I)

Solve the following systems of equations by elimination (see example 1 or 2).

1. $\begin{cases} x - y = 6 \\ x + y = 4 \end{cases}$

2. $\begin{cases} 3x - 2y = 1 \\ 2x + 2y = -6 \end{cases}$

3. $\begin{cases} 2x + 3y = 9 \\ -2x + 4y = 5 \end{cases}$

1. $(5, -1)$
2. $(-1, -2)$
3. $(\frac{3}{2}, 2)$

Part II

In many systems of equations, we are not always able to eliminate a variable by immediately adding the two equations. Frequently, it is necessary to multiply each term of either equation by an appropriate number that makes it possible to eliminate a variable. As an example, consider the following system of equations.

$$\begin{cases} 2x + 5y = 18 \\ x + 4y = 6 \end{cases}$$

Notice that neither the coefficients of the x terms nor the coefficients of the y terms are opposite in value. Thus, we cannot eliminate one of the variables by immediately adding the two equations.

However, if we multiply each term of the second equation by -2, we can get the coefficients of the x terms to be opposite in value.

$$2x + 5y = 18 \qquad\qquad 2x + 5y = 18$$
$$-2(x + 4y = 6) \;\Rightarrow\; -2x - 8y = -12$$

We can now solve this equivalent system as we have done previously.

- Add the equations to eliminate the x terms, and solve for y.

$$2x + 5y = 18$$
$$-2x - 8y = -12$$
$$\overline{0x - 3y = 6}$$
$$-3y = 6$$
$$y = \frac{6}{-3}$$
$$y = -2$$

- Substitute $y = -2$ into one of the given equations and solve for x. (We use the first equation.)

$$2x + 5y = 18$$
$$2x + 5(-2) = 18$$
$$2x - 10 = 18$$
$$2x = 18 + 10$$
$$2x = 28$$
$$x = \frac{28}{2}$$
$$x = 14$$

So,

the ordered pair $(14, -2)$ is the solution to the given system of equations.

The check is left to the student.

In summary, to solve a system of linear equations in two variables by the elimination method, we first make the coefficients of either the x terms or the y terms opposite in value. We next add the equations to eliminate that term and solve for the remaining variable. Once this value has been determined, we can solve for the other variable by substituting the value found previously into either one of the given equations. Finally, we check the solution for both equations.

PROCEDURE:

> To solve a system of linear equations by the elimination method:
>
> 1. Make the coefficients of either the x terms or y terms in each equation additive inverses of each other.
>
> 2. Add the equations, thereby eliminating one of the variables.
>
> 3. Solve the resulting equation for the remaining variable.
>
> 4. Substitute the answer found in (3) into either one of the given equations and solve for the other variable.
>
> 5. Check the solution.

EXAMPLE 1: Solve $\begin{cases} 2x + 2y = 0 \\ 3x - y = 2 \end{cases}$ by elimination.

SOLUTION: Multiply the second equation by 2 to make the coefficients of the y terms opposite in value.

$$2x + 2y = 0 \qquad\qquad 2x + 2y = 0$$
$$2(3x - y = 2) \;\Rightarrow\; 6x - 2y = 4$$

Add the two equations to eliminate the y terms and solve for x.

$$
\begin{aligned}
2x + 2y &= 0 \\
6x - 2y &= 4 \\
\hline
8x + 0y &= 4 \\
8x &= 4 \\
x &= \frac{4}{8} \\
x &= \frac{1}{2}
\end{aligned}
$$

Substitute $x = \frac{1}{2}$ into either equation and solve for y. (We use the first equation.)

$$
\begin{aligned}
2x + 2y &= 0 \\
2\left(\frac{1}{2}\right) + 2y &= 0 \\
1 + 2y &= 0 \\
2y &= 0 - 1 \\
2y &= -1 \\
y &= -\frac{1}{2}
\end{aligned}
$$

Final answer: $\left(\dfrac{1}{2},\ -\dfrac{1}{2}\right)$

The check is left for the student.

EXAMPLE 2: Solve $\begin{cases} 3x + 4y = 8 \\ 2x + 3y = 7 \end{cases}$ by elimination.

SOLUTION: Multiply the first equation by 3 and the second equation by -4 to get the y terms opposite in value.

$$3(3x + 4y = 8) \;\Rightarrow\; 9x + 12y = 24$$
$$-4(2x + 3y = 7) \;\Rightarrow\; -8x - 12y = -28$$

Add and solve for x,

$$9x + 12y = 24$$
$$\underline{-8x - 12y = -28}$$
$$x + 0y = -4$$
$$x = -4$$

Substitute $x = -4$ into either equation and solve for y. (We use the first equation.)

$$3x + 4y = 8$$
$$3(-4) + 4y = 8$$
$$-12 + 4y = 8$$
$$4y = 8 + 12$$
$$4y = 20$$
$$y = \frac{20}{4}$$
$$y = 5$$

Final answer: $(-4, 5)$

The check is left for the student.

EXAMPLE 3:   Solve the system $\begin{cases} y = 2x + 4 \\ 6x + 2y = 13 \end{cases}$ by elimination.

SOLUTION: Before we can solve this system by elimination we must rewrite the first equation so that the x terms, y terms, and constant terms are lined up under each other.

$$y = 2x + 4 \quad \Rightarrow \quad -2x + y = 4$$
$$6x + 2y = 13 \qquad\quad 6x + 2y = 13$$

We are now ready to solve this system.
Multiply the first equation by 3 to eliminate the x terms and solve for y.

$$3(-2x + y = 4) \quad \Rightarrow \quad -6x + 3y = 12$$
$$6x + 2y = 13 \qquad\qquad \underline{6x + 2y = 13}$$
$$0x + 5y = 25$$
$$5y = 25$$
$$y = \frac{25}{5}$$
$$y = 5$$

Example continued on next page.

Substitute 5 for y into one of the equations and solve for x. (We use the second equation.)

$$6x + 2y = 13$$
$$6x + 2(5) = 13$$
$$6x + 10 = 13$$
$$6x = 13 - 10$$
$$6x = 3$$
$$x = \frac{3}{6}$$
$$x = \frac{1}{2}$$

Final answer: $\left(\dfrac{1}{2}, 5\right)$

The check is left for the student.

EXAMPLE 4: Solve $\begin{cases} 2x + 10y = 1 \\ x + 5y = 6 \end{cases}$ by elimination.

SOLUTION: Multiply the second equation by -2 and add the equations.

$$
\begin{array}{ll}
2x + 10y = 1 & 2x + 10y = 1 \\
\underline{-2(x + 5y = 6)} \Rightarrow & \underline{-2x - 10y = -12} \\
& 0x + 0y = -11 \\
& 0 = -11
\end{array}
$$

Notice that both the x and y variables are eliminated and that the result, $0 = -11$, is a false statement. This means that there is no solution for this system of equations since there are no values for x and y that can make $0 = -11$ true. Thus, the equations of this system are inconsistent.

Final answer: No solution

EXAMPLE 5: Solve $\begin{cases} x + y = 4 \\ 2x + 2y = 8 \end{cases}$ by elimination.

SOLUTION: Multiply the first equation by -2 and add the equations.

$$
\begin{array}{ll}
\underline{-2(x + y = 4)} \Rightarrow & -2x - 2y = -8 \\
2x + 2y = 8 & \underline{2x + 2y = 8} \\
& 0x + 0y = 0 \\
& 0 = 0
\end{array}
$$

Notice that both the x and y variables and the constant have all been eliminated and that the result, $0 = 0$, is a true statement. This means that no matter what the values of x and y are, the result will always be true. There are an infinite number of solutions for this system. Thus, the equations of this system are dependent.

Final answer: Infinite solutions

NOTE:

> When solving a system of linear equations in two variables algebraically, we find that an inconsistent system will always result in a false statement and a dependent system will always result in a true statement.

Do the following practice set. Check your answers with the answers in the right-hand margin.

PRACTICE SET 11-3 (Part II)

Solve the following system of equations by elimination (see example 1 or 2).

1. $\begin{cases} x - 2y = 6 \\ 3x + y = 4 \end{cases}$ 2. $\begin{cases} 2x + y = 2 \\ x + 3y = 6 \end{cases}$ 3. $\begin{cases} 2x - 2y = 10 \\ 5x + 3y = 5 \end{cases}$

1. $(2, -2)$
2. $(0, 2)$
3. $\left(\frac{5}{2}, -\frac{5}{2}\right)$

Solve the following system of equations by elimination (see example 4).

4. $\begin{cases} y = 3x - 1 \\ 6x + y = 2 \end{cases}$ 5. $\begin{cases} 2x - y = -3 \\ x - y = -5 \end{cases}$ 6. $\begin{cases} 9x - 2y + 3 = 0 \\ 2y - 6x + 5 = 0 \end{cases}$

4. $\left(\frac{1}{3}, 0\right)$
5. $(2, 7)$
6. $\left(-\frac{8}{3}, -\frac{21}{2}\right)$

Solve the following system of equations by elimination. State if the system is inconsistent or dependent (see example 5 or 6).

7. $\begin{cases} 3x - 4y = 1 \\ 6x - 8y = -2 \end{cases}$ 8. $\begin{cases} x + y = 6 \\ 3x + 3y = 18 \end{cases}$ 9. $\begin{cases} 4x - 6y = -8 \\ 2x - 3y = -4 \end{cases}$

7.   inconsistent
8.   dependent
9.   dependent

EXERCISE 11-3

Solve the following system of equations by the elimination method.

1. $\begin{cases} x + y = 8 \\ x - y = 4 \end{cases}$ 2. $\begin{cases} 2x - y = 5 \\ x + y = 4 \end{cases}$ 3. $\begin{cases} x + 3y = 9 \\ -x - 2y = 6 \end{cases}$ 4. $\begin{cases} -x - 4y = 16 \\ x - 2y = -4 \end{cases}$

5. $\begin{cases} 3x + 2y = 9 \\ x - y = 8 \end{cases}$ 6. $\begin{cases} 4x - 3y = 4 \\ -2x - y = 8 \end{cases}$ 7. $\begin{cases} 2x - 3y = 2 \\ 4x + y = 4 \end{cases}$ 8. $\begin{cases} 2x - 6y = 24 \\ 3x - 2y = 1 \end{cases}$

9. $\begin{cases} 3x + 5y = -8 \\ -6x + 2y = -8 \end{cases}$ 10. $\begin{cases} 4x + 6y = 8 \\ 2x + 4y = -3 \end{cases}$ 11. $\begin{cases} 2x - 3y = -7 \\ 3x + 2y = 9 \end{cases}$ 12. $\begin{cases} 3x + 4y = -1 \\ -4x - 5y = -1 \end{cases}$

13. $\begin{cases} 2x - 3y = 9 \\ -3x + 5y = -17 \end{cases}$ 14. $\begin{cases} 6x - 2y = 8 \\ -3x + y = -4 \end{cases}$ 15. $\begin{cases} 2x - 6y = -2 \\ 3x + 5y = 4 \end{cases}$ 16. $\begin{cases} 9x + 3y = -1 \\ 7x + 2y = -1 \end{cases}$

17. $\begin{cases} 4x - 9y = 17 \\ 12x - 27y = -17 \end{cases}$ 18. $\begin{cases} 5x + 5y - 10 = 0 \\ -3x + 2y - 2 = 0 \end{cases}$ 19. $\begin{cases} 4y = -6x + 4 \\ 9x = 6y + 2 \end{cases}$ 20. $\begin{cases} y = 2x + 4 \\ y = 6x - 3 \end{cases}$

11-4   SOLUTION BY SUBSTITUTION

Another algebraic method that can be used to eliminate one of the variables when solving a system of equations is the substitution method. To use this method, we first solve one of the given equations for one variable in terms of the other variable. We then substitute this solution into the other given equation. As an example, consider the following system.

$$\begin{cases} x + y = 5 \\ x - y = 1 \end{cases}$$

To solve this system by substitution:

Solve the first equation for x in terms of y.

$$\begin{cases} x + y = 5 \\ x - y = 1 \end{cases} \Rightarrow \begin{matrix} x + y = 5 \\ x \quad = 5 - y \end{matrix}$$

Next, substitute this value of x into the second equation.

$$\begin{cases} x + y = 5 \\ x - y = 1 \end{cases} \Rightarrow \begin{matrix} x - y = 1 \\ (5 - y) - y = 1 \end{matrix}$$

Now, solve this equation for y.

$$(5 - y) - y = 1$$
$$5 - y - y = 1$$
$$5 - 2y = 1$$
$$-2y = 1 - 5$$
$$-2y = -4$$
$$y = \frac{-4}{-2}$$
$$y = 2$$

Finally, the corresponding x value can be found by substituting $y = 2$ into either one of the given equations and solving for x. (We use the first equation.)

$$x + y = 5$$
$$x + (2) = 5$$
$$x + 2 = 5$$
$$x = 5 - 2$$
$$x = 3$$

Thus,

the ordered pair $(3, 2)$ satisfies the given system.

The check is left to the student.

PROCEDURE:

To solve a system of linear equations in two variables by substitution:

1. Solve either one of the equations for one variable in terms of the other variable.

2. Substitute the expression into the second equation for its corresponding variable.

3. Solve this equivalent second equation.

4. Solve one of the given equations for the unknown variable, using the result from (3).

5. Check the solution.

EXAMPLE 1: Solve $\begin{cases} y = 2x \\ x + y = 15 \end{cases}$ by substitution.

SOLUTION: Since the first equation is already solved for y in terms of x, we can substitute $y = 2x$ into the second equation for y and solve for x.

$$x + y = 15$$
$$\Downarrow$$
$$x + (2x) = 15$$
$$3x = 15$$
$$x = \frac{15}{3}$$
$$x = 5$$

Now, solve for y by substituting $x = 5$ into either one of the given equations. (We use the first equation.)

$$y = 2\ x$$
$$\downarrow$$
$$y = 2(5)$$
$$y = 10$$

Final answer: (5, 10) consistent system

The check is left for the student.

EXAMPLE 2: Solve $\begin{cases} 3x + 2y = 7 \\ x = y + 4 \end{cases}$ by substitution.

Example continued on next page.

SOLUTION: Since the second equation is already solved for x in terms of y, we can substitute $x = y + 4$ into the first equation for x and solve for y.

$$3\,x + 2y = 7$$
$$3(y + 4) + 2y = 7$$
$$3y + 12 + 2y = 7$$
$$3y + 2y = 7 - 12$$
$$5y = -5$$
$$y = \frac{-5}{5}$$
$$y = -1$$

Now, solve for x by substituting $y = -1$ into either one of the given equations. (We use the second equation.)

$$x = y + 4$$
$$x = (-1) + 4$$
$$x = -1 + 4$$
$$x = 3$$

Final answer: $(3, -1)$ consistent system

The check is left for the student.

EXAMPLE 3: Solve $\begin{cases} 2x + 3y = 4 \\ x - 2y = -5 \end{cases}$ by substitution.

SOLUTION: Solve the second equation for x in terms of y.

$$\begin{cases} 2x + 3y = 4 \\ x - 2y = -5 \end{cases} \Rightarrow x - 2y = -5$$
$$x = -5 + 2y$$
$$\text{or}$$
$$x = 2y - 5$$

Substitute $x = 2y - 5$ into the first equation for x and solve for y.

$$2\,x + 3y = 4$$
$$2(2y - 5) + 3y = 4$$
$$4y - 10 + 3y = 4$$
$$4y + 3y = 4 + 10$$
$$7y = 14$$
$$y = \frac{14}{7}$$
$$y = 2$$

Substitute $y = 2$ into either one of the given equations and solve for x. (We use the second equation.)

$$x - 2\,y = -5$$
$$\downarrow$$
$$x - 2\,(2) = -5$$
$$x - 4 = -5$$
$$x = -5 + 4$$
$$x = -1$$

Final answer: $(-1, 2)$ consistent system

The check is left for the student.

EXAMPLE 4: Solve $\begin{cases} x - 2y = 3 \\ 3x - 6y = 2 \end{cases}$ by substitution.

SOLUTION: Solve the first equation for x in terms of y.

$$\begin{cases} \boxed{x - 2y = 3} \Rightarrow x - 2y = 3 \\ 3x - 6y = 2 \qquad\qquad x = 3 + 2y \end{cases}$$
$$\text{or}$$
$$x = 2y + 3$$

Substitute $x = 2y + 3$ into the second equation for x and solve for y.

$$3\,x - 6y = 2$$
$$3\,(2y + 3) - 6y = 2$$
$$6y + 9 - 6y = 2$$
$$6y - 6y = 2 - 9$$
$$0 = -7$$

Notice our result is a false statement. This indicates that the equations of this system are inconsistent and hence have no common solution.

Final answer: No solution (inconsistent system)

EXAMPLE 5: Solve $\begin{cases} 2x - 2y = 6 \\ x - y = 3 \end{cases}$ by substitution.

SOLUTION: Solve the second equation for x in terms of y.

$$\begin{cases} 2x - 2y = 6 \\ \boxed{x - y = 3} \Rightarrow x - y = 3 \\ \qquad\qquad\qquad x = 3 + y \end{cases}$$
$$\text{or}$$
$$x = y + 3$$

Example continued on next page.

Substitute $x = y + 3$ into the first equation for x and solve for y.

$$2x - 2y = 6$$
$$2(y + 3) - 2y = 6$$
$$2y + 6 - 2y = 6$$
$$2y - 2y = 6 - 6$$
$$0 = 0$$

Notice that our result is a true statement. This indicates that the equations of this system are dependent and hence have an infinite number of solutions.

Final answer: Infinite number of solutions (dependent system)

Do the following practice set. Check your answers with the answers in the right-hand margin.

PRACTICE SET 11-4

Solve the following system of equations by substitution (see example 1 or 2).

1. $\begin{cases} y = x \\ x + y = 16 \end{cases}$
2. $\begin{cases} x = 3y + 4 \\ x + 2y = 14 \end{cases}$
3. $\begin{cases} 3x + 2y = 8 \\ x - 6 = y \end{cases}$

1. $(8, 8)$
2. $(10, 2)$
3. $(4, -2)$

Solve the following system of equations by substitution (see example 3).

4. $\begin{cases} 3x = 2y + 2 \\ 4x - y = 6 \end{cases}$
5. $\begin{cases} x - 3y = 4 \\ 2x + 3y = 20 \end{cases}$
6. $\begin{cases} 2x + y = 3 \\ 2x + 3y = 7 \end{cases}$

4. $(2, 2)$
5. $(8, \frac{4}{3})$
6. $(\frac{1}{2}, 2)$

Solve the following system of equations by substitution. State if the system is inconsistent or dependent (see example 4 or 5).

7. $\begin{cases} 3x + 3y = 9 \\ x + y = 3 \end{cases}$
8. $\begin{cases} x - 3y = 4 \\ 3x - 9y = 2 \end{cases}$
9. $\begin{cases} y = 2x + 5 \\ 2y = 4x + 10 \end{cases}$

7. Dependent
8. Inconsistent
9. Dependent

EXERCISE 11-4

Solve the following systems of equations by substitution.

1. $\begin{cases} y = x \\ x + y = 16 \end{cases}$
2. $\begin{cases} x = y \\ x + y = 12 \end{cases}$
3. $\begin{cases} 2x + y = 8 \\ y = 2x \end{cases}$
4. $\begin{cases} x = 2y \\ x + 2y = -4 \end{cases}$

5. $\begin{cases} y = x + 2 \\ x + y = 6 \end{cases}$

6. $\begin{cases} y = x - 3 \\ 2x - y = -1 \end{cases}$

7. $\begin{cases} 2x - 3y = -12 \\ y = x + 6 \end{cases}$

8. $\begin{cases} x + 2y = 1 \\ y = 2x - 2 \end{cases}$

9. $\begin{cases} y = x - 6 \\ 3x - 2y = 0 \end{cases}$

10. $\begin{cases} x = 2y - 3 \\ 2x + y = -1 \end{cases}$

11. $\begin{cases} 2x - y = 8 \\ x - 2y = 1 \end{cases}$

12. $\begin{cases} x = 2y + 1 \\ 3y - 2x = 3 \end{cases}$

13. $\begin{cases} x - y = 6 \\ 2x + y = 0 \end{cases}$

14. $\begin{cases} y + x = 8 \\ 3x - 2y = -1 \end{cases}$

15. $\begin{cases} 2x + 4y = 8 \\ 2x - y = 8 \end{cases}$

16. $\begin{cases} x - 2y = -2 \\ 2x - y = 5 \end{cases}$

17. $\begin{cases} 2x + y = 0 \\ x - \dfrac{1}{2}y = 6 \end{cases}$

18. $\begin{cases} x - 2y = 0 \\ \dfrac{1}{4}x + \dfrac{3}{2}y = 6 \end{cases}$

19. $\begin{cases} 3x + 2y = 9 \\ x - 2y = 5 \end{cases}$

20. $\begin{cases} x - 3y = 4 \\ y = 3x \end{cases}$

11-5 USING SYSTEMS OF EQUATIONS IN TWO VARIABLES TO SOLVE VERBAL PROBLEMS

A. Solving Number Problems

In many instances, a problem can be solved more easily by using *two* variables rather than one variable. We demonstrate this by using number problems.

Solve the following problem by using two variables: The sum of two numbers is 18. Twice the larger number, decreased by 3 times the smaller number, is equal to 21. What are the numbers? Since we are dealing with *two* numbers:

$$\text{Let } x = \text{the larger number.}$$

and

$$\text{Let } y = \text{the smaller number,}$$

Now, translate the statements into equations.

The sum of two numbers is 18.

$$x + y \qquad = 18$$

Twice the larger, decreased by 3 times the smaller is 21.

$$2x \qquad\qquad - 3y \qquad = 21$$

Therefore,

the two equations we must solve are a *system of equations.* $\begin{cases} x + y = 18 \\ 2x - 3y = 21 \end{cases}$

To solve this system, we may use either the elimination method or the substitution method. (We choose the elimination method.)

Multiply the first equation by -2 to eliminate the x terms.

$\begin{cases} -2(x + y = 18) \Rightarrow -2x - 2y = -36 \\ \quad 2x - 3y = 21 \qquad\quad 2x - 3y = 21 \end{cases}$

Add the two equations and solve for y.

$$-2x - 2y = -36$$
$$\underline{2x - 3y = 21}$$
$$0x - 5y = -15$$
$$-5y = -15$$
$$y = \frac{-15}{-5}$$
$$y = 3$$

Substitute $y = 3$ into the first equation and solve for x.

$$x + y = 18$$
$$x + 3 = 18$$
$$x = 18 - 3$$
$$x = 15$$

Thus,

$$x = 15 \quad \text{and} \quad y = 3$$

Moreover, the larger number is 15 and the smaller number is 3.

In summary, to solve number problems by using two variables, we let x equal one of the numbers and y equal the other number. Once this has been decided, we translate the problem into two algebraic equations. We then solve this system of equations by using either the elimination or substitution method.

PROCEDURE:

To solve number problems by using two variables:

1. Let x equal one of the numbers and y equal the other number.

2. Translate the problem into two algebraic equations.

3. Solve this system of equations by using:
   a. The elimination method.

      or

   b. The substitution method.

EXAMPLE 1: Solve the following problem by using two variables.

Find two numbers whose sum is 43 and whose difference is 17.

SOLUTION: Assign the variables to a number.

Let $x =$ the first number

Let $y =$ the second number

Translate the problem into equations.

"... two numbers whose sum is 43 ..."

$$x + y \qquad = 43$$

"... two numbers ... whose difference is 17"

$$x - y \qquad = 17$$

Solve the system $\begin{cases} x + y = 43 \\ x - y = 17 \end{cases}$

Solve for x

$$x + y = 43$$
$$\underline{x - y = 17}$$
$$2x \quad\;\; = 60$$
$$2x = 60$$
$$x = \frac{60}{2}$$
$$x = 30$$

Solve for y

$$x + y = 43$$
$$(30) + y = 43$$
$$30 + y = 43$$
$$y = 43 - 30$$
$$y = 13$$

Final answer: The first number is 30; the second number is 13.
The check is left to the student.

EXAMPLE 2:   Solve the following problem by using two variables.

Separate 108 into two numbers such that 5 times the first number less twice the second number is equal to 85.

SOLUTION: Assign the variables to a number.

> Let x = the first number
> Let y = the second number

Translate the problem into equations.

"Separate 108 into two numbers ..."

$$x + y = 108$$

[The sum of the both numbers must equal 108.]

"... 5 times the first number less twice the second number is equal to 85"

$$5x \qquad\qquad - 2y \qquad\qquad = 85$$

Solve the system $\begin{cases} x + y = 108 \\ 5x - 2y = 85 \end{cases}$

Example continued on next page.

Solve for x		**Solve for y**

$$x + y = 108 \Rightarrow 2x + 2y = 216$$
$$5x - 2y = 85 \qquad 5x - 2y = 85$$
$$\overline{\qquad 7x \qquad = 301}$$
$$7x = 301$$
$$x = \frac{301}{7}$$
$$x = 43$$

$$x + y = 108$$
$$(43) + y = 108$$
$$43 + y = 108$$
$$y = 108 - 43$$
$$y = 65$$

Final answer: The first number is 43; the second number is 65.

The check is left to the student.

EXERCISE 11-5A

Solve the following problems by using two variables.

1. Find two numbers such that their sum is 73 and their difference is 23.

2. Find two numbers whose difference is 12 and whose sum is 46.

3. The sum of two numbers is 39. If twice the larger number is added to 3 times the smaller number, the sum is 97. Find the numbers.

4. The sum of two numbers is 39. If twice the larger number is added to 3 times the smaller number, the sum is 95. Find the numbers.

5. Find two numbers such that the sum of twice the larger number and twice the smaller number is 78. Also 3 times the larger, less twice the smaller, is 22.

6. Find two numbers such that the difference of 3 times the larger number and twice the smaller number is 16. Also twice the larger number, added to twice the smaller number, is 34.

7. Separate 113 into two parts such that 3 times the first part subtracted from twice the second part is equal to 126.

8. Separate 300 into two parts such that the difference between 5 times the first part and 3 times the second part is 132.

9. If 3 times the smaller of two numbers is added to 5 times the larger number, the result is 43. If twice the larger number is decreased by twice the smaller number, the result is 14. Find the numbers.

10. If 4 times the smaller of two numbers is subtracted from twice the larger number, the result is 22. Also, if 7 times the larger number is added to 3 times the smaller number, the result is 298. Find the numbers.

B. Strategies for Solving Word Problems by Using Systems of Equations

In Chapter 5, we discussed problem-solving techniques that we used to solve verbal problems such as coin, mixture, investment, and motion.

One of the techniques presented was first to identify the unknown in the problem and assign to it a variable (x). We then identified any other unknown and tried to represent it in terms of that variable (x).

Quite often, a problem may be solved more easily by using *two* variables, rather than by trying to represent each unknown in terms of a single variable. This means we will be using systems of equations, which is the basis of our discussion in this section—to use linear equations in two variables to solve word problems.

To do this, we use the problem-solving techniques discussed in Chapter 5, except we modify certain steps to reflect the fact that we are now using a system of equations to solve word problems.

TO SOLVE WORD PROBLEMS BY USING A SYSTEM OF EQUATIONS

1. Read the problem carefully.
2. Identify *each* unknown and represent it by using a *different* letter.
3. Construct a chart.
4. Determine the relationships in the problem and represent these relationships by algebraic equations.
5. Solve the system of equations.
6. Check the solution.

We now discuss solving coin, mixture, investment, and motion problems by using systems of equations.

C. Solving Coin Problems

EXAMPLE 1: José collected $25.05 in dimes and quarters from a pay telephone. The number of quarters was 5 more than 3 times the number of dimes. How many of each coin did he collect?

STEP 1: READ THE PROBLEM CAREFULLY.

STEP 2: IDENTIFY EACH UNKNOWN AND REPRESENT IT BY USING A DIFFERENT LETTER.

There are two unknowns in this problem.

- The number of dimes collected
- The number of quarters collected

> and
> Let d = the number of *dimes* collected
> Let q = the number of *quarters* collected

STEP 3: CONSTRUCT A CHART.

		$N \cdot V = T$	
Type of Coin	*Number of Coins*	*Value of Each Coin*	*Total Value*
Dime	d	10¢	$10d$
Quarter	q	25¢	$25q$

STEP 4: DETERMINE THE RELATIONSHIP IN THE PROBLEM.

There are *two* relationships in this problem.

1. The total value of dimes + The total value of quarters = \$25.05

$$10d \qquad\qquad + \qquad\qquad 25q \qquad\qquad\quad = 2505$$

2. The number of quarters was 5 more than 3 times the number of dimes.

$$q \qquad\qquad = \qquad 5 + \qquad\qquad 3d$$

Thus, we must solve the system $\begin{cases} 10d + 25q = 2505 \\ q = 5 + 3d \end{cases}$

STEP 5: SOLVE THE SYSTEM.

We choose to solve the system by substitution.

Solve for d | **Solve for q**

$\begin{cases} 10d + 25q = 2505 \\ q = 5 + 3d \quad \text{(substitute } 5 + 3d \text{ for } q) \end{cases}$

$$10d + \quad 25\,q \quad = 2505 \qquad\qquad q = 5 + 3d$$

$$10d + 25\,(\overbrace{5 + 3d}) = 2505 \qquad\qquad q = 5 + 3(28)$$

$$10d + 125 + 75d = 2505 \qquad\qquad q = 5 + 84$$

$$10d + 75d = 2505 - 125 \qquad\qquad q = 89$$

$$85d = 2380$$

$$d = \frac{2380}{85}$$

$$d = 28$$

Thus,

José collected 28 dimes and 89 quarters.

STEP 6: CHECK.

28 dimes has a total value of \$2.80. $\qquad\qquad 28 \times 10¢ = 280¢$

89 quarters has a total value of \$22.25. $\qquad\qquad 89 \times 25¢ = 2225¢$

The total value of dimes and quarters is \$25.05. $\qquad\qquad \$2.80 + \$22.25 = \$25.05$

Also,

The number of quarters is 5 more than
3 times the number of dimes.

$$q = 5 + 3d$$
$$89 = 5 + 3(28)$$
$$89 = 5 + 84$$
$$89 = 89$$

Final answer: 28 dimes; 89 quarters.

EXERCISE 11-5C

Solve the following problems by using systems of equations.

1. A grocer has a donation jar next to his cash register in which he collects donations for a local charity. Upon opening the jar, he found that it contained $23 in nickels and dimes. Furthermore, the number of nickels was twice the number of dimes. How many coins of each type were in the jar?

2. A vending machine that accepts only quarters and dimes had $33.45 when Darryl emptied it. If the number of quarters was 12 more than the number of dimes, how many of each coin were there?

3. A telephone repair man collected $15.65 in nickels and dimes from a pay telephone. All total, there were 165 coins. How many of each coin were there?

4. At the "Chuck-A-Luck" booth at a church festival, one of the booth operators collected $7.40 in quarters and dimes for one round. In all, she collected a total of 32 coins. How many of each coin were there?

5. A gambler walked away from the Blackjack table one night with $225 in $5 chips and $10 chips. There were 3 times as many $5 chips as $10 chips. How many chips of each kind did he have?

D.   Solving Mixture Problems

EXAMPLE 1:   The Portland Nut Shop sells a mixture of almonds and cashews for $1.75 per pound. If the almonds are worth $1.40 per pound and the cashews are worth $2.10 per pound, how many pounds of each kind must be used to make a mixture of 100 pounds?

STEP 1:   READ THE PROBLEM CAREFULLY.

STEP 2:   IDENTIFY EACH UNKNOWN AND REPRESENT IT BY USING A DIFFERENT LETTER.

There are two unknowns in this problem.

■   The number of pounds of almonds needed

■   The number of pounds of cashews needed

Example continued on next page.

> Let x = the number of pounds of almonds needed
>
> and Let y = the number of pounds of cashews needed

STEP 3: CONSTRUCT A CHART.

Type of Nut	Pounds Needed	Price per Pound	Total Value
Almonds	x	$1.40	$140x$ (in cents)
Cashews	y	$2.10	$210y$ (in cents)
Mixture	100	$1.75	$(175)(100)$ (in cents)

STEP 4: DETERMINE THE RELATIONSHIP IN THE PROBLEM.

There are two relationships in this problem.

1. | The total number of pounds of almonds | + | The total number of pounds of cashews | = | The total number of pounds of the mixture |

$$x \quad + \quad y \quad = \quad 100$$

2. | The total value of the almonds | + | The total value of the cashews | = | The total value of the mixture |

$$140x \quad + \quad 210y \quad = \quad (175)(100)$$

Thus, we must solve the system $\begin{cases} x + y = 100 \\ 140x + 210y = (175)(100) \end{cases}$

STEP 5: SOLVE THE SYSTEM.

We choose to solve the system by elimination.

Solve for y

$\begin{cases} x + y = 100 & \Rightarrow \quad -140x - 140y = -14000 \\ 140x + 210y = 17500 & \quad\quad 140x + 210y = \quad 17500 \end{cases}$

$$\begin{array}{r} -140x - 140y = -14000 \\ 140x + 210y = 17500 \\ \hline 70y = 3500 \end{array}$$

$$70y = 3500$$

$$y = \frac{3500}{70}$$

$$y = 50$$

Solve for x

$$x + y = 100$$

$$x + (50) = 100$$

$$x + 50 = 100$$

$$x = 100 - 50$$

$$x = 50$$

Thus,

50 pounds of almonds must be mixed with 50 pounds of cashews.

STEP 6: CHECK.

The number of pounds of almonds plus the number of pounds of cashews equals 100 pounds are the number of pounds that are needed for the mixture.

$$50 + 50 = 100$$

Also:

- The total value of the almonds is $70. $(\$1.40)(50) = \$70$
- The total value of the cashews is $105. $(\$2.10)(50) = \$105$
- The total value of the mixture is $175. $(\$70)(\$105) = \$175$

Final answer: 50 pounds almonds; 50 pounds cashews

EXERCISE 11-5D

Solve the following problems by using systems of equations.

1. The Wednesday special at the Deli-House is pickled bean salad which sells for $1.98 per pound. In making the salad, Ron had to mix waxed beans worth $1.80 per pound and kidney beans worth $2.20 per pound. How many pounds of each kind of bean did Ron use to make 50 pounds of salad?

2. Rubino's Italian Food Store sells dried *favés* for $1.60 per pound and dried *ceccis* (chick peas) for $1.90 per pound. How many pounds of each must be mixed to sell a 100-pound mixture for $1.72 per pound?

3. Leroy Crushed Stone sells #1CR (Crusher Run) stone for $12 per ton and No. 2 stone for $9 per ton. How many tons of each stone must be mixed so that Bennett Paving Company can purchase 12 tons for $10 per ton?

4. Hazel had to mix A1 grass seed worth $1.55 per pound and A19 grass seed worth 90¢ per pound in order to produce a 35-pound mixture worth $1.16 per pound. How many pounds of each type of seed did Hazel use for the mixture?

5. A roadway fruit stand was selling McIntosh apples for $1.10 per pound and Cortland apples for $1.50 per pound. A "mixed bag" would cost a customer $1.26 per pound. How many pounds of each kind would be included in a 15-pound mixed bag?

E. Solving Investment Problems

EXAMPLE 1: Mr. Fenby invested a sum of money from his pension. He invested part of it in a stock that yielded 18 percent and the remainder in a certificate of deposit that yielded 12 percent. The amount of money invested in the stock was $500 more than that invested in certificates. If the total annual interest Mr. Fenby received was $630, how much did he invest at each rate?

STEP 1: READ THE PROBLEM CAREFULLY.

STEP 2: IDENTIFY EACH UNKNOWN AND REPRESENT IT BY USING A DIFFERENT LETTER.

There are two unknowns in this problem.

■ The amount of money invested in stock at 18 percent

■ The amount of money invested in certificates at 12 percent

> and Let $x =$ amount of money invested in stock at 18 percent
> Let $y =$ amount of money invested in certificates at 12 percent

STEP 3: CONSTRUCT A CHART.

Investment	Principal	Rate	Interest
18% stock	x	0.18	$0.18x$
12% certificates	y	0.12	$0.12y$

STEP 4: DETERMINE THE RELATIONSHIP IN THE PROBLEM.

There are two relationships in this problem.

1. The amount invested in stock was $500 more than the amount invested in certificates

$$x \qquad = \qquad 500 + \qquad y$$

2. The total interest from stock $+$ The total interest from certificates $=$ The total amount of interest earned for the year

$$0.18x \quad + \quad 0.12y \quad = \quad \$630$$

Thus, we must solve the system $\begin{cases} x = 500 + y \\ 0.18x + 0.12y = 630 \end{cases}$

STEP 5: SOLVE THE SYSTEM.

We choose the substitution method.

Solve for y

$$\begin{cases} x = 500 + y & \text{(substitute } 500 + y \text{ for } x) \\ 0.18x + 0.12y = 630 \end{cases}$$

$$0.18\,x + 0.12y = 630$$

$$0.18\,(500 + y) + 0.12y = 630$$

$$90 + 0.18y + 0.12y = 630$$

$$0.18y + 0.12y = 630 - 90$$

$$0.3y = 540$$

$$y = \frac{540}{0.3}$$

$$y = 1800$$

Solve for x

$$x = 500 + y$$

$$x = 500 + (1800)$$

$$x = 500 + 1800$$

$$x = 2300$$

Thus, $2300 was invested in stocks at 18 percent and $1800 was invested in certificates of deposit at 12 percent.

STEP 6: CHECK.

The amount invested in stock (x) is equal to $500 more than the amount invested in certificates (y).

x	$500 + y$
2300	500 + 1800
	2300

Also:

■ The total interest from stock at 18 percent is $414.　$0.18(2300) = \$414$

■ The total interest from certificates at 12 percent is $216.　$0.12(1800) = \$216$

■ The total interest received is $630.　$\$414 + \$216 = \$630$

Final answer:　$2300 in stocks (at 18 percent); $1800 in certificates (at 12 percent)

EXERCISE 11-5E

Solve the following problems by using systems of equations.

1. Mrs. Dancer invested $2500, part at 8 percent and the rest at 11 percent. The total amount of interest she received was $251. How much did she invest at each rate?

2. Bob earned $336 from some money he invested at 7 percent and some at 5 percent. The total amount of money he invested was $5000. How much was invested at each rate?

3. Marilyn invested $12,600, part at 4 percent and the balance at $5\frac{1}{2}$ percent. The income on

4. Ronolog, Inc. invested $50,000, part at 6 percent and the rest at 10 percent. If the total

Exercise problems continued on next page.

the 4 percent investment was $29 more than the income on the $5\frac{1}{2}$ percent investment. How much did she invest at each rate?

5. Gloria invested $5000 at 12 percent. How much additional money must she invest at 6 percent in order to have her total annual interest equal 10 percent of her entire investment?

annual interest earned from both investments were equal, what was the amount invested at each rate?

F. Solving Motion Problems

Before we begin to solve motion problems by using systems, consider the following.

If an airplane can travel at a certain rate of speed in *still air*, then it stands to reason that it will travel at a higher rate if it travels *with the wind* and at a lower rate if it travels *against the wind*. For example, suppose an airplane can travel at a rate of 500 miles per hour in still air. If a wind is blowing at a rate of 20 miles per hour, then:

When the plane is flying with the wind, its speed is 520 miles per hour.

$$500 \quad + \quad 20 \quad = \quad 520$$

rate in still air   rate of wind   rate with wind

and

When the plane is flying against the wind, its speed is 480 miles per hour.

$$500 \quad - \quad 20 \quad = \quad 480$$

rate in still air   rate of wind   rate against wind

This leads to the following relationship.

> If r = the rate in still air
> and c = the rate of the wind,
> then $r + c$ = the rate with the wind
> and $r - c$ = the rate against the wind.

EXAMPLE 1: A weather plane left an airport and flew with the wind into a storm center for 2 hours. The distance traveled was 1200 miles. On the return trip back to the airport, it flew the same route against the same wind for 3 hours. Find the rate of speed of the plane in still air and the speed of the wind.

STEP 1: READ THE PROBLEM CAREFULLY.

STEP 2: IDENTIFY EACH UNKNOWN AND REPRESENT IT BY USING A DIFFERENT LETTER.

There are two unknowns in this problem.

■ The rate of speed of the airplane in still air.

■ The rate of the wind.

and
> Let r = the rate of speed of the airplane in still air
> Let c = the rate of the wind.

STEP 3: SET UP A CHART (SHOWING RATE · TIME = DISTANCE RELATIONSHIP).

| | $R \cdot T = D$ | | |
Direction	Rate	Time	Distance
With wind	$r + c$	2	$2(r + c)$
Against wind	$r - c$	3	$3(r - c)$

NOTE:

> The rate of speed of the plane when flying with the wind is given as $r + c$ (rate in still air plus rate of wind), and the rate of the plane when flying against the wind is given as $r - c$ (rate in still air minus rate of wind).

STEP 4: DETERMINE THE RELATIONSHIP IN THE PROBLEM.

There are two relationships in this problem.

1. The distance traveled with the wind is 1200 miles.

$$2(r + c) = 1200$$

or

$$2r + 2c = 1200$$

2. The distance traveled against the wind is 1200 miles.

$$3(r - c) = 1200$$

or

$$3r - 3c = 1200$$

Thus, the system we must solve is $\begin{cases} 2r + 2c = 1200 \\ 3r - 3c = 1200 \end{cases}$

STEP 5: SOLVE THE SYSTEM.

We choose elimination.

Solve for r

$$\begin{cases} 2r + 2c = 1200 \Rightarrow 6r + 6c = 3600 \\ 3r - 3c = 1200 \Rightarrow 6r - 6c = 2400 \end{cases}$$

$$6r + 6c = 3600$$
$$\underline{6r - 6c = 2400}$$
$$12r \quad\;\; = 6000$$
$$12r = 6000$$
$$r = \frac{6000}{12}$$
$$r = 500$$

Solve for c

$$2r + 2c = 1200$$
$$2(500) + 2c = 1200$$
$$1000 + 2c = 1200$$
$$2c = 1200 - 1000$$
$$2c = 200$$
$$c = \frac{200}{2}$$
$$c = 100$$

Thus, the rate of the airplane in still air is 500 miles per hour and the rate of the wind is 100 miles per hour.

STEP 6: CHECK.

With wind $\begin{cases} \text{The rate with the wind is} \\ \text{600 miles per hour.} \\\\ \text{The distance traveled with the wind} \\ \text{for 2 hours is 1200 miles.} \end{cases}$

$$r + c = \text{rate with wind}$$
$$500 + 100 = 600$$
$$d = r \cdot t$$
$$d = (600)(2)$$
$$d = 1200$$

Against wind $\begin{cases} \text{The rate against the wind is 400} \\ \text{miles per hour.} \\\\ \text{The distance traveled against the wind} \\ \text{for 3 hours is 1200 miles.} \end{cases}$

$$r - c = \text{rate against wind}$$
$$500 - 100 = 400$$
$$d = r \cdot t$$
$$d = (400)(3)$$
$$d = 1200$$

Final answer: The rate of speed of the airplane in still air is 500 miles per hour; the rate of speed of the wind is 100 miles per hour.

EXERCISE 11-5F

Solve the following problems by using systems of equations.

1. Milt has calculated that he can paddle his canoe in still water at an average rate of 6 miles per hour. One Sunday afternoon, Milt paddled

2. Calvin is able to jog at an average clip of 6 miles per hour (with no wind). One cold, winter day Calvin jogged $1\frac{1}{2}$ hours against a

upstream in 4 hours, whereas he took only 2 hours to return. Find the water current.

bitter cold wind and $\frac{1}{2}$ hour with the same wind. Find the speed of the wind.

NOTE:

> We may think of water current as wind current.
> Upstream is *against* current.
> Downstream is *with* current.

3. One normal working day, Superman was busy patrolling the skies. He spent 6 hours flying *against* a 25 miles per hour wind and 2 hours flying with the same wind. Find Superman's rate of speed in still air.

4. A Mountain Top Airway's plane left an airport and flew with the wind for 3 hours, covering 1800 miles. On the return trip, it flew against the same wind for 6 hours. Find the rate of the plane in still air and the speed of the wind.

5. Slim's favorite fishing hole is 30 miles downstream from his house. One Saturday morning, it took Slim 2 hours to travel by motorboat to get to the fishing hole, and it required 5 hours for him to get back home. Find the rate of speed of the motorboat and the rate of the current.

REVIEW EXERCISES

In problems 1–10, solve each system of equations graphically. State if the system is consistent, inconsistent, or dependent.

1. $\begin{cases} 2x - 3y = -7 \\ 3x + y = 17 \end{cases}$

2. $\begin{cases} x + y = -4 \\ 4x - 3y = -9 \end{cases}$

3. $\begin{cases} 4x + 2y = -6 \\ y = -2x + 3 \end{cases}$

4. $\begin{cases} x - y = 1 \\ 3x - 3y = 3 \end{cases}$

5. $\begin{cases} y = 3x \\ y = -2x \end{cases}$

6. $\begin{cases} y = 2x + 1 \\ y = 2x - 4 \end{cases}$

7. $\begin{cases} x = 4 \\ y = -6 \end{cases}$

8. $\begin{cases} y = x \\ y = -x - 3 \end{cases}$

9. $\begin{cases} 3x + y = 3 \\ 6x + 2y = 6 \end{cases}$

10. $\begin{cases} x + 2y = 5 \\ 2x + y = 5 \end{cases}$

In problems 11–20, solve each system of equations algebraically, either by elimination or substitution.

11. $\begin{cases} x + y = 8 \\ 2x - y = 4 \end{cases}$

12. $\begin{cases} y = -3x + 2 \\ 2x + y = 6 \end{cases}$

13. $\begin{cases} y = 2x + 4 \\ y = 3x + 6 \end{cases}$

14. $\begin{cases} 4x - 3y = -13 \\ y = 2x + 4 \end{cases}$

15. $\begin{cases} 2x - y = 0 \\ x + 2y = -5 \end{cases}$

16. $\begin{cases} 3x - 4y = 1 \\ 3x - 6y = -1 \end{cases}$

17. $\begin{cases} 3x + 2y = 11 \\ 5x + 7y = -11 \end{cases}$

18. $\begin{cases} y = 2x + 1 \\ y = -x - 1 \end{cases}$

19. $\begin{cases} 5x + 2y = 1 \\ 3x - y = 2 \end{cases}$

20. $\begin{cases} x - 8y = 14 \\ x - y = 0 \end{cases}$

Solve each of the following problems by using two variables.

21. The sum of two numbers is 80. If twice the larger number is increased by 3 times the smaller number, the result is 179. Find the numbers.

22. If 3 times the smaller of two numbers is subtracted from the larger number, the result is 9. Also if twice the larger number is added to twice the smaller number, the result is 122. Find the numbers.

23. A small boat can travel upstream in 2 hours. It takes only 1 hour to travel the same distance downstream. If the distance traveled is 24 miles, find the rate of speed of the boat in still water and the rate of the water current.

24. Neil invested $10,000, part at 8 percent and the balance at 12 percent. The income earned from both investments was $975. How much was invested at each rate?

25. Two types of coffee must be blended together in order to produce a special brand of coffee. The first type of coffee sells for $4.80 per pound and the second brand sells for $7.20 per pound. How many pounds of each must be mixed in order to produce 30 pounds of this "special" coffee that will sell for $5.80 per pound?

CHAPTER 12

Quadratic Equations

12-1 STANDARD FORM OF QUADRATIC EQUATIONS

A quadratic equation in one unknown is an equation that contains the second power as its highest power of the unknown. The following are examples of quadratic equations.

Each equation has 2 as its highest power of the unknown.

$$x^2 + 4x - 6 = 0$$
$$x^2 + 10 = 0$$
$$2x^2 - 7x = 0$$

The standard form of a quadratic equation is ordered in descending powers of x such as $ax^2 + bx + c = 0$.

The value of a is always the coefficient of the second-degree term; it cannot be equal to zero.

$$\boxed{a}\,x^2 + bx + c = 0$$

The value of b is always the coefficient of the first-degree term.

$$ax^2 + \boxed{b}\,x + c = 0$$

The value of c is always the constant term.

$$ax^2 + bx + \boxed{c} = 0$$

To find the value of a, b, or c in any quadratic equation, we must transform the equation so that one side (usually the right) is equal to zero. For example:

■ The equation $x^2 + 2x = 4$ must be written in standard form (set equal to zero) before identifying the values of a, b, or c.

$$x^2 + 2x = 4$$
$$x^2 + 2x - 4 = 0$$
$$\boxed{1}\,x^2 \boxed{+ 2}\,x \boxed{- 4} = 0$$
$$a = 1, \quad b = +2, \quad c = -4$$

■ The equation $2x^2 = 16$ must be written in standard form (set equal to zero) before identifying the value of a, b, and c.

$$2x^2 = 16$$
$$2x^2 - 16 = 0$$
$$\boxed{2}\,x^2 \boxed{- 16} = 0$$
$$a = 2, \quad c = -16$$

459

NOTE:

> If there is no 1st degree term, then $bx = 0$. The value of b is 0.
>
> We may attach a "dummy term" $0x$. $\quad 2x^2 + \boxed{0x} - 16 = 0$
>
> $\qquad\qquad\qquad\qquad\qquad\qquad\qquad a = 2, \quad b = 0, \quad c = -16$

We sometimes find it difficult to identify quadratic equations because of the form in which they are presented. As an easy solution to this problem, always transform equations into the standard form before attempting to identify or solve the equation. Equations containing like terms, parentheses, fractions, or radicals must be transformed into standard form. The following examples illustrate this.

EXAMPLE 1: Transform the equation $x^2 - 4x = 2x - 1$ into standard form.

SOLUTION: We must transform our equation so that one side is equal to zero.

$$x^2 - 4x = 2x - 1$$
$$x^2 - 4x - 2x + 1 = 0$$
$$x^2 - 6x + 1 = 0$$

Thus,

$$a = 1, \quad b = -6, \quad \text{and} \quad c = 1$$

Final answer: $\quad x^2 - 6x + 1 = 0$

EXAMPLE 2: Transform the equation $x(x - 1) = 6$ into standard form.

SOLUTION: We must first remove the parentheses and then set one side of the equation to zero.

$$x(x - 1) = 6$$
$$x^2 - x = 6$$
$$x^2 - x - 6 = 0$$

Thus,

$$a = 1, \quad b = -1, \quad \text{and} \quad c = -6$$

Final answer: $\quad x^2 - x - 6 = 0$

EXAMPLE 3: Express $\dfrac{2}{x} + x = 3$ in standard form.

SOLUTION: We first clear the fraction and then set one side to zero.

$$\frac{2}{x} + x = 3$$

$$x\left(\frac{2}{x}\right) + x(x) = (3)x$$

$$2 + x^2 = 3x$$

$$2 + x^2 - 3x = 0$$

or

$$x^2 - 3x + 2 = 0$$

Thus,

$$a = 1, \quad b = -3, \quad \text{and} \quad c = 2$$

Final answer: $x^2 - 3x + 2 = 0$

EXAMPLE 4: Express $\sqrt{x + 2} = 2x$ in standard form.

SOLUTION: We first eliminate the radical by squaring each side and then setting one side to zero. (Remember to isolate the radical before squaring.)

$$\sqrt{x + 2} = 2x$$

$$(\sqrt{x + 2})^2 = (2x)^2$$

$$x + 2 = 4x^2$$

$$0 = 4x^2 - x - 2$$

or

$$4x^2 - x - 2 = 0$$

Thus,

$$a = 4, \quad b = -1, \quad \text{and} \quad c = -2$$

Final answer: $4x^2 - x - 2 = 0$

Do the following practice set. Check your answers with the answers in the right-hand margin.

PRACTICE SET 12-1

Transform the following equations into the standard form, $ax^2 + bx + c = 0$ (see example 1, 2, 3, or 4).

1. $x^2 + 2x = 4x + 5$

2. $(x - 1)^2 = 2$

3. $x - 3 = \dfrac{7}{x}$

4. $\dfrac{2x + 5}{3} = \dfrac{4}{x}$

5. $\sqrt{x^2 - 1} = 2$

6. $\sqrt{x - 9} + 1 = x$

1. $x^2 - 2x - 5 = 0$
2. $x^2 - 2x - 1 = 0$
3. $x^2 - 3x - 7 = 0$
4. $2x^2 + 5x - 12 = 0$
5. $x^2 - 5 = 0$
6. $x^2 - 3x + 10 = 0$

EXERCISE 12-1

Transform the following equations into standard form.

1. $x(x + 6) = 40$
2. $2x(x - 4) = 11$
3. $x(x + 4) = 5$
4. $x^2 = 5(x + 4)$
5. $4(x + 2)^2 = 2x^2 + 82$
6. $2x - 27 = 2x(7 - 2x)$
7. $\dfrac{6 - 2x}{3} = \dfrac{1}{x}$
8. $\dfrac{x}{3} + \dfrac{5}{3x} = 2$
9. $\dfrac{x - 5}{2} = \dfrac{4}{x - 2}$
10. $\dfrac{x^2}{2} + 3 = \dfrac{x}{4}$
11. $x - 5 = \dfrac{7}{x}$
12. $\dfrac{x}{x + 1} - \dfrac{2}{x} = 6$
13. $\sqrt{x^2} = 2$
14. $\sqrt{x^2 - 5} = 3x$
15. $2\sqrt{x} = 6$
16. $x + 5 = \sqrt{7}$
17. $x = 6 + \sqrt{x}$
18. $\sqrt{2x^2 - 1} + 1 = x - 1$

12-2' SOLVING QUADRATIC EQUATIONS BY FACTORING

In Section 7-7, we learned how to solve quadratic equations by factoring. Although our problems have since become more involved (e.g., our equations now contain fractions or radicals), many quadratic equations can still be solved by using this method. Moreover, our approach remains the same.

Specifically, this approach involves expressing the equation in standard form ($ax^2 + bx + c = 0$), factoring the quadratic expression completely, setting each factor containing the unknown equal to zero, and then solving each of these equations accordingly. Our results should then be checked back into the original equation.

PROCEDURE:

> To solve quadratic equations by factoring:
>
> 1. Transform the equation into the standard form $ax^2 + bx + c = 0$.
>
> 2. Factor $ax^2 + bx + c$.
>
> 3. Set each factor equal to zero and solve each resulting equation.
>
> 4. Check the solution(s) in the original equation.

EXAMPLE 1: Solve $x^2 - x = 2x + 10$.

SOLUTION: First express the equation in standard form.

$$x^2 - x = 2x + 10$$
$$x^2 - x - 2x - 10 = 0$$
$$x^2 - 3x - 10 = 0$$

Factor the quadratic expression.

$$x^2 - 3x - 10 = 0$$
$$(x - 5)(x + 2) = 0$$

Set each factor to zero and solve the equations accordingly.

$$(x - 5)(x + 2) = 0$$

$x - 5 = 0$	$x + 2 = 0$
$x = 5$	$x = -2$

Check:

	$x = 5$			$x = -2$	
$x^2 - x$		$2x + 10$	$x^2 - x$		$2x + 10$
$(5)^2 - (5)$		$2(5) + 10$	$(-2)^2 - (-2)$		$2(-2) + 10$
$25 - 5$		$10 + 10$	$4 + 2$		$-4 + 10$
20		20	6		6

Final answer: $x = 5, x = -2$

EXAMPLE 2:   Solve $x^2 = 7x$.

SOLUTION: Express the equation in standard form.

$$x^2 = 7x$$
$$x^2 - 7x = 0$$

Factor.

$$x^2 - 7x = 0$$
$$x(x - 7) = 0$$

Set each factor to zero and solve accordingly.

$$x(x - 7) = 0$$

$x = 0$	$x - 7 = 0$
$x = 0$	$x = 7$

Check:

	$x = 0$			$x = 7$	
x^2		$7x$	x^2		$7x$
$(0)^2$		$7(0)$	$(7)^2$		$7(7)$
0		0	49		49

Final answer: $x = 0, x = 7$

EXAMPLE 3: Solve $x^2 = 16$.

SOLUTION: Express in standard form.

$$x^2 = 16$$
$$x^2 - 16 = 0$$

Factor, set each factor to zero, and solve accordingly.

$$x^2 - 16 = 0$$
$$(x + 4)(x - 4) = 0$$

$x + 4 = 0$	$x - 4 = 0$
$x = -4$	$x = 4$

Check:

$x = -4$		$x = -4$	
x^2	16	x^2	16
$(-4)^2$	16	$(4)^2$	16
16		16	

Final answer: $x = -4, x = 4$

EXAMPLE 4: Solve $x(x - 1) = 6$.

SOLUTION: Express in standard form.

$$x(x - 1) = 6$$
$$x^2 - x = 6$$
$$x^2 - x - 6 = 0$$

Factor, set each factor to zero, and solve accordingly.

$$x^2 - x - 6 = 0$$
$$(x - 3)(x + 2) = 0$$

$x - 3 = 0$	$x + 2 = 0$
$x = 3$	$x = -2$

Check:

$x = 3$		$x = -2$	
$x(x - 1)$	6	$x(x - 1)$	6
$3(3 - 1)$	6	$-2((-2) - 1)$	6
$3(2)$		$-2(-3)$	
6		6	

Final answer: $x = 3, x = -2$

EXAMPLE 5:   Solve $\dfrac{3x-2}{5} = \dfrac{8}{x}$.

SOLUTION: Express in standard form.

$$\frac{3x-2}{5} = \frac{8}{x}$$

$$5x\left(\frac{3x-2}{5}\right) = \left(\frac{8}{x}\right)5x$$

$$x(3x-2) = 8(5)$$

$$3x^2 - 2x = 40$$

$$3x^2 - 2x - 40 = 0$$

Solve the equation by factoring.

$$3x^2 - 2x - 40 = 0$$

$$(3x+10)(x-4) = 0$$

$$
\begin{array}{c|c}
3x + 10 = 0 & x - 4 = 0 \\
3x = -10 & x = 4 \\
x = \dfrac{-10}{3} &
\end{array}
$$

Check:

$$x = \frac{-10}{3} \hspace{5cm} x = 4$$

$$
\begin{array}{c|c}
\dfrac{3x-2}{5} & \dfrac{8}{x} \\[2mm]
\hline
\dfrac{3\left(\dfrac{-10}{3}\right) - 2}{5} & \dfrac{8}{\left(\dfrac{-10}{3}\right)} \\[4mm]
\dfrac{-10 - 2}{5} & \left(\dfrac{8}{1}\right)\left(\dfrac{3}{-10}\right) \\[2mm]
\boxed{\dfrac{-12}{5}} & \dfrac{24}{-10} \\[2mm]
& \boxed{\dfrac{-12}{5}}
\end{array}
$$

$$
\begin{array}{c|c}
\dfrac{3x-2}{5} & \dfrac{8}{x} \\[2mm]
\hline
\dfrac{3(4)-2}{5} & \dfrac{8}{4} \\[2mm]
\dfrac{12-2}{5} & \boxed{2} \\[2mm]
\dfrac{10}{5} & \\[2mm]
\boxed{2} &
\end{array}
$$

Final answer: $x = \dfrac{-10}{3}, \quad x = 4$

EXAMPLE 6: Solve $\sqrt{x} + 6 = x$.

SOLUTION: Express in standard form.

$$\sqrt{x} + 6 = x$$
$$\sqrt{x} = x - 6$$
$$(\sqrt{x})^2 = (x - 6)^2$$
$$x = x^2 - 12x + 36$$
$$0 = x^2 - 12x - x + 36$$
$$0 = x^2 - 13x + 36$$

or

$$x^2 - 13x + 36 = 0$$

Solve the equation by factoring.

$$x^2 - 13x + 36 = 0$$
$$(x - 4)(x - 9) = 0$$

$$x - 4 = 0 \qquad \mid \qquad x - 9 = 0$$
$$x = 4 \qquad \mid \qquad x = 9$$

Check:

$x = 4$		$x = 9$	
$\sqrt{x} + 6$	x	$\sqrt{x} + 6$	x
$\sqrt{4} + 6$	4	$\sqrt{9} + 6$	9
$2 + 6$		$3 + 6$	
8		9	

[$x = 4$ is an extraneous solution.]

Final answer: $x = 9$

Do the following practice set. Check your answers with the answers in the right-hand margin.

PRACTICE SET 12-2

Solve the following equations by factoring (see example 1, 2, or 3).

1. $x^2 - 3x = 10$ 2. $x^2 = 7x$ 3. $2x^2 = 50$

Solve the following equations by factoring (see example 4, 5, or 6).

4. $x(x - 3) = 4$ 5. $x = \dfrac{2x}{x + 6}$ 6. $x = \sqrt{15 - 2x}$

1. $x = 5, \quad x = -2$
2. $x = 0, \quad x = 7$
3. $x = +5, \quad x = -5$

4. $x = 4, \quad x = -1$
5. $x = 0, \quad x = -4$
6. $x = 3, \quad x = -5$

EXERCISE 12-2

Solve the following equations by factoring and check the roots.

1. $x^2 - 6x - 7 = 0$

2. $x^2 + 14 = 9x$

3. $20 + 6x = 2x^2$

4. $2x^2 - 7x + 6 = 0$

5. $x^2 - 25 = 0$

6. $2x^2 = 32$

7. $3x^2 = 27$

8. $5x^2 - 125 = 0$

9. $x^2 - 9x = 0$

10. $2x^2 + 16x = 0$

11. $3x^2 = 12x$

12. $2x^2 = -5x$

13. $3(x^2 - 4) = 4x + 3$

14. $x(x + 9) = 10$

15. $x(x - 3) = 10$

16. $6x(x - 3) = 4x + 28$

17. $x(x - 6) = -9$

18. $x(2x - 5) = 3$

19. $\dfrac{x}{3} = \dfrac{9}{x - 6}$

20. $\dfrac{x^2}{2} - 3 = \dfrac{5x}{4}$

21. $\dfrac{4}{x} = \dfrac{x}{9}$

22. $\dfrac{x}{3} + \dfrac{9}{x} = 4$

23. $\dfrac{10}{x} + 3 = 4x$

24. $x = \dfrac{40}{x - 3}$

25. $\sqrt{6x} = x$

26. $\sqrt{x^2} = 3$

27. $\sqrt{5x + 1} - 1 = x$

28. $\sqrt{x} = x - 6$

29. $2\sqrt{x^2 - 5x} = 6x$

30. $\sqrt{4x^2 - 7} = x + 1$

12-3 SOLVING INCOMPLETE QUADRATIC EQUATIONS BY THE SQUARE ROOT METHOD

Another method we can use to solve a quadratic equation is the square root method. This method is quite useful for solving quadratic equations that are incomplete (i.e., missing the first-degree, or b term) and which cannot be factored. The following are examples of typical incomplete quadratic equations that cannot be factored.

$$\left. \begin{array}{r} x^2 - 7 = 0 \\ 3x^2 - 16 = 0 \\ x^2 + 6 = 0 \\ 2x^2 = 0 \end{array} \right\} \text{ incomplete quadratics that cannot be factored}$$

To develop the square root method for solving quadratic equations, consider for a moment the incomplete quadratic equation, $x^2 - 16 = 0$. Notice that:

1. This incomplete quadratic equation can be factored and solved accordingly.

$$x^2 - 16 = 0$$
$$(x + 4)(x - 4) = 0$$
$$x + 4 = 0 \quad | \quad x - 4 = 0$$
$$x = -4 \quad | \quad x = 4$$

2. This equation can be solved by first setting the two terms equal to each other (i.e., by solving the equation for x^2), and then, taking the square root of each side.

$$x^2 - 16 = 0$$
$$x^2 = 16$$
$$\sqrt{x^2} = \pm\sqrt{16}$$
$$x = \pm 4$$

NOTE:

Recall that what we do to one side of an equation, we must also do to the other side to maintain the equality. Thus, if we take the square root of one side, then we must take the square root of the other side.

Further, recall (from Section 9-1A) that every positive number has two roots—one positive and one negative. So, whenever we take the square root of a positive quantity, we must account for both the positive and negative values. We indicate this by writing "plus or minus" ($\pm$) before the square root sign.

Using this method, we are now able to solve an incomplete quadratic equation that cannot be factored. For example, to solve $x^2 - 8 = 0$:

First, set the terms equal to each other (i.e., solve for x^2).

$$x^2 - 8 = 0$$
$$x^2 = 8$$

And then, take the square root of each side.

$$x^2 = 8$$
$$\sqrt{x^2} = \pm\sqrt{8}$$
$$x = \pm\sqrt{8}$$

If the solution contains radicals, simplify the radicals.

$$x = \pm\sqrt{(4)(2)}$$
$$x = \pm 2\sqrt{2}$$

Thus,

the solution to the equation $x^2 - 8 = 0$ is $x = 2\sqrt{2}$ and $x = -2\sqrt{2}$.

The check is left to the student.

In summary, then, to use the square root method for solving incomplete quadratic equations, we first solve the equation for x^2 and then take the square root of both sides, accounting for both the positive and negative values of the solution. Furthermore, if our solution contains a radical, we must simplify the radical. The solutions should then be checked.

PROCEDURE:

To solve an incomplete quadratic equation by using the square root method:

1. Solve the equation for x^2.

2. Take the square root of each side of the equation.

3. Simplify the radicals.

4. Check the solution(s) in the original equation.

EXAMPLE 1: Solve $x^2 - 25 = 0$.

SOLUTION: Solve the equation for x^2.

$$x^2 - 25 = 0$$
$$x^2 = 25$$

Now, take the square root of both sides.

$$x^2 = 25$$
$$\sqrt{x^2} = \pm\sqrt{25}$$
$$x = \pm\sqrt{25}$$
$$x = \pm 5$$

Final answer: $x = 5$, $x = -5$

The check is left to the student.

EXAMPLE 2: Solve $3x^2 - 27 = 0$.

SOLUTION: Solve the equation for x^2.

$$3x^2 - 27 = 0$$
$$3x^2 = 27$$
$$x^2 = \frac{27}{3}$$
$$x^2 = 9$$

Take the square root of both sides.

$$x^2 = 9$$
$$\sqrt{x^2} = \pm\sqrt{9}$$
$$x = \pm\sqrt{9}$$
$$x = \pm 3$$

Final answer: $x = 3$, $x = -3$

The check is left to the student.

EXAMPLE 3: Solve $4x^2 + 1 = 4$.

SOLUTION: Solve for x^2.

$$4x^2 + 1 = 4$$
$$4x^2 = 4 - 1$$
$$4x^2 = 3$$
$$x^2 = \frac{3}{4}$$

Example continued on next page.

Take the square root of each side.

$$x^2 = \frac{3}{4}$$

$$\sqrt{x^2} = \pm\sqrt{\frac{3}{4}}$$

$$x = \pm\sqrt{\frac{3}{4}}$$

Simplify the radical.

$$x = \pm\sqrt{\frac{3}{4}}$$

$$x = \pm\frac{\sqrt{3}}{\sqrt{4}}$$

$$x = \frac{\pm\sqrt{3}}{2}$$

Final answer: $x = \dfrac{\sqrt{3}}{2}, \quad x = \dfrac{-\sqrt{3}}{2}$

The check is left to the student.

EXAMPLE 4: Solve $2x^2 + 8 = 0$

SOLUTION: Solve for x^2.

$$2x^2 + 8 = 0$$

$$2x^2 = -8$$

$$x^2 = \frac{-8}{2}$$

$$x^2 = -4$$

Take the square root of each side.

$$x^2 = -4$$

$$\sqrt{x^2} = \pm\sqrt{-4}$$

$$x = \pm\sqrt{-4}$$

NOTE:

We cannot find the square root of a negative number in our number system. Thus, there is no solution.

Final answer: No solution (in the real number system).

Do the following practice set. Check your answers with the answers in the right-hand margin.

PRACTICE SET 12-3

Solve by the square root method (see example 1, 2, 3, or 4).

1. $x^2 - 64 = 0$
2. $3x^2 - 24 = 3$
3. $2x^2 = 24$
4. $4x^2 = 25$
5. $5x^2 + 30 = 5$
6. $2x^2 = 3$

1. $x = 8, \quad x = -8$
2. $x = 3, \quad x = -3$
3. $x = 2\sqrt{3}, \quad x = -2\sqrt{3}$
4. $x = \frac{5}{2}, \quad x = -\frac{5}{2}$
5. No solution
6. $x = \sqrt{6}/2, \quad x = -\sqrt{6}/2$

EXERCISE 12-3

Solve the following quadratic equations by the square root method.

1. $6x^2 = 54$
2. $2x^2 = 75 - x^2$
3. $18 = 2x^2$
4. $2x^2 - 50 = 0$
5. $7x^2 - 7 = 0$
6. $2x^2 - 200 = 0$
7. $3x^2 + 3 = 30$
8. $6x^2 - x^2 = 45$
9. $x^2 = 12$
10. $3x^2 = 6$
11. $x^2 + x^2 = 16$
12. $5x^2 - 3x^2 = 54$
13. $4x^2 - 17 = x^2 + 67$
14. $5x^2 = 3x^2 + 4$
15. $\dfrac{x}{8} = \dfrac{4}{x}$
16. $3x^2 - 10 = -x^2 + 30$

12-4 SOLVING QUADRATIC EQUATIONS BY COMPLETING THE SQUARE

We now extend our square root method for solving incomplete quadratic equations to solving any type of quadratic equations.

It is implicit in our square root method for solving incomplete quadratic equations that we have a perfect square on one side of the equation and a constant on the other side. Thus, when taking the square root of each side of the equation, we, in effect, are eliminating the "squared" term, since the square root of a perfect square is simply the quantity itself. For example:

$$\text{If } \underbrace{x^2}_{\substack{\text{perfect} \\ \text{square}}} = \underbrace{16}_{\text{constant}}$$

$$\text{If } \underbrace{(x + 2)^2}_{\substack{\text{perfect} \\ \text{square}}} = \underbrace{4}_{\text{constant}}$$

$$\text{then } \sqrt{x^2} = \pm\sqrt{16}$$
$$\text{and } \quad x = \pm\sqrt{16}$$

$$\text{then } \sqrt{(x + 2)^2} = \pm\sqrt{4}$$
$$\text{and } \quad x + 2 = \pm\sqrt{4}$$

Notice in the first example, $x^2 = 16$, that we could also have solved this equation by factoring.

$$x^2 = 16$$
$$x^2 - 16 = 0$$
$$(x + 4)(x - 4) = 0$$

Notice in the second example, $(x + 2)^2 = 4$, that $(x + 2)^2$ is really the trinomial, $x^2 + 4x + 4$, which was factored as a binomial squared. Such trinomials are called perfect square trinomials.

$$x^2 + 4x + 4 = (x + 2)(x + 2)$$
$$= (x + 2)^2$$

Thus,

each of these quadratic equations is *factorable*.

Unfortunately, not all quadratic equations are factorable. However, we can make any quadratic equation factorable by employing a method called completing the square. The general principle behind this method is to put one side of the equation in the form $(x + d)^2$, where d is a real number. This side of the equation would then be a perfect square and hence we would be able to apply our square root method. To do this,

first, note that, in general, the constant term of a perfect square trinomial, (d^2), can be obtained by taking one half the coefficient of the x term, and then squaring this value.

$$(x + d)^2 = (x + d)(x + d)$$
$$= x^2 \underbrace{+ dx + dx} + d^2$$
$$= x^2 + 2dx + d^2$$
$$x^2 + \boxed{2d\,x} + d^2$$
$$\left[\left(\frac{1}{2}\right)(2d)\right]^2 = d^2$$

For example, in the expression, $x^2 + 4x$:

If we take one half the coefficient of the x term,

$$x^2 + \boxed{4}\,x \qquad \frac{1}{2}(4) = 2$$

Square this value,

$$\left[\frac{1}{2}(4)\right]^2 = (2)^2 = 4$$

And add it to the expression, we then have a perfect square trinomial that can be factored as a binomial squared.

$$x^2 + 4x \qquad x^2 + 4x \boxed{+4}$$
$$\boxed{x^2 + 4x + 4} = (x + 2)(x + 2)$$
$$= (x + 2)^2$$

As another example, consider the expression, $x^2 - 9x$.

If we take one half the coefficient of the x term,

$$x^2 - \boxed{9}\,x \qquad \frac{1}{2}(-9) = \frac{-9}{2}$$

Square it,

$$\left[\frac{1}{2}(-9)\right]^2 = \left(\frac{-9}{2}\right)^2 = \frac{81}{4}$$

And add it to the expression, we get a perfect square trinomial that can be factored as a binomial squared.

$$x^2 - 9x \qquad x^2 - 9x + \boxed{\frac{81}{4}}$$

$$x^2 - 9x + \frac{81}{4}$$
$$= \left(x - \frac{9}{2}\right)\left(x - \frac{9}{2}\right)$$
$$= \left(x - \frac{9}{2}\right)^2$$

Knowing this technique, we can now solve any quadratic equation. However, before applying the technique of completing the square, we must be sensitive to two things: First, isolate the constant term, and second, make certain the coefficient of the squared term is one. Once this has been done, we can then complete the square and solve the resulting equation by using our square root method accordingly. For example, to solve $3x^2 - 3x - 18 = 0$:

First, isolate the constant term.

$$3x^2 - 3x - 18 = 0$$
$$3x^2 - 3x = 18$$

Next, make the coefficient of the squared term equal to one.

$$3x^2 - 3x = 18$$
$$\frac{3x^2}{3} - \frac{3x}{3} = \frac{18}{3}$$
$$x^2 - x = 6$$

Now, complete the square.

$$\left[\begin{array}{c} \frac{1}{2}(-1) = -\frac{1}{2} \\ \text{and} \\ \left(-\frac{1}{2}\right)^2 = \frac{1}{4} \end{array} \right.$$

$$x^2 - x = 6$$
$$x^2 - x + \frac{1}{4} = 6 + \frac{1}{4}$$

Factor the perfect square trinomial as a binomial squared and combine like terms.

$$x^2 - x + \frac{1}{4} = 6 + \frac{1}{4}$$
$$\left(x - \frac{1}{2}\right)^2 = \frac{25}{4}$$

Finally, solve this equation by our square root method.

$$\left(x - \frac{1}{2}\right)^2 = \frac{25}{4}$$
$$\sqrt{\left(x - \frac{1}{2}\right)^2} = \pm\sqrt{\frac{25}{4}}$$
$$x - \frac{1}{2} = \pm\frac{\sqrt{25}}{\sqrt{4}}$$
$$x - \frac{1}{2} = \pm\frac{5}{2}$$

NOTE:

$x - \frac{1}{2} = \pm\frac{5}{2}$ is broken into two separate equations:

$$x - \frac{1}{2} = \frac{5}{2} \quad \text{and} \quad x - \frac{1}{2} = \frac{-5}{2}$$

$$x - \frac{1}{2} = \frac{5}{2} \qquad\qquad x - \frac{1}{2} = -\frac{5}{2}$$
$$x = \frac{5}{2} + \frac{1}{2} \qquad\qquad x = \frac{-5}{2} + \frac{1}{2}$$
$$x = \frac{6}{2} \qquad\qquad x = \frac{-4}{2}$$
$$x = 3 \qquad\qquad x = -2$$

Thus,

$$x = 3, \quad \text{and} \quad x = -2.$$

The check is left to the student.

PROCEDURE:

To solve quadratic equations by completing the square:

1. Transform the equation into the form $x^2 + bx = c$, where the coefficient of x^2 is one.

2. Complete the square.
   a. Take one half the coefficient of x.
   b. Square this number.
   c. Add this value to each side of the equation.

3. Factor the perfect square trinomial into a binomial squared and combine like terms.

4. Solve by using the square root method.

5. Check the solution(s) in the original equations.

EXAMPLE 1: Solve $x^2 + 10x - 7 = 0$ by completing the square.

SOLUTION: Isolate the constant term.

$$x^2 + 10x - 7 = 0$$
$$x^2 + 10x = 7$$

Complete the square by adding $[\frac{1}{2}(10)]^2 = (5)^2 = 25$ to both sides of the equation.

$$x^2 + 10x = 7$$
$$x^2 + 10x + 25 = 7 + 25$$
$$(x + 5)^2 = 32$$

Solve by using the square root method.

$$(x + 5)^2 = 32$$
$$\sqrt{(x + 5)^2} = \pm\sqrt{32}$$
$$x + 5 = \pm\sqrt{16}\sqrt{2}$$
$$x + 5 = \pm 4\sqrt{2}$$

$x + 5 = 4\sqrt{2}$	$x + 5 = -4\sqrt{2}$
$x = 4\sqrt{2} - 5$	$x = -4\sqrt{2} - 5$

Final answer: $x = -5 + 4\sqrt{2}$; $x = -5 - 4\sqrt{2}$

The check is left to the student.

EXAMPLE 2:   Solve $3x^2 - 2x - 4 = 0$ by completing the square.

SOLUTION: Isolate the constant term and make the coefficient of the squared term equal to one.

$$3x^2 - 2x - 4 = 0$$
$$3x^2 - 2x = 4$$
$$x^2 - \frac{2}{3}x = \frac{4}{3}$$

Complete the square by adding $[\frac{1}{2}(-\frac{2}{3})]^2 = [-\frac{1}{3}]^2 = \frac{1}{9}$ to both sides of the equation.

$$x^2 - \frac{2}{3}x = \frac{4}{3}$$
$$x^2 - \frac{2}{3}x + \frac{1}{9} = \frac{4}{3} + \frac{1}{9}$$
$$\left(x - \frac{1}{3}\right)^2 = \frac{12}{9} + \frac{1}{9}$$
$$\left(x - \frac{1}{3}\right)^2 = \frac{13}{9}$$

Solve by using the square root method.

$$\left(x - \frac{1}{3}\right)^2 = \frac{13}{9}$$
$$\sqrt{\left(x - \frac{1}{3}\right)^2} = \pm\sqrt{\frac{13}{9}}$$
$$x - \frac{1}{3} = \frac{\pm\sqrt{13}}{\sqrt{9}}$$
$$x - \frac{1}{3} = \frac{\pm\sqrt{13}}{3}$$

$$x - \frac{1}{3} = \frac{\sqrt{13}}{3} \qquad \bigg| \qquad x - \frac{1}{3} = \frac{-\sqrt{13}}{3}$$

$$x = \frac{\sqrt{13}}{3} + \frac{1}{3} \qquad \bigg| \qquad x = \frac{-\sqrt{13}}{3} + \frac{1}{3}$$

$$x = \frac{\sqrt{13} + 1}{3} \qquad \bigg| \qquad x = \frac{-\sqrt{13} + 1}{3}$$

Final answer: $x = \dfrac{\sqrt{13} + 1}{3}$, $x = \dfrac{-\sqrt{13} + 1}{3}$

The check is left to the student.

Do the following practice set. Check your answers with the answers in the right-hand margin.

PRACTICE SET 12-4

Solve each of the following quadratic equations by completing the square (see example 1 or 2).

1. $x^2 + 6x - 1 = 0$
2. $x^2 + 12x + 24 = 0$
3. $4x^2 + 8x + 3 = 0$
4. $3x^2 - 4x - 9 = 0$

1. $x = -3 + \sqrt{10}$
    $x = -3 - \sqrt{10}$
2. $x = -6 + 2\sqrt{3}$
    $x = -6 - 2\sqrt{3}$
3. $x = -\frac{1}{2}, \quad x = -\frac{3}{2}$
4. $x = (2 + \sqrt{31})/3$
    $x = (2 - \sqrt{31})/3$

EXERCISE 12-4

Solve and check the following equations by completing the square. Answers may be left in radical form.

1. $x^2 - 2x - 3 = 0$
2. $x^2 - 10x + 25 = 0$
3. $x^2 + 3x - 4 = 0$
4. $x^2 + 5x - 14 = 0$
5. $x^2 - 5x + 4 = 0$
6. $2x^2 + 6x = 0$
7. $2x^2 + 9x + 10 = 0$
8. $2x^2 + x - 36 = 0$
9. $x^2 + 2x - 9 = 0$
10. $x^2 - 6x + 1 = 0$
11. $x^2 - 6x - 8 = 0$
12. $x^2 - 24x - 10 = 0$
13. $x^2 = 6x + 3$
14. $x^2 = 4x + 6$
15. $x^2 = 4x - 2$
16. $3x^2 - 6x - 2 = 0$
17. $2x^2 - 6x - 3 = 0$
18. $4x^2 + 4x - 5 = 0$
19. $5x^2 + 6x + 1 = 0$
20. $3x^2 + 2x - 7 = 0$

12-5 SOLVING QUADRATIC EQUATIONS BY THE QUADRATIC FORMULA

The method of completing the square can be used to develop a formula that will permit us to find the solution of any quadratic equation. The derivation of this formula is as follows.

Starting with the standard form of a quadratic equation, we first manipulate the equation in order to complete the square. (The constant must be isolated, and the coefficient of the squared term must be one.)

$$ax^2 + bx + c = 0$$
$$ax^2 + bx = -c$$
$$x^2 + \frac{b}{a}x = \frac{-c}{a}$$

Now, complete the square by adding

$$\left[\frac{1}{2}\left(\frac{b}{a}\right)\right]^2 = \left(\frac{b}{2a}\right)^2 = \frac{b^2}{4a^2}$$

to both sides of the equation; factor the resulting perfect square trinomial and combine the fractions.

Solve for x by using the square root method.

$$x^2 + \frac{b}{a}x = \frac{-c}{a}$$

$$x^2 + \frac{b}{a}x + \frac{b^2}{4a^2} = \frac{-c}{a} + \frac{b^2}{4a^2}$$

$$\left(x + \frac{b}{2a}\right)^2 = \frac{b^2}{4a^2} - \frac{c}{a}$$

$$\left(x + \frac{b}{2a}\right)^2 = \frac{b^2 - 4ac}{4a^2}$$

$$\left(x + \frac{b}{2a}\right)^2 = \frac{b^2 - 4ac}{4a^2}$$

$$\sqrt{\left(x + \frac{b}{2a}\right)^2} = \pm\sqrt{\frac{b^2 - 4ac}{4a^2}}$$

$$x + \frac{b}{2a} = \pm\sqrt{\frac{b^2 - 4ac}{4a^2}}$$

$$x + \frac{b}{2a} = \pm\frac{\sqrt{b^2 - 4ac}}{2a}$$

$$x = \frac{-b}{2a} \pm \frac{\sqrt{b^2 - 4ac}}{2a}$$

$$x = \frac{-b \pm \sqrt{b^2 - 4ac}}{2a}$$

Thus, the quadratic formula is

$$x = \frac{-b \pm \sqrt{b^2 - 4ac}}{2a}$$

Note from this formula that the solution consists of

$$x = \frac{-b + \sqrt{b^2 - 4ac}}{2a} \qquad \text{and} \qquad x = \frac{-b - \sqrt{b^2 - 4ac}}{2a}$$

Further, note that the entire quantity, $-b + \sqrt{b^2 - 4ac}$ and $-b - \sqrt{b^2 - 4ac}$ is being divided by $2a$.

This formula is very useful to us because it will allow us to solve any quadratic equation. Therefore, it is important to memorize this formula carefully.

When using this formula, we must first make sure that our quadratic equation is in standard form. Once this has been done, we identify the values for a, b, and c, and substitute these values into the formula accordingly. We then evaluate the formula. As an example, consider the equation $3x^2 + 6x + 1 = 0$.

With the equation in standard form, we can easily identify the values for a, b, and c, respectively.

$$3x^2 + 6x + 1 = 0$$

$$\boxed{3}\,x^2 \; \boxed{+\,6}\,x \; \boxed{+\,1} = 0$$

$$a = 3 \quad b = 6 \quad c = 1$$

We now substitute these values into the quadratic formula directly,

$$x = \frac{-b \pm \sqrt{b^2 - 4ac}}{2a}$$

$$x = \frac{-(6) \pm \sqrt{(6)^2 - 4(3)(1)}}{2(3)}$$

And evaluate the formula.

$$x = \frac{-6 \pm \sqrt{36 - 12}}{6}$$

$$x = \frac{-6 \pm \sqrt{24}}{6}$$

$$x = \frac{-6 \pm \sqrt{(4)(6)}}{6}$$

$$x = \frac{-6 \pm 2\sqrt{6}}{6}$$

Thus,

$$x = \frac{-6 + 2\sqrt{6}}{6} \qquad \text{and} \qquad x = \frac{-6 - 2\sqrt{6}}{6}$$

Reducing these fractions, by canceling out a common factor of 2, we get

$$x = \frac{-3 + \sqrt{6}}{3} \qquad \text{and} \qquad x = \frac{-3 - \sqrt{6}}{3}$$

We then may either leave our answers expressed in radical form (as previously shown) or evaluate the radical and express the answers in decimal form.

PROCEDURE:

> To solve quadratic equations by the quadratic formula:
>
> 1. Express the quadratic equation in standard form.
>
> 2. Identify the values for a, b, and c.
>
> 3. Replace the values for a, b, and c directly into the quadratic formula. Use parentheses to maintain form.
>
> 4. Perform the indicated operations.
>
> 5. Simplify any resulting radicals.
>
> 6. Check the solution(s) in the original equation.

EXAMPLE 1: Solve $x^2 - 4x + 4 = 0$ by the quadratic formula.

SOLUTION: With the equation in standard form, we identify the values for a, b, and c.

$$x^2 - 4x + 4 = 0$$
$$a = 1, \quad b = -4, \quad c = 4$$

Substitute these values directly into the quadratic formula.

$$x = \frac{-b \pm \sqrt{b^2 - 4ac}}{2a}$$

$$x = \frac{-(-4) \pm \sqrt{(-4)^2 - 4(1)(4)}}{2(1)}$$

Evaluate the formula.

$$x = \frac{-(-4) \pm \sqrt{(-4)^2 - 4(1)(4)}}{2(1)}$$

$$x = \frac{4 \pm \sqrt{16 - 16}}{2}$$

$$x = \frac{4 \pm \sqrt{0}}{2}$$

$$x = \frac{4 \pm 0}{2}$$

$$x = \frac{4}{2}$$

$$x = 2$$

Final answer: $x = 2$

The check is left to the student.

EXAMPLE 2: Solve $2x^2 = 40$.

SOLUTION: With the equation in standard form, we identify the values for a, b, and c.

$$2x^2 = 40$$
$$2x^2 - 40 = 0$$
$$2x^2 + 0x - 40 = 0$$
$$a = 2, \quad b = 0, \quad c = -40$$

Substitute these values directly into the formula.

$$x = \frac{-b \pm \sqrt{b^2 - 4ac}}{2a}$$

$$x = \frac{-(0) \pm \sqrt{(0)^2 - 4(2)(-40)}}{2(2)}$$

Example continued on next page.

Evaluate the formula.

$$x = \frac{-(0) \pm \sqrt{(0)^2 - 4(2)(-40)}}{2(2)}$$

$$x = \frac{0 \pm \sqrt{0 + 320}}{4}$$

$$x = \frac{\pm\sqrt{320}}{4}$$

$$x = \frac{\pm\sqrt{(64)(5)}}{4}$$

$$x = \frac{\pm 8\sqrt{5}}{4}$$

$$x = \pm 2\sqrt{5}$$

Final answer: $x = 2\sqrt{5}, \quad x = -2\sqrt{5}$

The check is left to the student.

EXAMPLE 3:   Solve $2x^2 + 6x = 9$ by the quadratic formula. Express your answer to the nearest tenth.

SOLUTION: With the equation in standard form, we identify values for a, b, and c.

$$2x^2 + 6x = 9$$
$$2x^2 + 6x - 9 = 0$$
$$a = 2, \quad b = 6, \quad c = -9$$

Substitute these values directly into the formula and evaluate the formula.

$$x = \frac{-b \pm \sqrt{b^2 - 4ac}}{2a}$$

$$x = \frac{-(6) \pm \sqrt{(6)^2 - 4(2)(-9)}}{2(2)}$$

$$x = \frac{-6 \pm \sqrt{36 + 72}}{4}$$

$$x = \frac{-6 \pm \sqrt{108}}{4}$$

Using $\sqrt{108} \approx 10.39$, we now complete our evaluation.

$$x = \frac{-6 \pm 10.39}{4}$$

$$x = \frac{-6 + 10.39}{4} \quad \text{and} \quad x = \frac{-6 - 10.39}{4}$$

$$x = \frac{4.39}{4} \qquad\qquad\qquad x = \frac{-16.39}{4}$$

$$x = 1.09 \qquad\qquad\qquad x = -4.09$$

$x = 1.1$ (to the nearest tenth) $\qquad x = -4.1$ (to the nearest tenth)

Final answer: $x = 1.1$ and $x = -4.1$

Do the following practice set. Check your answers with the answers in the right-hand margin.

PRACTICE SET 12-5

Solve by using the quadratic formula (see example 1, 2, or 3).

1. $2x^2 - x - 6 = 0$
2. $x^2 - 10x - 2 = 0$
3. $x^2 + x = 1$
4. $3x^2 + 6x = -2$
5. $x^2 - 3x - 3 = 0$
   (Give answers to the nearest tenth)
6. $3x^2 + 6x + 1 = 0$
   (Give answers to the nearest tenth)

1. $x = 2, \quad x = -\frac{3}{2}$
2. $x = 5 + 3\sqrt{3}$
   $x = 5 - 3\sqrt{3}$
3. $x = (-1 + \sqrt{5})/2$
   $x = (-1 - \sqrt{5})/2$
4. $x = (-3 + \sqrt{3})/3$
   $x = (-3 - \sqrt{3})/3$
5. $x = 3.8, \quad x = -0.8$
6. $x = -0.2, \quad x = -1.8$

EXERCISE 12-5

Solve the following quadratic equations by using the quadratic formula. Express radicals in the simplest form.

1. $x^2 - 2x - 15 = 0$
2. $x^2 - 8x + 7 = 0$
3. $x^2 - 2x - 35 = 0$
4. $x^2 + 2x - 8 = 0$
5. $3x^2 + 5x + 2 = 0$
6. $2x^2 + 5x + 2 = 0$
7. $4x^2 + 7x - 2 = 0$
8. $3x^2 - 10x + 3 = 0$
9. $x^2 - 4 = 0$
10. $x^2 - 6x = 0$
11. $x^2 - 3 = 0$
12. $3x^2 = 0$
13. $x^2 + 6x + 3 = 0$
14. $x^2 - 10x - 5 = 0$
15. $x^2 + 2x - 1 = 0$
16. $x^2 - 4x + 1 = 0$
17. $2x^2 - 13x + 4 = 0$
18. $5x^2 + 8x + 3 = 0$
19. $2x^2 + x - 4 = 0$
20. $3x^2 - 17 = 0$
21. $5x^2 - 8x + 2 = 0$
22. $x^2 + 12x + 9 = 0$
23. $4x^2 - 12x + 3 = 0$
24. $3x^2 - 9x - 1 = 0$
25. $\dfrac{x - 5}{2} = \dfrac{4}{x - 2}$
26. $x(2x - 3) = 1$
27. $x(x - 2) = 2$
28. $x = 4 + \dfrac{2}{x}$
29. $x + \dfrac{1}{x} = \dfrac{10}{3}$
30. $(x + 4)^2 = 13x + 15$

Solve the following quadratic equations by using the quadratic formula. Express radicals correct to the nearest tenth.

31. $x^2 - 3x - 3 = 0$

32. $x^2 + 7x - 3 = 0$

33. $x^2 - x - 1 = 0$

34. $4x^2 - 6x + 1 = 0$

35. $3x^2 + 4x - 2 = 0$

36. $7x^2 - 6x - 4 = 0$

Summary of Algebraic Techniques for Solving Quadratic Equations

We have developed a variety of algebraic techniques to solve quadratic equations. These techniques include factoring a binomial, factoring a trinomial, the difference of two squares, the square root method, completing the square, and the quadratic formula. When solving a quadratic equation, our technique should be the easiest and most effective for the given quadratic equation.

The following summary of techniques is presented to aid us in our selection.

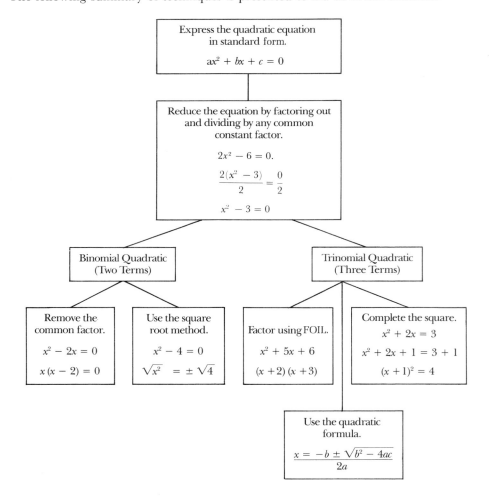

EXERCISE 12-5 (SUPPLEMENTARY)

Solve each of the following equations by using the most convenient method. Express all radical answers in the simplest radical form.

1. $x^2 = 6x$

2. $x^2 - 25 = 0$

3. $x^2 + 4x = 21$

4. $2x^2 = 72$

5. $2x^2 + x = 8 - 4x$

6. $x^2 - 6x + 7 = 0$

7. $3x^2 + 9x = 0$

8. $3x^2 - 24 = 0$

9. $5x^2 + 5 = 11x$

10. $2x^2 = 9$

11. $2(x^2 - 14) = x$

12. $2(4x^2 - 5x) = -3$

13. $3(5x - 2) = (x + 5)(2x - 1)$

14. $x + \dfrac{1}{x} = \dfrac{10}{3}$

15. $\sqrt{2x + 1} - 1 = 2$

16. $x^2 + (x + 1)^2 = 85$

12-6 GRAPHING QUADRATIC EQUATIONS

In Chapter 10, we learned how to graph a linear equation such as $y = -3x + 10$ by generating ordered pairs that satisfy the given equation. We now extend this technique to include quadratic (second-degree) equations. For example, to graph the quadratic equation $y = x^2 + 6x + 8$, we can find ordered pairs that satisfy the equation by simply choosing values for x and then finding corresponding y values by substituting the x values into the equation and solving for y.

If $x = 0$

then $y = x^2 - 6x + 8$
$= 0^2 - 6(0) + 8$
$= 0 - 0 + 8$
$= 8$

Thus,

$(0, 8)$ is an ordered pair.

If $x = 1$

then $y = x^2 - 6x + 8$
$y = (1)^2 - 6(1) + 8$
$y = 1 - 6 + 8$
$y = -5 + 8$
$y = 3$

Thus,

$(1, 3)$ is an ordered pair.

If $x = 2$

then $y = x^2 - 6x + 8$
$y = (2)^2 - 6(2) + 8$
$y = 4 - 12 + 8$
$y = -8 + 8$
$y = 0$

Thus,

$(2, 0)$ is an ordered pair.

If $x = 3$

then $y = x^2 - 6x + 8$
$y = (3)^2 - 6(3) + 8$
$y = 9 - 18 + 8$
$y = -9 + 8$
$y = -1$

Thus,

$(3, -1)$ is an ordered pair.

If $x = 4$

then $y = x^2 - 6x + 8$
$y = (4)^2 - 6(4) + 8$
$y = 16 - 24 + 8$
$y = -8 + 8$
$y = 0$

Thus,

$(4, 0)$ is an ordered pair.

If $x = 5$

then $y = x^2 - 6x + 8$
$y = (5)^2 - 6(5) + 8$
$y = 25 - 30 + 8$
$y = -5 + 8$
$y = 3$

Thus,

$(5, 3)$ is an ordered pair.

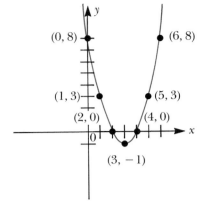

Plotting these points and connecting them with a smooth curve, we get the following graph.

This curve is called a parabola.

All quadratic equations have the shape of the parabola. Parabolas may open up, down, to the right, or to the left.

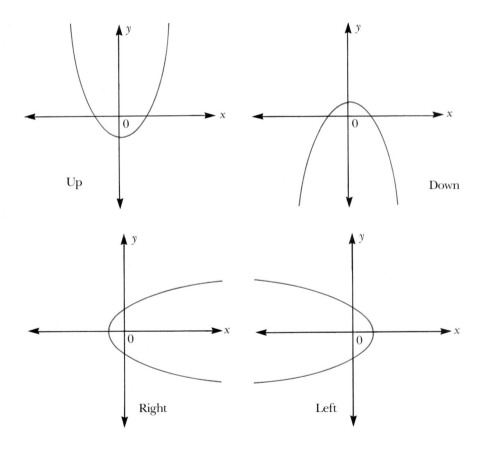

The quadratics (parabolas) we graph will have the form $y = ax^2 + bx + c$. When $a > 0$, this parabola opens "up."

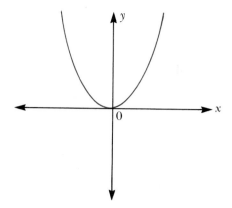

The critical part of every quadratic is where it turns. Knowing where it *turns* and plotting a few points on either side of this turn will produce our curve satisfactorily.

The axis of symmetry is a line that splits our curve into two equal parts. For example, in the parabola at the right:

We are able to identify its turning point (called the vertex) and axis of symmetry.

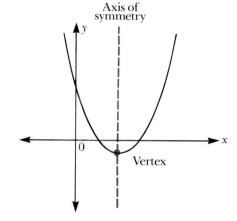

By knowing the axis of symmetry, we have the *x* value of our turning point. The equation $x = -b/2a$ gives us the axis of symmetry. (The values for *a* and *b* are taken from our quadratic equation, expressed in standard form.)

For our original problem $y = x^2 - 6x + 8$:

$$y = x^2 - 6x + 8$$
$$a = 1 \quad b = -6$$

The axis of symmetry is found to be $x = 3$.

$$x = \frac{-b}{2a}$$

$$x = \frac{-(-6)}{2(1)}$$

$$x = \frac{6}{2}$$

$$x = 3$$

This tells us half of our curve is found to the left of this line, and half is found to the right of this line.

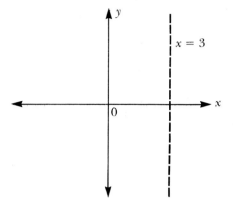

Using the ordered pairs generated earlier (where the x values are 3, less than 3 such as 0, 1, 2, and greater than 3 such as 4, 5, 6), we graph the critical area of our curve.

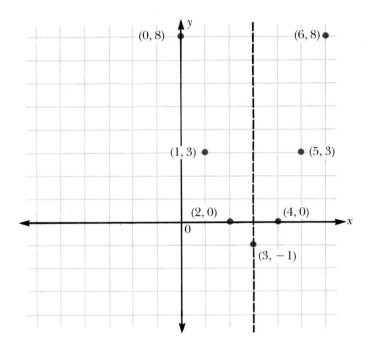

We can see how the axis of symmetry "splits" our parabola into two equal parts.

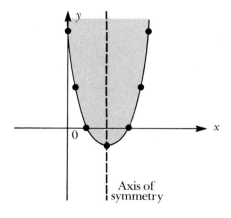

Thus, to graph a quadratic equation in the form $ax^2 + bx + c = y$, we first determine the axis of symmetry by $x = -b/2a$. Next, we find the ordered pair for the turning point (using the x value obtained for the axis of symmetry), and generate a minimum of three ordered pairs on either side of the axis of symmetry. Finally, we plot these points and trace the curve through these points.

PROCEDURE:

To graph a quadratic equation in the form $y = ax^2 + bx + c$:

1. Determine the axis of symmetry by $x = \dfrac{-b}{2a}$.

2. Find the ordered pair for the turning point by substituting the value for x found in (1) into the given equation and solving for y.

3. Generate a minimum of three ordered pairs on either side of the axis of symmetry.

4. Plot all the points generated on the same set of axes.

5. Trace the curve through these points.

EXAMPLE 1: Sketch the graph of $y = x^2 + 4x$.

SOLUTION: With the equation in standard form, we first find the axis of symmetry.

$$x = \frac{-b}{2a}$$

$$x = \frac{-4}{2(1)}$$

$$x = \frac{-4}{2}$$

$$x = -2$$

We now find the ordered pair for the turning point by using $x = -2$

If
$$x = -2$$

then
$$y = x^2 + 4x$$
$$y = (-2)^2 + 4(-2)$$
$$y = 4 - 8$$
$$y = -4$$

Example continued on next page.

Thus,

$$(-2, -4) \text{ are the coordinates of the turning point.}$$

Generate a minimum of three ordered pairs on either side of the turning point; we get the following table.

	x	y
	−5	5
	−4	0
	−3	−3
Turning point	−2	−4
	−1	−3
	0	0
	1	5

Plot these points and trace the curve through them.

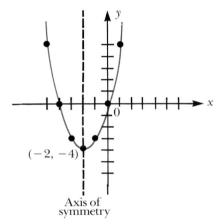

$(-2, -4)$

Axis of symmetry

EXAMPLE 2: Sketch the graph of $y = x^2 - 3x - 4$.

SOLUTION: With the equation in standard form, we first find the axis of symmetry.

$$x = \frac{-b}{2a}$$

$$x = \frac{-(-3)}{2(1)}$$

$$x = \frac{3}{2}$$

Using $x = \frac{3}{2}$, we find the coordinates of the turning point.

If

$$x = \frac{3}{2}$$

then

$$y = x^2 - 3x - 4$$

$$y = \left(\frac{3}{2}\right)^2 - 3\left(\frac{3}{2}\right) - 4$$

$$y = \frac{9}{4} - \frac{9}{2} - 4$$

$$y = \frac{9}{4} - \frac{18}{4} - \frac{16}{4}$$

$$y = \frac{-9}{4} - \frac{16}{4}$$

$$y = \frac{-25}{4}$$

Thus,

$$\left(\frac{3}{2}, \frac{-25}{4}\right)$$ are the coordinates of the turning point.

Generate a minimum of three ordered pairs on either side of the turning point; we get the following table.

	x	y
	-1	0
	0	-4
	1	-6
Turning point	$\dfrac{3}{2}$	$\dfrac{-25}{4}$
	2	-6
	3	-4
	4	0

Plot these points and trace the curve through them.

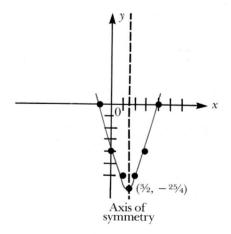

$(\sfrac{3}{2}, -\sfrac{25}{4})$

Axis of symmetry

Do the following practice set. Check your answers with the answers that follow the problems.

PRACTICE SET 12-6

Graph the following quadratic equations (see example 1, 2, or 3).

1. $y = x^2 - 6x$ 2. $y = x^2 + 4x + 3$ 3. $y = x^2 - 4x + 1$

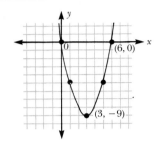 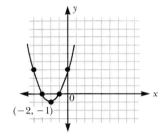 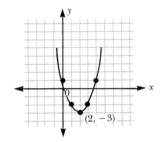

EXERCISE 12-6

Graph the following quadratic equations.

1. $y = x^2$ 2. $y = 4x^2$ 3. $y = 3x^2$
4. $y = x^2 + 1$ 5. $y = x^2 - 1$ 6. $y = x^2 + 4$
7. $y = x^2 - 2x$ 8. $y = x^2 + 6x$ 9. $y = 2x^2 - x$
10. $y = 3x^2 - 4x$ 11. $y = x^2 - 2x + 2$ 12. $y = 2x^2 - 3x + 2$
13. $y = x^2 - 6x + 8$ 14. $y = x^2 - 2x - 3$ 15. $y = 2x^2 + 4x - 5$
16. $y = 3x^2 - x - 6$

12-7 SOLVING QUADRATIC EQUATIONS BY GRAPHING

To solve a quadratic equation graphically:

First, write our quadratic equation in standard form and then replace the zero with y.

$$x^2 - 2x - 3 = 0$$

$$x^2 - 2x - 3 = \boxed{0}$$

$$x^2 - 2x - 3 = y$$

or

$$y = x^2 - 2x - 3$$

Now, plot the graph of this equation as developed in Section 12-6. We plot the following points, $(-2, 5)$, $(-1, 0)$, $(0, -3)$, $(1, -4)$, $(2, -3)$, $(3, 0)$, $(4, 5)$ and connect the points with a smooth curve.

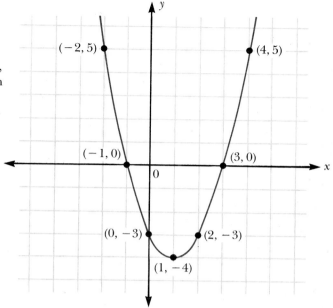

The value of x is determined where the graph crosses the x axis. At these points y equals 0.

$$y = x^2 - 2x - 3$$
$$0 = x^2 - 2x - 3$$

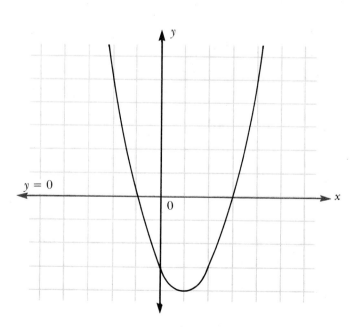

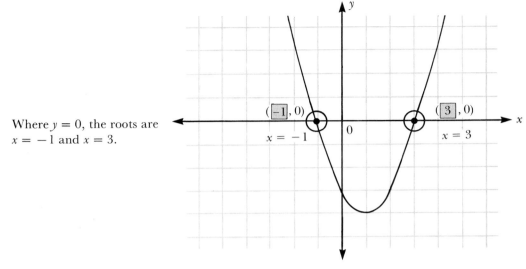

Where $y = 0$, the roots are $x = -1$ and $x = 3$.

If our curve crosses our x axis between two points, then we give an approximate value (to the nearest tenth). For example, in the graph at the right:

We approximate our values for x as $x \approx -1.4$ and $x \approx 1.4$.

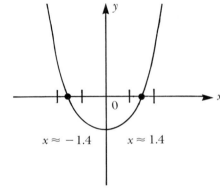

Thus, to solve a quadratic equation by graphing, we simply graph the equation $y = ax^2 + bx + c$ and estimate the values of x where the graph crosses the x axis. These value(s) will then be our solution.

PROCEDURE:

To solve a quadratic equation by graphing:

1. Graph the equation.

2. The solution(s) to the equation can be found by estimating at what point(s) the graph crosses the x axis.

EXAMPLE 1: Solve $x^2 - x - 6 = 0$ by graphing.

SOLUTION: First, graph the equation $y = x^2 - x - 6$.

x	y
-3	6
-2	0
-1	-4
0	-6
Turning point $\frac{1}{2}$	$-6\frac{1}{4}$
1	-6
2	-4
3	0
4	6

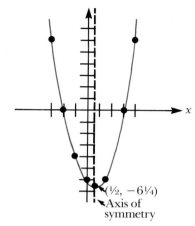

The graph crosses the x axis at $x = -2$ and $x = 3$.

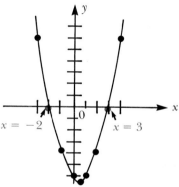

Final answer: $x = -2$ and $x = 3$

EXAMPLE 2:   Solve $x^2 - 2x - 1 = 0$ by graphing.

SOLUTION: First, graph the equation $y = x^2 - 2x - 1$.

x	y
-2	7
-1	2
0	-1
Turning point 1	-2
2	-1
3	2
4	7

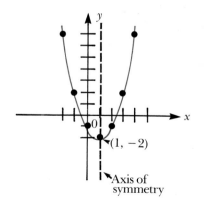

Example continued on next page.

The graph crosses the x axis at approximately
$$x \approx -0.4 \quad \text{and} \quad x \approx 2.4$$

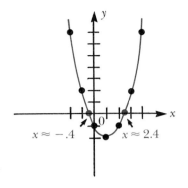

Final answer: $x = -0.4$ and $x = 2.4$

Do the following practice set. Check your answers with the answers in the right-hand margin.

PRACTICE SET 12-7

Solve the following quadratic equations graphically. Find answers to the nearest tenth.

1. $x^2 - 6x + 8 = 0$

2. $x^2 - 9 = 0$

3. $x^2 - 5x + 2 = 0$

1. $x = 2, \quad x = 4$

2. $x = 3, \quad x = -3$

3. $x = 0.4, \quad x = 4.6$

EXERCISE 12-7

Solve the following quadratic equations graphically. Find answers to the nearest tenth.

1. $x^2 - 25 = 0$ 2. $x^2 - 12 = 0$ 3. $x^2 - 4x = 0$ 4. $2x^2 - 3x = 0$

5. $x^2 - 2x - 8 = 0$ 6. $x^2 + x - 6 = 0$ 7. $x^2 - x - 1 = 0$ 8. $x^2 - 2x - 1 = 0$

9. $2x^2 + 5x - 3 = 0$ 10. $2x^2 + 4x - 5 = 0$

REVIEW EXERCISES

Transform the following equations into standard form.

1. $2(x^2 - 3) = 5x$ 2. $\dfrac{2}{x} - 3 = \dfrac{x}{2}$ 3. $\sqrt{x - 1} = x - 1$

Solve the following equations by factoring.

4. $2x^2 - 6x = 0$ 5. $3x^2 = 48$ 6. $x^2 + 6x = -8$ 7. $x^2 + 14 = 9x$

8. $x^2 + 4x = 12$ 9. $x^2 + 8x + 16 = 0$ 10. $x(x - 4) = 21$ 11. $1 = \dfrac{3}{x} + \dfrac{10}{x^2}$

Solve the following quadratic equations by the square root method.

12. $3x^2 = 0$ 13. $2x^2 = 32$ 14. $2x^2 = 16$ 15. $\dfrac{x}{6} = \dfrac{3}{x}$

Solve the following quadratic equations by completing the square.

16. $x^2 + 4x + 3 = 0$ 17. $x^2 - 3x + 2 = 0$ 18. $2x^2 + 5x - 3 = 0$ 19. $x^2 + 6x - 1 = 0$

Solve the following quadratic equations by the quadratic formula.

20. $x^2 - 5x + 6 = 0$ 21. $x^2 + 4x - 2 = 0$ (answer in radical form)

22. $x^2 + 10x - 10 = 0$ (answer to the nearest tenth)

Graph the following quadratic equations.

23. $y = x^2 - 3x - 4$ 24. $y = x^2 - 9$ 25. $y = x^2 - 3x - 1$

Solve the following quadratic equations by graphing. Find the answer to the nearest tenth.

26. $x^2 + x - 6 = 0$ 27. $5x^2 - 8x - 1 = 0$ 28. $x^2 + 2x - 7 = 0$

Appendix: Sets

A-1   SET NOTATION AND DESCRIPTION

A beginning student of algebra should be familiar with the concept and properties of sets, for the ideology of sets allows us to describe many mathematical theories in a most simplified manner.

A **set** may be thought of as a collection of objects. Every object that belongs to a set is called a *member* or an *element* of the set. We define a set as a well-defined collection of objects such that we will always be able to determine if an object is or is not a member of the set. For example, all the teachers in your school form a set and your algebra teacher is indeed a member of this set. However, objects such as the secretary of a department, or a school maintenance worker are not members of the set.

In specifying a set, we make use of special symbols and notation. As an illustration, consider the set whose elements are the months of the year that begin with the letter "J." One way of indicating this set is to list the names of its elements between braces, { }, separating each element by a comma. Thus, this set would be written as

$$\{\text{January, June, July}\}$$

This method of specifying a set is commonly referred to as the *roster method*, *tabulation method*, or the *listing method*.

Sets are also given a name. This name is usually in the form of a capital letter. Hence, using the preceding example, we may write

$$A = \{\text{January, June, July}\}$$

to mean that set A is the set whose elements are the months January, June, and July.

When a set is specified by the roster method, the order in which the elements are listed is not important. For example,

$$A = \{\text{January, June, July}\}$$

is identical (or equal) to all of the following listings.

$$\{\text{January, July, June}\}$$
$$\{\text{June, January, July}\}$$

$$\{\text{June, July, January}\}$$
$$\{\text{July, January, June}\}$$
$$\{\text{July, June, January}\}$$

Frequently, it is impractical to list all the elements of a set. For example, in mathematics, the set of natural numbers between 1 and 500, inclusive, consists of 500 elements. To specify this particular set by roster would be inconvenient. However, we may utilize an abbreviated roster form to list these elements. This form employs the use of three dots, . . . , called an *ellipsis*, to indicate that a previously established pattern is to continue. For example, the set of natural numbers between 1 and 500, inclusive, may be written as

$$A = \{1, 2, 3, \ldots, 498, 499, 500\}$$

The use of an ellipsis, in effect, tells us that there are elements in the set that follow the established pattern but for convenience sake have not been written down. Thus, in the previous example, 300, 19, and 297 are elements of A, but 0 and 648 are not elements of A.

To indicate whether or not an object is indeed a member of a set, we use the symbol $\in$, to mean "is an element of" and the symbol, $\notin$, to mean "is not an element of." Using our example,

$$A = \{\text{January, June, July}\}$$

we could write June $\in A$, or March $\notin A$.

EXAMPLE 1: Write the set of months with 31 days by roster.

SOLUTION: $A = \{\text{January, March, May, July, August, October, December}\}$

EXAMPLE 2: State whether each of the following is true or false.

      a. $P \in \{A, B, C, D\}$

      b. $p \notin \{P, A, C, E\}$

      c. $12 \in \{1, 2, 3, \ldots, 23, 24, 25\}$

SOLUTION: a. False. The set contains only four elements, none of which are the letter P.

      b. True. The lower-case letter, p, is not an element of the set with the upper-case letters, P, A, C, and E.

      c. True. The use of the ellipsis indicates that the number 12 is indeed included in the set.

Often it is inconvenient to specify a set by listing all its elements in roster form. For this reason, we introduce another method for specifying a set—the *rule method*.

In using the rule method, we specify a particular set by writing within braces a **description** or rule that describes or identifies the members of the set. Moreover, we use a variable (such as "x") to represent any one of the members of the set.

As an example, consider once again the set of months of the year that begin with the letter J. Using the rule method, we write

$$A = \{x \mid x \text{ is a month of the year that begins with the letter J}\}$$

This notation is read as

$$
\begin{array}{ccccl}
\text{set} & \text{is equal} & \text{all} & \text{such} & \\
A & \text{to} & x & \text{that} & \text{(a written description of the set)} \\
\downarrow & \downarrow & \downarrow & \swarrow & \\
A & = & \{x & \mid & x \text{ is a month of the year that begins with the letter J}\}
\end{array}
$$

EXAMPLE 3:   Specify the following set by the rule method.

$$A = \{\text{Sunday, Monday, Tuesday, Wednesday, Thursday, Friday, Saturday}\}$$

SOLUTION:
$$A = \{x \mid x \text{ is a day of the week}\}$$

EXAMPLE 4:   List the members of the following set by roster.

$$A = \{x \mid x \text{ is a month of the year with 31 days}\}$$

SOLUTION:
$$A = \{\text{January, March, May, July, August, October, December}\}$$

Do the following practice set. Check your answers with the answers in the right-hand margin.

PRACTICE SET A-1

In problems 1–3, write the sets indicated by roster (see example 1 or 4).

1. The set of days of the week that begin with the letter T.
2. $\{x \mid x \text{ is a vowel in the English language}\}$
3. $\{x \mid x \text{ is a digit in the decimal number system}\}$

1. {Tuesday, Thursday}
2. $\{a, e, i, o, u\}$
3. $\{0, 1, 2, 3, 4, 5, 6, 7, 8, 9,\}$

In problems 4–6, specify the given set by the rule method (see example 3).

4. {Sunday, Saturday}
5. {New York, New Mexico, New Jersey, New Hamsphire}
6. {2, 3, 5, 7, 11, 13, 17, 19}

In problems 7–12, state true or false.

7. $5 \in \{1, 2, 3, 4, 5\}$
8. $2 \notin \{123\}$
9. $d \in \{a, b, c, \ldots, x, y, z\}$
10. black $\in \{x \mid x$ is a color of the rainbow$\}$
11. $10 \notin \{x \mid x$ is a multiple of 3$\}$
12. spring $\in \{x \mid x$ is a season of the year$\}$

4. $\{x \mid x$ is a day of the week beginning with the letter "S"$\}$
5. $\{x \mid x$ is a state beginning with the word "New"$\}$
6. $\{x \mid x$ is a prime number less than 20$\}$

7. True
8. True
9. True
10. False
11. True
12. True

EXERCISE A-1

In problems 1–15, list the elements of each set in roster form.

1. The set of letters in the word "musk."
2. The set of states that begin with the letter A.
3. The set of U.S. presidents who have been assassinated while in office.
4. The set whose elements are the digits of the number 2358.
5. The set whose elements are the odd numbers that can divide evenly into 18.
6. $\{x \mid x$ is a month with 30 days$\}$
7. $\{x \mid x$ is the capital of New York State$\}$
8. $\{x \mid x$ is the name of all the "bases" on a baseball diamond$\}$
9. $\{x \mid x$ is a color of the rainbow$\}$
10. $\{x \mid x$ is a Japanese-made automobile$\}$
11. $\{x \mid x + 8 = 12\}$
12. $\{x \mid x - 6 = 9\}$
13. $\{x \mid x + 5 = 5\}$
14. $\{x \mid x + 0 = 2\}$
15. $\{x \mid \frac{x}{4} = 2\}$

In problems 16–20, specify the given set by using the rule method.

16. $\{a, b, c, \ldots, x, y, z\}$
17. $\{5, 10, 15, 20, \ldots, 100\}$
18. {April, June, September, November}
19. {Arctic, Indian, Atlantic, Pacific}
20. {reading, 'riting, 'rithmetic}

In problems 21–30, answer true or false.

21. $6 \in \{1, 2, 3, \ldots, 99, 100\}$

22. $6 \in \{1, 1, 2, 3, 5, \ldots\}$

23. $! \in \{x \mid x$ is a punctuation mark$\}$

24. Liz $\in \{x \mid x$ is a girl's name beginning with the letter "E"$\}$

25. $3 \in \{345\}$

26. New York $\notin \{x \mid x$ is a city in the United States$\}$

27. diamonds $\in \{x \mid x$ is a suit of cards$\}$

28. $L \notin \{a, b, c, \ldots, x, y, z\}$

29. $0 \in \{\ \}$

30. $21 \in \{x \mid x$ is divisible by 3$\}$

A-2 TYPES OF SETS

A set may contain a finite number of elements, an infinite number of elements or no elements at all. Respectively, these sets are called finite sets, infinite sets, and empty sets.

Finite Sets

A finite set is a set that contains a definite number of elements. For example, each of the following are finite sets.

$$A = \{1, 2, 3, 4\}$$
$$B = \{1, 2, 3, \ldots, 48, 49, 50\}$$
$$C = \{1, 3, 5, \ldots, 99\}$$

Notice that in each set listed it is actually possible to count the number of elements in the set and that there is an end to the listing of the elements of the set. These are the characteristics of a finite set.

Infinite Sets

Unlike finite sets, infinite sets have an unlimited number of elements. For example, each of the following are infinite sets.

$$A = \{1, 3, 5, 7, 9, \ldots\}$$
$$B = \{2, 4, 6, 8, \ldots\}$$
$$C = \{1, \frac{1}{2}, \frac{1}{4}, \frac{1}{8}, \frac{1}{16}, \ldots\}$$

Notice that in each of these sets we use an ellipsis to indicate that the established pattern will continue indefinitely. The listing of the elements of the set, then, is an endless process, thus making it impossible to count the number of elements of the set.

Some of the special sets of numbers that are discussed in this textbook are infinite sets. We list these sets here.

THE SET OF NATURAL NUMBERS

$$\mathcal{N} = \{1, 2, 3, 4, \ldots\}$$

THE SET OF WHOLE NUMBERS

$$W = \{0, 1, 2, 3, \ldots\}$$

THE SET OF INTEGERS

$$J = \{\ldots, -3, -2, -1, 0, 1, 2, 3, \ldots\}$$

THE SET OF RATIONAL NUMBERS

$$Q = \left\{ \frac{p}{q} \middle| p \in J, q \in J \text{ and } q \neq 0 \right\}$$

(A rational number is a fraction in the form of p/q where both p and q are integers and $q \neq 0$.)

THE SET OF IRRATIONAL NUMBERS

(Any number that is not rational is irrational.)

THE SET OF REAL NUMBERS

$$R = \{x \mid x \in Q \text{ or } x \in \text{irrationals}\}$$

If the elements of a set are numbers, then we may describe the set by a graph whereby each element of the set is represented as a point on the number line. In doing so, we use a closed circle, ●, to represent a particular element of the set,, an open circle, ○, to represent a point that is not part of the graph, and a shaded line, ▬, to indicate that every point on the line is a member of the set. Consider the following examples.

SET	**GRAPH**
{All natural numbers less than or equal to 10}	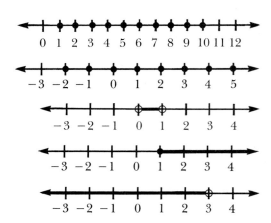
{All integers greater than −3 but less than or equal to 5}	
{All real numbers between 0 and 1}	
{All real numbers greater than or equal to 1}	
{All real numbers less than 3}	

EMPTY SETS

It is also possible for a set to contain zero elements. Such a set is called the *empty set* or *null set*. The empty set is designated by placing nothing between braces { } or by the symbol, ∅, which is written without braces. We must exercise caution not to write {∅} to indicate the empty set. This notation does not mean the empty set; it represents the set with the element, ∅.

EQUAL SETS AND EQUIVALENT SETS

If two sets contain the exact same elements, we say that the two sets are *equal*. For example, if $A = \{1, 3, 5\}$ and $B = \{3, 1, 5\}$, then set A is equal to set B, denoted $A = B$. If, however, $Q = \{2, 4, 6\}$ and $R = \{2, 4\}$, then set Q and set R are not equal to each other, denoted $Q \neq R$.

Consider now the two sets $A = \{1, 3, 5\}$ and $B = \{a, b, c\}$. Clearly, $A \neq B$. Notice, though, that both A and B have the exact same number of elements—each set contains three elements. Two sets, such as A and B, which contain exactly the same number of elements, are said to be *equivalent* sets. Observe that if two sets are equal, then they are also equivalent. However, if two sets are equivalent, they are not necessarily equal.

Do the following practice set. Check your answers with the answers in the right-hand margin.

PRACTICE SET A-2

In problems 1–6, state whether the set described is finite, infinite, or empty.

1.	The set of real numbers less than 5.	1. Infinite
2.	The set of books in print in the United States.	2. Finite
3.	The set of dogs who can climb trees.	3. Empty
4.	The set of natural numbers between 0 and 2.	4. Finite
5.	The set of real numbers between 0 and 2.	5. Infinite
6.	The set of real numbers greater than 3 but less than 0.	6. Empty

In problems 7–9, state whether the given sets are equivalent. If so, state if they are also equal.

7.	$A = \{1, 3, 9\}, \quad B = \{x, y, z\}$	7. Equivalent
8.	$A = \{M, I, C, R, O, S\}, \quad B = \{S, O, R, C, I, M\}$	8. Equivalent and equal
9.	$A = \{125\}, \quad B = \{5, 2, 1\}$	9. Not Equivalent

EXERCISE A-2

In problems 1–10, state if the given set is infinite, finite, or empty.

1. The set of all natural numbers less than one billion.

2. The set of all natural numbers greater than one billion.

3. The set of real numbers between one half and seven tenths.

4. The set of the number of miles of roadways in the world.

5. The set of two-digit numbers greater than one hundred.

6. The set of female presidents of the United States.

7. The set of six-digit numbers on the odometer of a car.

8. $\{\varnothing\}$

9. $\{10, 20, 30, 40, \ldots\}$

10. $\{1, 3, 5, \ldots, 99\}$

In problems 11–15, state whether the given sets are equivalent. If so, state if they are also equal.

11. $\{2, 4, 6\}$ and $\{1, 3, 5\}$

12. $\{b, 1, u, e\}$ and $\{b, 1, e, w\}$

13. $\{0, 2, 4, 6, \ldots, 98, 100\}$ and $\{1, 3, 5, \ldots, 97, 99\}$

14. $\{t, a, r\}$ and $\{tar\}$

15. $\{t, a, r\}$ and $\{r, a, t\}$

A-3   SUBSETS

Often we find two sets that are neither equal nor equivalent. For example, $A = \{2, 4\}$ and $B = \{1, 2, 3, 4\}$, are two such sets. However, notice that every element in set A is also contained in set B. In cases such as this, we say that set A is a *proper subset* of set B and denote this as $A \subset B$. In more precise terms, if every element in A is also an element in B, and $A \neq B$, then set A is a proper subset of set B.

EXAMPLE 1:   a.   Given $A = \{1, 3, 5\}$ and $B = \{1, 2, 3, \ldots, 9, 10\}$, then $A \subset B$ since every element in A is contained in B and $A \neq B$.

b.   Given $N = \{1, 2, 3, \ldots\}$ and $E = \{2, 4, 6, \ldots\}$ then $E \subset N$ since every element in E (the set of even natural numbers) is contained in N (the set of natural numbers) and $E \neq N$.

If we remove the restriction that $A \neq B$ in our preceding definition of a proper subset, then we have the definition of a *subset*. That is, when a set A is contained in another set B, then A is a subset of B, denoted $A \subseteq B$.

It is necessary to make a distinction between a subset and a proper subset because of the fact that every set is a subset of itself. For example, consider the two sets

$$A = \{1, 2, 3\} \quad \text{and} \quad B = \{1, 2, 3, 4, 5\}.$$

Clearly, A is a subset of itself $(A \subseteq A)$, since every element in A is also contained in A. But A is not a proper subset of itself $(A \not\subset A)$, since $A = A$. Rather, A is a proper subset of B $(A \subset B)$ since every element in A is also contained in B and $A \neq B$.

When writing all subsets of a given set, we make two observations: 2^n, where n is the number of elements in the set, represents the number of subsets possible for the given set, and the empty set is a subset of every set.

EXAMPLE 2:   State true or false for each of the following.

    a. $\{2\} \subset \{1, 2, 3\}$

    b. $\{\emptyset\} \subset \{\ \}$

    c. $\{1, 3, 5, \ldots\} \subseteq \{1, 2, 3, \ldots\}$

SOLUTION: a.   True. All the elements of the first set are contained in the second set, and the two sets are not equal.

    b.   False. All the elements of the first set are not contained in the second set.

    c.   False. Although all the elements of the first set are contained in the second set, the two sets must also be equal in order to use the symbol $\subseteq$.

EXAMPLE 3:   Given the set, $\{1, 2, 3\}$:

    a.   State how many subsets the set has.

    b.   List all of the subsets.

SOLUTION: a.   Since the set $\{1, 2, 3\}$ has three elements, there are 2^3 or 8 subsets.

    b.   Listing the subsets in the order of zero elements at a time, one element at a time, two elements at a time, and three elements at a time, we have the following subsets.

$$\underbrace{\{\ \},}_{\substack{\text{zero at a time}}} \quad \underbrace{\{1\}, \{2\}, \{3\},}_{\text{one at a time}} \underbrace{\{1, 2\}, \{1, 3\}, \{2, 3\},}_{\text{two at a time}} \quad \underbrace{\{1, 2, 3\}}_{\text{three at a time}}$$

(the empty set is a   (every set is a

subset of any set)   subset of itself)

Do the following practice set. Check your answers with the answers in the right-hand margin.

PRACTICE SET A-3

In problems 1–3, state true or false (see example 2 or 3).

1. $\{2, 4, 6, \ldots\} \subseteq \{1, 2, 3, \ldots\}$

2. $\{1, 3, 5\} \subseteq \{135\}$

3. $\{1, 2, 3, \ldots, 98, 99, 100\} \not\subset \{1, 2, 3, \ldots\}$

4.   Given the set $\{a, b\}$:
    a. State the number of subsets the set has.
    b. List all such subsets.

1.   True

2.   False

3.   False

4.   a. 2^2 or 4 subsets
    b. $\{\ \}, \{a\}, \{b\}, \{a, b\}$

EXERCISE A-3

Given $A = \{1, 3, 5\}$, $B = \{2, 3, 4\}$, and $C = \{1, 2, 3, 4, 5\}$, answer true or false in problems $1-10$.

1. $A \subset B$
2. $A \subseteq A$
3. $C \not\subset A$
4. $A \subset C$
5. $B = B$
6. $\emptyset \subseteq B$
7. $B \subset C$
8. $A \not\subset C$
9. $A \subset \emptyset$
10. $C \subset C$

Using your knowledge of set notation, answer true or false in problems $11-20$.

11. $\emptyset \subset \{a\}$
12. $\{a\} \subseteq \{a\}$
13. $1 = \{1\}$
14. $\{0\} \subseteq \{\emptyset\}$
15. $8 \notin \{1, 2, 3\}$
16. $\{8\} \in \{1, 2, 3, \ldots, 10\}$
17. $\{5\} \subset \{x \mid x \text{ is a prime number}\}$
18. $3 \subset \{x \mid x \text{ is a single digit number}\}$
19. $X \subseteq X$
20. $4 \in (1, 2, 3, 4)$
21. List all subsets of $\{2, 8, 9\}$.
22. List all subsets of $\{a, b, c, d\}$.

A-4 SET OPERATIONS

There are three distinct set operations that result in the generation of new sets: complement, intersection, and union.

COMPLEMENTS

Whenever we consider problems dealing with sets, we frequently employ the use of a *universal set*. This universal set, denoted by U, is a general set and contains all the elements being discussed in a particular problem. With this being the case, every set that is being considered in a problem would then be a subset of the universal set. For example, if we were to discuss sets of numbers, then our universal set may be the set of real numbers, or the set of natural numbers. The selection of a universal set is usually determined by the nature of the discussion and may change from problem to problem.

If, in a particular problem, we are given a universal set and an appropriate subset, then this subset will have a corresponding *complement*. This complement set would contain all elements in the universal set that are not contained in the subset. For example, if

$$U = \{1, 2, 3, 4, 5, 6, 7, 8, 9, 10\}$$

and

$$A = \{3, 4, 5, 9\}$$

then the complement of A, denoted $\bar{A}$ would contain every element in U except those contained in A. That is,

$$\bar{A} = \{1, 2, 6, 7, 8, 10\}$$

Thus, the complement of set A, is the set of all elements that are found in the universal set, U, except for those elements that are in set A.

EXAMPLE 1:   Given $U = \{1, 2, 3, 4, 5, 6, 7, 8, 9, 10\}$, $A = \{1, 3, 7, 9\}$, $B = \{2, 9, 10\}$,   and $C = \{1, 2, 3, 4, 5, 6, 7, 8, 9, 10\}$, find:
    a. $\bar{A}$
    b. $\bar{B}$
    c. $\bar{C}$

SOLUTION: By crossing out the elements of sets A, B, and C, respectively, from the universal set, we can determine the complements of each set.

    a. If U $= \{1, 2, 3, 4, 5, 6, 7, 8, 9, 10\}$
        and $A = \{1, 3, 7, 9\}$
        then $\bar{A} = \{\cancel{1}, 2, \cancel{3}, 4, 5, 6, \cancel{7}, 8, \cancel{9}, 10\}$
        or $\bar{A}$ $= \{2, 4, 5, 6, 8, 10\}$

    b. If U $= \{1, 2, 3, 4, 5, 6, 7, 8, 9, 10\}$
        and $B = \{2, 9, 10\}$
        then $\bar{B} = \{1, \cancel{2}, 3, 4, 5, 6, 7, 8, \cancel{9}, \cancel{10}\}$
        or $\bar{B}$ $= \{1, 3, 4, 5, 6, 7, 8\}$

    c. If U $= \{1, 2, 3, 4, 5, 6, 7, 8, 9, 10\}$
        and $C = \{1, 2, 3, 4, 5, 6, 7, 8, 9, 10\}$
        then $\bar{C} = \{\cancel{1}, \cancel{2}, \cancel{3}, \cancel{4}, \cancel{5}, \cancel{6}, \cancel{7}, \cancel{8}, \cancel{9}, \cancel{10}\}$
        or $\bar{C}$ $= \{\ \}$ (the empty set)

INTERSECTION

Given two sets A and B, the intersection of A and B, denoted $A \cap B$, is a new set that contains all the elements common to both A and B. For example, if $A = \{1, 2, 3, 4, 5\}$ and $B = \{2, 4, 6, 8, 10\}$, then $A \cap B = \{2, 4\}$ since the elements 2 and 4 are the only elements contained in both A and B.

EXAMPLE 2:   Given $U = \{1, 2, 3, 4, 5, 6, 7, 8, 9, 10\}$, $A = \{2, 3, 6, 8, 9, 10\}$,
    $B = \{1, 2, 3, 5, 9\}$, and $C = \{4, 7\}$, find:
    a. $A \cap B$

    b. $A \cap C$

    c. $\bar{B} \cap C$

Example continued on next page.

SOLUTION: a. If A $= \{2, 3, 6, 8, 9, 10\}$
and $B = \{1, 2, 3, 5, 9\}$
then $A \cap B = \{2, 3, 9\}$

b. If A $= \{2, 3, 6, 8, 9, 10\}$
and $C = \{4, 7\}$
then $A \cap C = \{\ \}$

NOTE:

> Two sets that have nothing in common, that is, their intersection yields the empty set, are said to be *disjoint sets*.

c. Before we can find $\bar{B} \cap C$, we must first find $\bar{B}$.
If $U = \{1, 2, 3, 4, 5, 6, 7, 8, 9, 10\}$
and $B = \{1, 2, 3, 5, 9\}$
then $\bar{B} = \{4, 6, 7, 8, 10\}$

Now, we can find $\bar{B} \cap C$.

$$\bar{B} = \{4, 6, 7, 8, 10\}$$
$$C = \{4, 7\}$$

Thus, $\bar{B} \cap C = \{4, 7\}$, or simply, C.

UNION

Given two sets A and B, the union of A and B, denoted $A \cup B$, is a new set that contains all the elements that are contained in either A and B or both. For example, if $A = \{1, 2, 3, 4, 5\}$ and $B = \{2, 4, 6, 8, 10\}$, then $A \cup B = \{1, 2, 3, 4, 5, 6, 8, 10\}$.

EXAMPLE 3: Given $U = \{1, 2, 3, 4, 5, 6, 7, 8, 9, 10\}$, $A = \{1, 2, 6, 9\}$, $B = \{2, 4, 5, 8\}$, and $C = \{5, 9, 10\}$, find
a. $A \cup B$
b. $A \cup C$
c. $\bar{A} \cup B$

SOLUTION: a. $A \cup B = \{1, 2, 4, 5, 6, 8, 9\}$

b. $A \cup C = \{1, 2, 5, 6, 9, 10\}$

c. $\bar{A} = \{3, 4, 5, 7, 8, 10\}$

$\bar{A} \cup B = \{2, 3, 4, 5, 7, 8, 10\}$

Do the following practice set. Check your answers with the answers in the right-hand margin.

PRACTICE SET A-4

Given $U = \{1, 2, 3, 4, 5, 6, 7, 8, 9, 10\}$, $A = \{1, 3, 8, 9\}$, $B = \{2, 4, 5, 6, 7, 8\}$, and $C = \{3, 7, 9\}$. In problems 1–3, find the complement (see example 1).

1. $\bar{A}$
2. $\bar{B}$
3. $\bar{C}$

  1. $\{2, 4, 5, 6, 7, 10\}$
  2. $\{1, 3, 9, 10\}$
  3. $\{1, 2, 4, 5, 6, 8, 10\}$

In problems 4–6, find the intersection (see example 2).

4. $A \cap B$
5. $B \cap C$
6. $A \cap \bar{C}$

  4. $\{8\}$
  5. $\{7\}$
  6. $\{1, 8\}$

In problems 7–9, find the union (see example 3).

7. $A \cup B$
8. $A \cup C$
9. $\bar{B} \cup C$

  7. $\{1, 2, 3, 4, 5, 6, 7, 8, 9\}$
  8. $\{1, 3, 7, 8, 9\}$
  9. $\{1, 3, 7, 9, 10\}$

EXERCISE A-4

Given $U = \{1, 2, 3, 4, 5, 6, 7, 8, 9, 10\}$, $A = \{1, 2, 4, 8\}$, $B = \{1, 3, 7, 9, 10\}$, $C = \{3, 4, 5, 6, 7, 8\}$, and $D = \{4, 6, 8\}$, find the sets indicated in problems 1–20.

1. $A \cap B$
2. $A \cap C$
3. $B \cap C$
4. $B \cap D$
5. $A \cup C$
6. $B \cup C$
7. $\bar{A}$
8. $\bar{D}$
9. $\bar{B} \cap C$
10. $\bar{C} \cap A$
11. $\bar{A} \cup \bar{D}$
12. $B \cup \bar{B}$
13. $B \cap \varnothing$
14. $\bar{A} \cup \varnothing$
15. $A \cap (B \cup C)$
16. $A \cup (B \cap D)$
17. $C \cup (B \cup D)$
18. $B \cap (A \cup C)$
19. $\bar{A} \cap (\bar{B} \cup D)$
20. $(\overline{(B \cap C) \cup D})$
21. If $P = \{1, 3, 5, 7, 8\}$ and $Q = \{2, 3, 4, 7, 9\}$, write $P \cap Q$ and $P \cup Q$.
22. If $R = \{x \mid x$ is a real number less than $8\}$ and $S = \{x \mid x$ is a real number greater than or equal to $4\}$, graph $R \cap S$ and $R \cup S$.

Answers to Odd-Numbered Exercises

CHAPTER 1

EXERCISE 1-1 (Page 7)

1. a. $1\frac{1}{2}$ (or 1.5)

    b. 2

    c. $3\frac{1}{2}$ (or 3.5)

    d. $\frac{1}{2}$ (or 0.5)

3. a. $\frac{2}{10}$ (or 0.2)

    b. $\frac{5}{10}$ (or 0.5)

    c. $\frac{8}{10}$ (or 0.8)

    d. $1\frac{4}{10}$ (or 1.4)

5.

```
 ──┼───┼───┼───┼──────▶
   0   1   2   3
```

7.

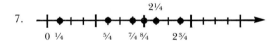

9.

```
      ¼
──┼─◆─◆◆◆─┼─◆─┼┼┼┼─┼──▶
  0 ¹⁄₁₀ ⅕ ⅓   ½
```

11. $F, K, P, U, \mathcal{Z}$

13. All points A–$\mathcal{Z}$

EXERCISE 1-2 (Page 17)

1. $\frac{1}{3}$

3. $\frac{1}{4}$

5. $\frac{3}{5}$

7. $\frac{3}{2}$

9. 5

11. $\frac{33}{40}$

13. $\frac{3}{4}$

15. $\frac{17}{20}$

17. $5\frac{3}{25}$

19. 0.75

21. 0.1

23. 0.35

25. 0.375

27. $0.66\bar{6}$

29. $\frac{31}{6}$

31. $\frac{14}{3}$

33. $\frac{83}{7}$

35. $\frac{247}{10}$

37. $2\frac{1}{2}$

39. 1

41. $2\dfrac{1}{5}$ 43. $3\dfrac{3}{7}$ 45. $5\dfrac{3}{7}$

47. The decimal 3.6 is not greater than 4.2.   49. The fraction $\dfrac{3}{8}$ does not equal $\dfrac{2}{3}$.

51. The fraction $\dfrac{3}{4}$ is greater than $\dfrac{1}{2}$ and less than $\dfrac{2}{3}$.   53. $3 > 0$

55. $\dfrac{5}{6} < \dfrac{7}{6}$ 57. $\dfrac{3}{8} < 1\dfrac{3}{4}$ 59. $0.9 > 0.256$

EXERCISE 1-3 (Page 27)

1. $\dfrac{46}{35}$ 3. $\dfrac{41}{24}$ 5. $\dfrac{5}{36}$ 7. $\dfrac{15}{28}$

9. $\dfrac{1}{20}$ 11. $\dfrac{15}{4}$ 13. $3\dfrac{11}{12}$ 15. $40\dfrac{5}{24}$

17. $2\dfrac{11}{12}$ 19. $\dfrac{51}{8}$ 21. 12   23. $\dfrac{9}{10}$

25. 49.9538   27. 2.3   29. 0.3990   31. 0.04

EXERCISE 1-4 (Page 31)

1. 10^2 3. 7^5 5. 6^0 7. 2^8 9. 9^0
11. 2^3 13. 10^3 15. y^3 17. $2^3 \cdot 3^4$ 19. $9^1 \cdot x^3$
21. 16   23. 10,000   25. 125   27. 48   29. 75
31. 40   33. 120   35. 168   37. 900   39. 96
41. 3240   43. 1440   45. 6   47. 1000

EXERCISE 1-5 (Page 37)

1. 28   3. 6   5. 0   7. 48   9. 69
11. 14   13. 24   15. 36   17. 8   19. 120
21. 40   23. 1   25. 6   27. 109   29. 42.875

31. 729   33. 125.81   35. 369   37. $3\dfrac{4}{9}$ 39. $\dfrac{1369}{3600}$

REVIEW EXERCISES (CHAPTER 1) (Page 38)

1.

2. 0

3. There are many fractions equal to the given numbers; we list three.

    a. $2 = \dfrac{2}{1}$ $\dfrac{2 \times 2}{1 \times 2} = \boxed{\dfrac{4}{2}}$ $\dfrac{2 \times 3}{1 \times 3} = \boxed{\dfrac{6}{3}}$ $\dfrac{2 \times 4}{1 \times 4} = \boxed{\dfrac{8}{4}}$

    b. $6 = \dfrac{6}{1}$ $\dfrac{6 \times 2}{1 \times 2} = \boxed{\dfrac{12}{2}}$ $\dfrac{6 \times 3}{1 \times 3} = \boxed{\dfrac{18}{3}}$

       $\dfrac{6 \times 4}{1 \times 4} = \boxed{\dfrac{24}{4}}$

    c. $0 = \dfrac{0}{1}$ $\dfrac{0 \times 2}{1 \times 2} = \boxed{\dfrac{0}{2}}$ $\dfrac{0 \times 3}{1 \times 3} = \boxed{\dfrac{0}{3}}$ $\dfrac{0 \times 4}{1 \times 4} = \boxed{\dfrac{0}{4}}$

4. There are many fractions equivalent to the given fractions; we list three.

    a. $\dfrac{2}{3} = \dfrac{2 \times 2}{3 \times 2} = \boxed{\dfrac{4}{6}}$ $\dfrac{2 \times 3}{3 \times 3} = \boxed{\dfrac{6}{9}}$ $\dfrac{2 \times 4}{3 \times 4} = \boxed{\dfrac{8}{12}}$

    b. $\dfrac{3}{5} = \dfrac{3 \times 2}{5 \times 2} = \boxed{\dfrac{6}{10}}$ $\dfrac{3 \times 3}{5 \times 3} = \boxed{\dfrac{9}{15}}$ $\dfrac{3 \times 4}{5 \times 4} = \boxed{\dfrac{12}{20}}$

    c. $\dfrac{2}{4} = \dfrac{2 \div 2}{4 \div 2} = \boxed{\dfrac{1}{2}}$ $\dfrac{2 \times 2}{4 \times 2} = \boxed{\dfrac{4}{8}}$ $\dfrac{2 \times 3}{4 \times 3} = \boxed{\dfrac{6}{12}}$

5.　a.　$\dfrac{3}{10}$

　　b.　$\dfrac{7}{10}$

　　c.　$2\dfrac{3}{5}$

6.　a.　$\dfrac{5}{3}$

　　b.　$\dfrac{7}{2}$

　　c.　$\dfrac{21}{5}$

7.　a.　Point F
　　b.　Point C
　　c.　Point B

8.　12.96

9.　53.21

10.　165.51

11.　15.3954

12.　259.12

13.　533.19

14.　5.714

15.　99.57

16.　28.2009

17.　$\dfrac{5}{6}$

18.　$\dfrac{5}{7}$

19.　1

20.　$\dfrac{5}{8}$

21.　$\dfrac{7}{12}$

22.　$\dfrac{23}{36}$

23.　$9\dfrac{5}{8}$

24.　$9\dfrac{5}{12}$

25.　$11\dfrac{2}{5}$

26.　1.53

27.　3.9

28.　200.064

29.　1.92

30.　5.39

31.　2.441

32.　6.13

33.　709.36

34.　4.487

35.　$\dfrac{1}{3}$

36.　$\dfrac{1}{9}$

37.　$\dfrac{6}{13}$

38.　$\dfrac{1}{2}$

39.　$\dfrac{5}{24}$

40.　$\dfrac{11}{18}$

41.　$1\dfrac{1}{2}$

42.　$1\dfrac{3}{4}$

43.　$2\dfrac{1}{6}$

44.　3.64

45.　0.0792

46.　0.00372

47.　5.26

48.　58.7664

49.　11.2336

50.　0.0009

51.　65.12

52.　51

53.　$\dfrac{15}{28}$

54.　$\dfrac{3}{10}$

55.　$\dfrac{3}{8}$

56.　$\dfrac{7}{16}$

57.　$\dfrac{1}{4}$

58.　$\dfrac{2}{9}$

59.　$26\dfrac{2}{3}$

60.　6

61.　$21\dfrac{1}{3}$

62.　57.5

63.　60

64.　12.3

65.　1.4

66.　65.4

67.　3.6

68.　320

69.　10.8

70.　0.6

71.　2

72.　$\dfrac{9}{2}$

73.　$\dfrac{7}{6}$

74.　$\dfrac{21}{4}$

75.　$\dfrac{9}{7}$

76.　$\dfrac{160}{9}$

77.　$\dfrac{7}{9}$

78.　$\dfrac{5}{9}$

79.　$\dfrac{6}{11}$

80.　a.　$13 > 5$
　　b.　$5.6 > 2.007$

　　c.　$\dfrac{3}{4} > \dfrac{2}{3}$

　　d.　$0.2 < \dfrac{3}{5}$

81.　a.　144
　　b.　1800
　　c.　1

82.　a.　32　　b.　45
　　c.　75　　d.　54

　　e.　69　　f.　$\dfrac{1}{6}$

　　g.　30　　h.　1
　　i.　73　　j.　3000
　　k.　63　　l.　900

CHAPTER 2

EXERCISE 2-1 (Page 46)

1.

3.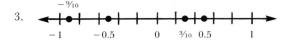

5.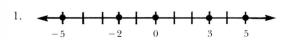

7. $0 < +6$

9. $+1\frac{1}{2} > -\frac{3}{2}$

11. $+1.2 > -\frac{8}{4}$

13. $-9 < +0.9$

15. $-1\frac{3}{4} > \frac{-15}{8}$

17. $-2\frac{7}{8} < -2\frac{3}{4}$

19. $-9.1 < -0.6 < 8\frac{1}{8}$

21. $+\$300$

23. $+\$10$

25. -20

27. $-\$100$

29. -12 yards

31. $+6$

33. -0.5

35. 0

37. Negative

39. Positive

EXERCISE 2-2 (Page 50)

1. 6

3. 8.5

5. 9.42

7. $4\frac{1}{4}$

9. 5

11. $|6| > -6$

13. $|-5| > -5$

15. $|-3| = |3|$

17. $|-5| > 2$

19. $14 > -36$

21. 11

23. 1

25. 50

27. 9

29. 9

EXERCISE 2-3 (Page 60)

1. $+11$

3. $+4$

5. -28

7. -4

9. $+2$

11. $+2$

13. $+8$

15. -9

17. -156

19. $+2.287$

21. -3.92

23. -1.7

25. $+8.85$

27. $+1.24$

29. $-\frac{8}{9}$

31. -1

33. $+\frac{1}{4}$

35. $-\frac{1}{2}$

37. $-\frac{5}{6}$

39. $-\frac{73}{15}$

41. $+1$

43. -46

45. -14

47. 0

49. -4

51. $+12$

53. $+30$

55. -10

57. $-\frac{1}{2}$

59. $+20$; 4 yards per play average

61. Balance: $99.99; no, she is not overdrawn.

63. Weight of freight is 54,296 pounds.

65. $-87°F$

EXERCISE 2-4 (Page 68)

1. $+72$

3. -200

5. $+30$

7. -72

9. -200

11. -5.4

13. -118.3

15. $+461$

17. 0

19. $+\frac{3}{25}$

21. $-\frac{96}{5}$

23. -1

25. -240

27. $+100$

29. -60

31. 0

33. -25

35. 0

37. $+8$

39. -8

41. -11

43. $+64$

45. $+625$

47. $+180$

49. -27

51. -36

53. -1

55. -540

57. $+324$

EXERCISE 2-5 (Page 73)

1. $+2$	3. -20	5. -7	7. $+2$	9. $+1$
11. -5	13. $+\dfrac{7}{9}$	15. $-\dfrac{1}{2}$	17. $+2.8$	19. $-\dfrac{8}{9}$
21. $+\dfrac{1}{18}$	23. $+29$	25. -4	27. $+\dfrac{2}{3}$	29. $+189$

EXERCISE 2-6 (Page 76)

1. -4	3. $+6$	5. -36	7. -178	9. $+2$
11. 8	13. -14	15. $+16$	17. -66	19. -5
21. 0	23. $+9$	25. 10	27. -5	29. $+541$

EXERCISE 2-7 (Page 88)

1. True; commutative property of multiplication
3. False
5. True; distributive property
7. True; associative property of multiplication
9. True; identity property of addition
11. False
13. False
15. False
17. True; distributive property
19. False
21. Commutative property of addition
23. Additive inverse
25. Associative property of addition
27. Zero property of multiplication
29. Commutative property of multiplication
31. $(8 + 2)x$
33. $(4 + 1)y$
35. $5(3) + 5(4)$
37. $12(x) - 12(y)$
39. $8(6) + 1(6)$
41. 30
43. 2
45. -6
47. 8

REVIEW EXERCISES (CHAPTER 2) (Page 89)

1. g	2. i	3. f	4. h	5. b
6. e	7. a	8. c	9. d	10. $+2$
11. -14	12. $+10$	13. -16	14. -2	15. $+17$
16. $+12$	17. $+33$	18. 22	19. -112	20. 24
21. -35	22. 15	23. 19	24. -4	25. -18
26. $+54$	27. $+10$	28. -216	29. -1728	30. $+144$
31. $+9$	32. -1	33. -5	34. -2	35. -16
36. -64	37. 62	38. -5	39. -54	40. 56
41. 60	42. 192	43. -150	44. -36	45. 0

CHAPTER 3

EXERCISE 3-1 (Page 97)

1. a. Variable is x
   b. Constant is $+6$

3. a. Variable is p
   b. Constant is $+15$

5. a. Variables: x and y
   b. Constant is -16

7. a. Variables: x and y
   b. Constant is $-\dfrac{3}{4}$

9. a. Variables: x and y
   b. Constant is -8

11. a. Two terms
    b. First term: $4y$
       Second term: $-6z$
    c. Coefficient of first term: 4
       Coefficient of second term: -6

13. a. Three terms
    b. First term: x
       Second term: $-4z$
       Third term: $+16$
    c. Coefficient of first term: 1
       Coefficient of second term: -4
       Constant term is $+16$

15. a. Three terms
     b. First term: x
         Second term: $-y$
         Third term: $-z$
     c. Coefficient of first term: 1
         Coefficient of second term: -1
         Coefficient of third term: -1

17. a. Two terms
     b. First term: $4x$
         Second term: $(3y + 6z)$
     c. Coefficient of first term: 4
         Coefficient of second term: 1

19. a. One term
     b. Term is: $-4x(3y + 6z)$
     c. Coefficient is -4

21. a. Three terms
     b. First term: $3ab^2$
         Second term: $\left(+\dfrac{2x - y}{3xy} \right)$
         Third term: $2(x^2 - 6y)$
     c. Coefficient of first term: 3
         Coefficient of second term: $+1$
         Coefficient of third term: $+2$

23. $x^3 y^1$ 25. $5x^2 y$ 27. $-16x^2 y^2$ 29. $3(x^3 + y^2)$

31. $x \cdot x$ 33. $-5 \cdot y \cdot y \cdot y$ 35. $(-5x)(-5x)$

EXERCISE 3-2 (Page 102)

1. 15   3. 1   5. -6 7. 9   9. $\dfrac{5}{2}$

11. 8   13. 0   15. -18 17. -14 19. -14

21. -420 23. $+21$ 25. 100   27. 164   29. 36

31. -52 33. -16 35. 29   37. $c = 20$ 39. $A = 24$

41. $c = 27.2$ 43. $s = 104$ 45. $s = 185.9$ 47. $A = 6655$ 49. $l = 0.3125$

51. a. 0.4 amps
     b. 14 amps
     c. 240 amps

53. a. 2300°F
     b. 1940°F

55. 1800 horsepower

57. a. 90.7 cubic inches
     b. 1397 cc

59. a. 400 candlepower
     b. Yes, the intensity level for
         Setek's office is 70.9
         candlepower.

61. Approximately 62.5%

63. $27,000

65. a. 158 beats per second
     b. 86.6 beats per second
     c. between 6 and 22 minutes

EXERCISE 3-3 (Page 111)

1. $11x$ 3. $-94t$ 5. $14m$ 7. $17x$ 9. $4x + 12y$

11. $9c + 8d$ 13. $-2w - 3x$ 15. $5(x + y)$ 17. $6xy - 3x$ 19. $3x^2 + 12x$

21. $-3x^2 y$ 23. $-3x^2 y + 8xy^2$ 25. $-a^2 b + 3ab^2$

27. $-5 + 5ab + 5a - b$ 29. $6xy^2 - 3x^2 y^2 + 2x^2 y - 6x$ 31. $3x - y$

33. $-x + 24$ 35. $2x - 7$ 37. $33x + 48$

39. $6a - 11b$ 41. $5x^2 - 13x + 84$ 43. $7x^2 - 24x + 8$

45. $19x^2 - 6x - 35$ 47. $45x - 100$ 49. $39x - 130$

51. $13x + 87y$ 53. $-46x + 59y$ 55. $-6x + 9a$

57. $-20x - 20y$ 59. $230x - 285$

REVIEW EXERCISES (CHAPTER 3) (Page 113)

1. Variable 2. Variable 3. Term 4. Term 5. Constant
6. Constant 7. Coefficient 8. Coefficient 9. Factor 10. Factor
11. Coefficient 12. Factor 13. -1
14. a. Five terms b. First term: $8x^4$ c. Coefficient of first term: 8
    Second term: $+5x^3$ Coefficient of second term: $+5$
    Third term: $-x^2$ Coefficient of third term: -1
    Fourth term: $+7x$ Coefficient of fourth term: $+7$
    Fifth term: -6 Constant term is -6

15. $5x^4y^3$ 16. $8xxxyyyy$ 17. 15
18. -15 19. -6 20. 9
21. -270 22. $I = 25{,}000$ 23. $l = \dfrac{3}{16}$
24. $A = 50.24$ 25. $F = 59$ 26. $5x - y$
27. $6a - 12b$ 28. $21x^2 - 30x - 4$ 29. $-10x + 23y$

CHAPTER 4

EXERCISE 4-2 (Page 120)

1. a. Addition 3. a. Subtraction 5. a. Subtraction
   b. Subtract 6 from both sides b. Add 5 to both sides of b. Add 5 to both sides of
      of equation. equation. equation.
7. a. Addition 9. a. Addition and multiplication
   b. Subtract 23 from both sides of equation. b. First subtract 5 from both sides of equation,
      then divide both sides of equation by 2.

11. $x = 3$ 13. $x = 17$ 15. $x = 25$ 17. $x = -25$ 19. $x = -7$
21. $x = 19$ 23. $x = 15$ 25. $x = 22$ 27. $x = 2$ 29. $x = -28$
31. $x = 8$ 33. $x = 5$ 35. $x = 27$ 37. $x = -9$ 39. $x = -3$
41. $x = -24$ 43. $x = -8$ 45. $x = -2$ 47. $x = 60$ 49. $x = 1$
51. $x = -9$

EXERCISE 4-3 (Page 125)

1. $x = 3$ 3. $x = 7$ 5. $x = 1$ 7. $x = -5$ 9. $x = -11\dfrac{1}{3}$
11. $x = -3$ 13. $x = -2$ 15. $x = -25$ 17. $x = -22$ 19. $x = 10$
21. $x = 15$ 23. $x = -30$ 25. $x = 15$ 27. $x = -28$ 29. $x = -54$
31. $x = 56$ 33. $x = 80$ 35. $x = 21$ 37. $x = -9$ 39. $x = 5$
41. $x = 12$ 43. $x = -5$ 45. $x = -13$ 47. $x = -30$ 49. $x = -12$
51. $x = -20$

EXERCISE 4-4 (Page 128)

1. $x = 3$ 3. $x = 3$ 5. $x = -72$ 7. $x = 31$ 9. $x = -60$
11. $x = 2$ 13. $x = -8$ 15. $x = 2$ 17. $x = 100$ 19. $x = \dfrac{275}{2}$

EXERCISE 4-5 (Page 132)

1. $x = 5$ 3. $x = 7$ 5. $x = -5$ 7. $x = 7$ 9. $x = -2$
11. $x = -18$ 13. $x = 1$ 15. $x = \dfrac{6}{11}$ 17. $x = 3$ 19. $x = 3$

EXERCISE 4-6 (Page 136)

1. $x = 3$ 3. $x = 8$ 5. $x = 1$ 7. $x = 2$ 9. $x = 9$

11. $x = -7$ 13. $x = -10$ 15. $x = 23$ 17. $x = -11$ 19. $x = 2$

21. $x = -5$ 23. $x = 2$ 25. $x = 4$ 27. $x = -12$ 29. $x = 3$

31. $x = -\dfrac{40}{19}$ 33. $x = -7$ 35. $x = \dfrac{28}{5}$ 37. $x = \dfrac{11}{9}$ 39. $x = 2$

41. $x = \dfrac{24}{5}$ 43. $x = \dfrac{30}{7}$ 45. $x = 3$ 47. $x = -5$ 49. $x = 12$

51. $x = -\dfrac{19}{6}$ 53. $x = 80$

EXERCISE 4-7 (Page 139)

1. $H = 19 - l$ 3. $\dfrac{F}{A} = M$ 5. $\dfrac{C}{D} = \pi$ 7. $W = \dfrac{V}{LH}$

9. $b = y - mx$ 11. $b = \dfrac{2A}{H}$ 13. $C = \dfrac{5}{9}(F - 32)$ 15. $a = \dfrac{2S}{N} - l$

17. $F = \dfrac{9}{5}C + 32$ 19. $t = \dfrac{A}{pr} - \dfrac{1}{r}$

EXERCISE 4-8 (Page 149)

1.
$x < 7$

3.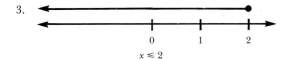
$x \leq 2$

5.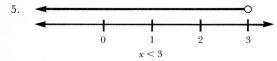
$x < 3$

7.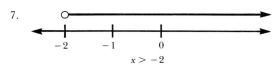
$x > -2$

9.
$x > 6$

11.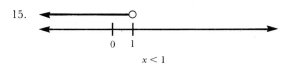
$x < 5$

13.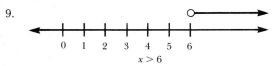
$x \leq 7$

15.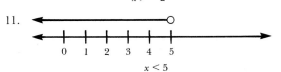
$x < 1$

17.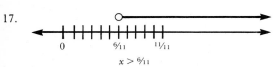
$x > \frac{6}{11}$

19.
$x \geq 8$

21.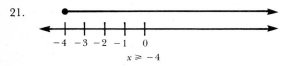
$x \geq -4$

23.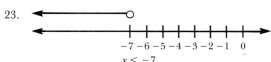
$x < -7$

25.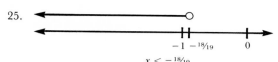

$x < -{}^{18}\!/_{19}$

27.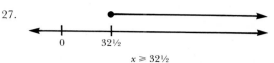

$x \geqslant 32\frac{1}{2}$

29.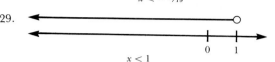

$x < 1$

31.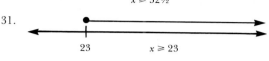

$x \geqslant 23$

33.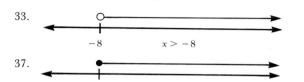

$x > -8$

35.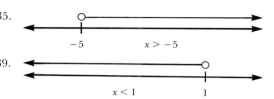

$x > -5$

37.

$x \geqslant {}^{5}\!/_{3}$

39.

$x < 1$

REVIEW EXERCISES (CHAPTER 4) (Page 150)

1. $x = 3$
2. $x = 9$
3. $x = 1$
4. $x = 5$
5. $x = -14$

6. $x = -18$
7. $x = 5$
8. $x = \dfrac{16}{3}$
9. $x = 6$
10. $x = 19$

11. $x = -\dfrac{14}{5}$
12. $x = 6$
13. $x = 80$
14. $x = 36$
15. $x = 32$

16. $x = 14$
17. $x = -\dfrac{9}{2}$
18. $x = -\dfrac{66}{5}$
19. $x = -1$
20. $x = -\dfrac{5}{2}$

21. $x = -1$
22. $x = -\dfrac{14}{3}$
23. $x = 3$
24. $x = 3$
25. $x = -\dfrac{5}{3}$

26. $x = 1$
27. $x = 6$
28. $x = \dfrac{14}{3}$
29. $x = \dfrac{17}{2}$
30. $x = -7$

31. $x = 5$
32. $x = \dfrac{3}{2}$
33. $x = \dfrac{13}{5}$
34. $x = \dfrac{45}{2}$
35. $x = 10$

36. $x = 36$
37. $x = -5$
38. $x = -\dfrac{80}{3}$
39. $x = -7$
40. $x = 28$

41. $x = 5$
42. $x = -\dfrac{19}{4}$
43. $x = \dfrac{24}{5}$
44. $x = -\dfrac{11}{15}$
45. $x = -\dfrac{1}{5}$

46. $x = 2$
47. $x = \dfrac{7}{8}$
48. $x = 1$
49. $x = -8$
50. $x = \dfrac{1}{11}$

51. $x = \dfrac{19}{7}$
52. $x = 4$
53. $x = -\dfrac{3}{2}$
54. $x = -3$
55. $x = -2$

56. $x = -2$
57. $x = \dfrac{5}{3}$
58. $x = 1$
59. $x = 6$
60. $x = \dfrac{10}{3}$

61. $x = 4$
62. $x = -\dfrac{19}{4}$
63. $x = 6$
64. $x = -1$
65. $x = 4$

66. $x = \dfrac{5}{13}$
67. $x = -\dfrac{1}{3}$
68. $x = -\dfrac{5}{7}$
69. $x = \dfrac{9}{5}$
70. $x = \dfrac{58}{5}$

71. $x = -5$

72.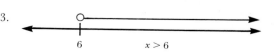

$x > 9$

73.

$x > 6$

74.

$x \geqslant -2$

75.

$x > -4$

76.

$x \leqslant -4$

77.

$x > 4$

78.

$x < -4$

79.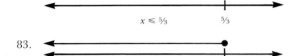

$x \geqslant 3$

80.

$x > 3$

81.

$x \leqslant {}^5\!/_3$ ${}^5\!/_3$

82.

$x \geqslant -27$

83.

$x \leqslant -{}^{12}\!/_5$ $-{}^{12}\!/_5$

84.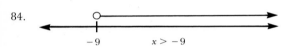

$x > -9$

85.

$x > 2$

86. $V = \dfrac{375H}{D}$

87. $g = \dfrac{2S}{t^2}$

88. $l = \dfrac{2S}{V} - a$

89. $N = \dfrac{2.5HP}{D^2}$

CHAPTER 5

EXERCISE 5-1 (Page 157)

1. $3 + x$

3. $x + y$

5. $3x - 2$

7. $12 \div (2x)$ or $\dfrac{12}{2x}$

9. $9x$

11. $2(3 + x)$

13. $x + y - (2x + 4)$

15. $x \div y$ or $\dfrac{x}{y}$

17. $(x - y) \div 2$ or $\dfrac{(x - y)}{2}$

19. $5[(k - 3) \div (x - y)]$ or $5\left(\dfrac{k - 3}{x - y}\right)$

21. $x^2 \div y^2$ or $\dfrac{x^2}{y^2}$

23. $3 + x = 23$

25. $6x - 5 = 41$

27. $3x - 19 = 8$

29. $9 + 6x = 63 + 2x$

31. $x - y = 5x$

33. $18 + 2x = \dfrac{1}{3}x$

35. $\dfrac{x}{9} = 3x - 5$

37. $2x - 13 = 29$

39. $x - 5 = 23$

41. $2x - 9 = 31$

43. $2(x + 5) = 80$

45. $\dfrac{x}{6} + 23 = 7$

47. $\dfrac{4 + 3x}{8} = 15$

49. $\dfrac{8 + 3x}{4} = 20$

51. $4(12 + 3x) = 96$

EXERCISE 5-2 (Page 163)

1. $x = 2$

3. $x = 25$

5. $x = 30$

7. $x = 13$

9. $x = \dfrac{44}{3}$

11. $x = -6$

13. First number is 12.
     Second number is 24.

15. First number is 6.
     Second number is 24.

17. $x = 3$

19. $x = 15$

21. $x = 13\dfrac{1}{3}$

23. $x = 5$

25. First Number is 4.
Second Number is 16.

27. Mike's Age is 28.
Milt's Age is 40.

EXERCISE 5-4 (Page 170)

1. 6 nickels
6 dimes

3. 8 dimes
24 quarters

5. 32 nickels
48 dimes
16 quarters

7. 37 nickels
49 dimes
72 quarters

9. 2500 tickets at $8.50 each
2500 tickets at $6.50 each
5000 tickets at $4.50 each

11. 23 fifteen cent stamps
37 twenty-five cent stamps

13. 28 twenty-eight cent stamps
58 thirty-five cent stamps
14 forty cent stamps

EXERCISE 5-5 (Page 175)

1. 25 pounds of sunflower seeds
25 pounds of wheat germ nuggets

3. 260 gallons of regular unleaded
140 gallons of super unleaded

5. a. 40 pounds of cashews;   b. yes

7. 16 ounces of Hazel's special grass seed mix
32 ounces of Crossman's grass seed mix

9. 3.2 ounces of Wheatmate
8 ounces of Cornhusk

11. 8 ounces of water

13. 4 quarts

EXERCISE 5-6 (Page 181)

1. $350 at 6 percent
$650 at 10 percent

3. $320 at 5 percent
$820 at 7 percent
A total of $1140 was invested.

5. $3000 at 6 percent (savings account)
$500 at 18 percent (Stock)

7. a. $1300 at 8 percent
$2300 at 15 percent
$1000 at 12 percent
b. $4600 was invested.

9. $2400 at 6 percent
$3600 at 11 percent

11. $2250 at 14 percent
$750 at 6 percent

13. $350 at 8 percent
$450 at 6 percent

EXERCISE 5-7 (Page 188)

1. 4 hours

3. First truck: 60 miles per hour
Second truck: 85 miles per hour

5. a. 6 hours
b. 55 miles from Howard's house

7. a. 3 minutes
b. 2.2 miles

9. First part: 65 miles per hour
Second part: 20 miles per hour

11. Start: 4 A.M.
End: 2 P.M.

13. 66 miles

15. 4994 feet

REVIEW EXERCISES (CHAPTER 5)(Page 189)

1. $x = 13$

2. $x = -3$

3. $x = -256$

4. $x = 3$

5. $x = 29$

6. $x = -6$

7. $x = -8$

8. $x = 5$

9. First number is 100.
Second number is 25.

10. $x = 50$

11. $x = -1$

12. $x = -7$

13. $x = 2$

14. Land: $34,500
House: $103,500

15. 12 five dollar bills
36 ten dollar bills
60 twenty dollar bills

16. 28 pounds of apricots
22 pounds of blanched nuts

17. $61,250 at 16 percent
$38,750 at 12 percent

18. a. 10 hours
b. 3 A.M.

19. 3.5 gallons

20. 34 fifteen cent stamps
1360 eighteen cent stamps
3400 three cent stamps

21. Insulation: $2300
Siding: $5750

22. $450 at 3 percent
$1650 at 8 percent
$2100 total

23. Ken: 45 miles per hour
Jacki: 60 miles per hour

CHAPTER 6

EXERCISE 6-1 (Page 193)

1. Two terms; first-degree polynomial
3. Three terms; second-degree polynomial
5. Two terms; second-degree polynomial
7. Three terms; first-degree polynomial
9. Two terms; first-degree polynomial

EXERCISE 6-2 (Page 197)

1. $17x$
3. $-5ab$
5. $-7xy$
7. $-32x^2y$
9. $-19y$
11. $14c$
13. $-3x^2$
15. $-3(x + y)$
17. $-12y$
19. $+6x^2$
21. 0
23. $-5xy$
25. $-4c^2$
27. $7x + 14y$
29. $5a$
31. $2x^3 - 1$
33. $2x^2 + 2xy + y^2$
35. $2a + 17b + 4c$
37. $8x^2 - 3x + 1$
39. $3z^2 + 2z$
41. $3x - 6y$
43. $7c - 3d$
45. $3x^2 + x$
47. $10x^2 - 3x + 1$
49. $-y^2 + 10y + 13$
51. $-6x^2 - 4x + 8$
53. $+3y^2 - 3y + 2$
55. 0

EXERCISE 6-3 (Page 202)

1. $7x$
3. $+3x$
5. $0z = 0$
7. $3y$
9. $1.0x^2$
11. $-7a$
13. $3y^2$
15. $-6ab$
17. $19y$
19. $0ab$ or 0
21. $-4z$
23. $-11y^2$
25. $-4a$
27. $10(a + b)$
29. $-2.2x$
31. $12.0d$
33. $-3ab$
35. $-25x^2$
37. 0
39. $8x$
41. $-8x^2$
43. $-4(a + b)$
45. $-7ab$
47. $13xy$
49. $-6x^2$
51. $-12cd$
53. $-4\frac{1}{2}x$
55. $13x^2 - 3x$
57. $-3a + 6b + 7$
59. $a - 5b$
61. $-16x + 23y$
63. $4y - 5x - 3$
65. $4x^3 + 10x^2 + 12$
67. $7a + 2b$
69. $-3x$
71. $-3x^2$
73. $4x^2 + 2x$
75. $8y - 4$
77. $6x^2 - 18x + 5$
79. $a + 2b$
81. $-3a + 2b$
83. $-17a + 13b$
85. $2x^2 - 6x$
87. $3x^2 - 24$
89. $9x^2 + 6x + y^2 + 5y - 8$
91. $-3y + 2$
93. 0
95. $14r - 18s + 9t$
97. $2x^2 - 3x + 1$
99. $-3y^2 - 8y + 11$

EXERCISE 6-4 (Page 209)

1. a^5
3. y^6
5. a^6
7. a^4b^5
9. c^6d^4
11. 4^5 or 1024
13. 3^3x^3 or $27x^3$
15. x^{2+b}
17. y^8
19. y^{10}
21. 2^{15} or $32,768$
23. $3^{m \cdot n}$
25. z^{14}
27. a^{5b}
29. x^8y^{12}
31. $49a^2$
33. $8a^6$
35. $4x^7$
37. $-27y^3z^6$

EXERCISE 6-5 (Page 217)

1. $-24abc$
3. $24x^3y^6$
5. $-4x^3y^3$
7. $4a$
9. $-8y^2$
11. $+48ab$
13. $-6xyz$
15. $+1x^6$
17. $-18x^7y^5$
19. $-20x^2y^4$
21. $4x^2$
23. $9a^4b^6$
25. $27y^2$
27. $2y^{11}$
29. $4a + 4b$
31. $3x + 6y$
33. $-3a^3 + 3a^2$
35. $2x^4 - 3x^3 + 4x^2$
37. $3a^3b + 4a^2b^2 + 5ab^3$
39. $3y - 6y^2 + 12y^3$
41. $10x^4 + 25x^3y^2 - 20x^2y^3$
43. $2x^2 - 4x - 7$
45. $6x^3 - 12x^2 + 15x$
47. $-12a^2bc + 6ab^2c + 3abc^2 - 12abc$
49. $x^2 + 7x + 12$
51. $z^2 + 4z + 4$
53. $x^2 - 9x + 14$
55. $x^2 + 4x - 21$
57. $z^2 - 7z - 18$
59. $y^2 - 4$
61. $2a^2 + 5a + 3$
63. $3x^2 + 2a - 8$
65. $2y^2 - y - 1$
67. $3 + 2z - z^2$
69. $2x^2 - 11x - 6$
71. $6a^2 + 21a + 9$
73. $10x^2 - 6x - 4$
75. $10x^2 - 25x + 15$
77. $-x^2 + 6x - 8$
79. $2x^2 + 3xy + y^2$
81. $a^4 + a^2 - 2$
83. $x^2 + 2xy + y^2$
85. $x^2 - 2xy + y^2$
87. $6x^2 + 7xy - 20y^2$
89. $a^2 + 2a + 1$
91. $4a^2 + 4ab + b^2$
93. $9a^2 + 12ab + 4b^2$
95. $y^3 - 9y^2 + 27y - 27$
97. $6a^3 + 23a^2 + 25a + 6$
99. $6x^3 + 13x^2 - 19x - 12$
101. $2x^3 - 10x^2 + 6x + 12$
103. $4x^3 - 13xy^2 + 6y^3$
105. $3z^3 - 13z^2 + 15z - 4$
107. $-x^3 + 3x^2 + 13x - 15$
109. $x^4 + 2x^2 + 9$

EXERCISE 6-6 (Page 226)

1. x^7
3. x^5
5. z^{2b}
7. 10^3
9. y
11. y^3
13. 10^2
15. a^2b
17. x^2yz^4
19. a^2b^2
21. x^2yz^7
23. 1
25. 1
27. 1
29. 1
31. x
33. b
35. ac^2
37. $\dfrac{1}{y^6}$
39. $\dfrac{1}{z}$
41. $\dfrac{1}{b^2}$
43. $\dfrac{1}{abc}$
45. $\dfrac{a}{b^2}$
47. $\dfrac{x^4}{z^4}$

EXERCISE 6-7 (Page 235)

1. $3x^3$
3. $-18ab$
5. $2x$
7. $-2y$
9. b
11. $-3xy^3$
13. $2b$
15. -1
17. $4d$
19. $-2x^4yz^3$
21. $-z$
23. $\dfrac{2}{r^2}$
25. $\dfrac{-6x^3}{y^3}$
27. $\dfrac{4}{y}$
29. $-5(a + b)$
31. $2x + 3y$
33. $3x - 1$
35. $2d^2 + d$
37. $y + z$
39. $x + 6$
41. $4y^2 - 3y$
43. $y - 2$
45. $-3a + 1$
47. $3c - d$
49. $-x^3 + 3x^2 + 2$
51. $3x - 1 + 2y$
53. $x - 2$
55. $x + 3$
57. $x + 3$
59. $x - 5$
61. $x - 7$
63. $x - 5$
65. $5x - 7$
67. $x^2 + 11 R\ 27$
69. $3y^2 + 2y - 7$
71. $3y^2 + y + 3$
73. $4x^2 - 3x + 5 R - 6$
75. $3x - 2R - 9$
77. $2y^2 - 2y + 1$
79. $x^2 - 2$

REVIEW EXERCISES (CHAPTER 6) (Page 237)

1. 4
2. -2
3. $2, -5$
4. One
5. Two
6. One
7. $-x^3 + x^2 + x$
8. $-2x^2 + 4xy + 7y^2$ $7y^2 + 4xy - 2x^2$
9. $r^3s - 3r^2s^2 - 3rs^3 + s^4$
10. $12a^2$
11. $3a$
12. 0
13. $4x^2 - 2x$
14. $2x^2 + 5x + 3$
15. $2x^3 + 6x^2 + x + 1$
16. $4y^4 + 6y^3 - 9y^2 - 17$
17. $2x$
18. $+4x^2$
19. 0
20. $5a + 10$

21. $x^2 - 2x - 1$

22. $-2x^2 + x - 5$

23. $-2x^2 + 7x - 17$

24. $-3x$

25. $3x^2 - 4x + 5$

26. $2x + 3$

27. x^5

28. 2^{11}

29. $a^4 b^6 c^3$

30. $-10a^2$

31. $12x^5$

32. $-12a^3 b^4 c^4$

33. $-8x + 12$

34. $2a^3 - 6a^2 + 12a$

35. $+3x^4 - 9x^3 + 12x^2$

36. $2x^3 - 12x^2 + 16x$

37. $5z^3 - 17z^2 + 31z - 10$

38. $a^3 - 6a^2 + 12a - 8$

39. y^7

40. xy^2

41. $\dfrac{x}{z}$

42. $\dfrac{c^3}{b^3}$

43. $4a^2$

44. $-2b$

45. $-25xz^2$

46. $x + 3$

47. $-x^2 + 3y + 4$

48. $-3xy + 2$

49. $2x - 3$

50. $3a - 4b$

CHAPTER 7

EXERCISE 7-2 (Page 245)

1. $1 - 2x^2$

3. $x + y$

5. $x^2 + 2x - 5$

7. $x - 2y$

9. $x^2 - 5x + 6$

11. $z + x - y$

13. 3

15. $7x$

17. $2x$

19. $5xy^2$

21. $5xyz$

23. xyz

25. $7(x + y)$

27. $10a$

29. $3a(b - 2c)$

31. $6x^2$

33. $6(x^2 - 2xy + y^2)$

35. $3b^2(1 - ab - b^2)$

37. $(x + 3)(a + b)$

39. $(x + 2)(x + 3)$

EXERCISE 7-3 (Page 250)

1. $x^2 + 8x + 15$

3. $x^2 - 5x - 14$

5. $z^2 - 5z - 14$

7. $x^2 - 5x + 4$

9. $x^2 - 1$

11. $2x^2 + 11x + 5$

13. $3x^2 + 10x - 8$

15. $4x^2 - 1$

17. $15x^2 + 4x - 35$

19. $x^4 + 4x^2 + 3$

21. $18 + 3x - x^2$

23. $x^2 + 12x + 36$

25. $y^2 - 14y + 49$

27. $x^4 + 12x^2 + 36$

29. $4a^2 - 12ab + 9b^2$

EXERCISE 7-4 (Page 257)

1. $(x + 2)(x + 1)$

3. $(x + 2)(x + 3)$

5. $(z + 2)(z + 2)$

7. $(x + 1)(x + 9)$

9. $(x + 3)(x + 6)$

11. $(x - 3)(x - 3)$

13. $(x - 2)(x - 5)$

15. $(x - 1)(x - 7)$

17. $(x - 2)(x - 4)$

19. $(y - 1)(y - 6)$

21. $(y + 2)(y - 1)$

23. $(x + 8)(x - 3)$

25. $(x + 6)(x - 3)$

27. $(x + 4)(x - 2)$

29. $(y + 3)(y - 2)$

31. $(y - 5)(y + 2)$

33. $(x - 9)(x + 2)$

35. $(x - 4)(x + 2)$

37. $(z - 8)(z + 3)$

39. $(x - 5)(x + 3)$

41. $(y + 8)(y - 8)$

43. $(c + 10)(c - 10)$

45. $(3 + x)(3 - x)$

47. $(2x + 5)(2x - 5)$

49. $(11 + x)(11 - x)$

51. $(6x + 5y)(6x - 5y)$

53. $(x + 1)(x - 1)$

55. $(1 + x)(1 - x)$

57. $\left(x + \dfrac{1}{2}\right)\left(x - \dfrac{1}{2}\right)$

EXERCISE 7-5 (Page 264)

1. $(2x + 1)(x + 2)$

3. $(2x + 3)(x + 1)$

5. $(2x + 3)(x + 4)$

7. $(6x + 1)(x + 4)$

9. $(2x - 3)(2x - 1)$

11. $(4x - 1)(2x - 3)$

13. $(2y - 5)(2y - 1)$

15. $(3y - 7)(y - 2)$

17. $(2x - 3)(x + 5)$

19. $(3y - 4)(y + 3)$

21. $(3x - 5)(x + 2)$

23. $(3x - 4)(2x + 1)$

25. $(2x + 3)(x - 5)$

27. $(3x + 4)(x - 3)$

29. $(5x + 1)(3x - 1)$

31. $(3x - 4)(5x + 2)$

33. $(4x + 5)(4x - 5)$

35. $(5x + 7)(5x - 7)$

EXERCISE 7-6 (Page 267)

1. $3(x + 3)(x - 3)$

3. $(z^2 + 4)(z + 2)(z - 2)$

5. $a(x + y)(x - y)$

7. $z(z + 1)(z - 1)$

9. $4(a + 3)(a - 3)$

11. $x(x + 5)(x - 5)$

13. $2(x-2)(x-2)$

15. $a(x+2)(x+1)$

17. $2y(2x-3)(x-1)$

19. $x(x+5)(x+2)$

21. $(x^2+2)(x+1)(x-1)$

23. $xy(x+5y)(x+5y)$

EXERCISE 7-7 (Page 271)

1. $x=0, \quad x=7$

3. $x=0, \quad x=4$

5. $x=0, \quad x=\dfrac{3}{2}$

7. $x=3, \quad x=-3$

9. $z=3, \quad z=-3$

11. $x=\dfrac{3}{2}, \quad x=-\dfrac{3}{2}$

13. $x=1, \quad x=3$

15. $x=2, \quad x=5$

17. $x=1, \quad x=-6$

19. $y=1, \quad y=2$

21. $z=2, \quad z=9$

23. $y=-1, \quad y=7$

25. $x=-\dfrac{1}{2}, \quad x=-2$

27. $x=-\dfrac{4}{3}, \quad x=1$

29. $x=-\dfrac{1}{3}, \quad x=-3$

31. $y=-\dfrac{3}{2}, \quad y=-2$

33. $x=\dfrac{3}{4}, \quad x=-4$

35. $y=\dfrac{1}{2}, \quad y=\dfrac{1}{2}$

37. $x=\dfrac{5}{2}, \quad x=-\dfrac{4}{3}$

39. $x=-\dfrac{1}{2}, \quad x=10$

REVIEW EXERCISES (CHAPTER 7) (Page 272)

1. $(6x^3)$

2. (x^2-4)

3. $(x+2)$

4. $(x-3)$

5. 5

6. $3x^2$

7. $2xy$

8. 5

9. $3x(x+2)$

10. $ab(b^2+a)$

11. $6(-2x^2+5x+1)$

12. $4x(2x-3-6x^2)$

13. $(x-4)(x+3)$

14. $x^2-4x-21$

15. $y^2+7y-18$

16. $2x^2-5x-12$

17. $6y^2-y-15$

18. x^2-4

19. $9x^2-16$

20. a^2-b^2

21. x^2-9

22. $y^2+10y+25$

23. z^2-6z+9

24. $4x^2-12x+9$

25. $(x+5)(x-3)$

26. $(x-7)(x+1)$

27. $(x+2)(x+6)$

28. $(y-7)(y-5)$

29. $(2x+1)(x-4)$

30. $(3x-1)(2x+3)$

31. $(2x-1)(2x-1)$

32. $(2x-3)(3x+2)$

33. $(x+5)(x-5)$

34. $(2y+5)(2y-5)$

35. $(4+3x)(4-3x)$

36. $3(x+1)(x-1)$

37. $z^3(z+1)(z-1)$

38. $(y^2+4)(y+2)(y-2)$

39. $3(x-4)(x+3)$

40. $2x(x-1)(x-1)$

41. $y=0, \quad y=7$

42. $z=0, \quad z=4$

43. $y=4, \quad y=-4$

44. $x=5, \quad x=-5$

45. $x=5, x=-3$

46. $x=-7, \quad x=1$

47. $x=6, \quad x=2$

48. $y=-7, \quad y=-5$

49. $x=\dfrac{1}{2}, \quad x=-4$

50. $x=\dfrac{3}{2}, \quad x=-\dfrac{1}{3}$

CHAPTER 8

EXERCISE 8-1 (Page 279)

1. $\dfrac{(3)(x)}{(4)(y)}$

3. $\dfrac{x(x+2)}{2(x+2)}$

5. $\dfrac{(x+3)(x-3)}{(x+3)(x+3)}$

7. $\dfrac{(x+3)(x-2)}{(x-2)(x-2)}$

9. $\dfrac{2(x+2)(x-1)}{4(x+1)(x-1)}$

11. $+\dfrac{3x}{4}$

13. $-\dfrac{x-2}{-2+x}$

15. $+\dfrac{2(-1+y)}{3(y-1)}$

17. $+\dfrac{2x(-1+x)}{3(x-1)}$

EXERCISE 8-2 (Page 284)

1. $\dfrac{x}{y}$

3. $\dfrac{x}{4}$

5. $\dfrac{1}{2}$

7. $\dfrac{x}{4y^2}$

9. $-\dfrac{2x}{3y}$

11. $\dfrac{3}{4}$

13. $\dfrac{1}{2x}$

15. $\dfrac{3}{5}$

17. $\dfrac{1}{4}$

19. $\dfrac{2(x-y)}{3}$

21. $\dfrac{2}{x-2}$ 23. $\dfrac{1}{x+1}$ 25. $\dfrac{7}{x+1}$ 27. $\dfrac{x-2}{x+2}$ 29. $\dfrac{4}{x+3}$

31. $3(x-1)$ 33. $\dfrac{x-1}{x+4}$ 35. $\dfrac{y-5}{y+2}$ 37. $-\dfrac{2}{3}$ 39. $\dfrac{x+5}{x-8}$

EXERCISE 8-3 (Page 290)

1. $\dfrac{3}{5}$ 3. $\dfrac{5}{3a}$ 5. $\dfrac{3}{xy}$ 7. $\dfrac{x}{6}$ 9. $\dfrac{2x^3}{y^4}$

11. $\dfrac{2}{3}$ 13. $\dfrac{3}{2}$ 15. $\dfrac{3}{4}$ 17. $\dfrac{15x}{2(x-2)(x-3)}$

19. 2 21. $\dfrac{1+3x}{2(x-2)}$ 23. $\dfrac{x(x+1)}{x+4}$ 25. 1

EXERCISE 8-4 (Page 293)

1. $\dfrac{1}{x}$ 3. x 5. 12 7. $6x$ 9. $\dfrac{y}{21x}$

11. $\dfrac{x}{14}$ 13. $\dfrac{3}{2}$ 15. $\dfrac{2(x-y)}{3}$ 17. $\dfrac{2}{x+y}$ 19. 1

21. $3(x+3)$

EXERCISE 8-5 (Page 298)

1. x 3. $\dfrac{5x}{12}$ 5. $\dfrac{5}{x+1}$ 7. 1

9. $x+y$ 11. $\dfrac{6x+1}{3}$ 13. 9 15. $\dfrac{2x+1}{x+3}$

EXERCISE 8-6 (Page 311)

1. $6x$ 3. $2x$ 5. $6x$

7. $4(x+1)$ or $4x+4$ 9. $(x-2)(x-2)$ or x^2-4x+4

11. $12x$ 13. x^2 15. $50x^2$ 17. $x(x+1)$

19. $(x+y)(x-y)$ 21. $(x-2)(x-3)$ 23. $\dfrac{19x}{30}$ 25. $\dfrac{19}{10x}$

27. $\dfrac{3y^2-2x^2}{18x^2y^2}$ 29. $\dfrac{-3(2x-1)}{4}$ 31. $\dfrac{-y-16}{3(y+4)(y-4)}$ 33. $\dfrac{x}{3(y+2)}$

35. $\dfrac{24}{(x+6)(x-6)}$ 37. $\dfrac{3x-2}{(x-2)(x-1)}$ 39. $\dfrac{-x^2+3x+2}{(x+6)(x+1)}$ 41. $\dfrac{9x^2+7x+5}{(x+3)(x-1)(x+2)}$

43. $\dfrac{x^2+3}{x}$ 45. $\dfrac{x^2+x+1}{x+1}$ 47. $\dfrac{4(-x+3)}{x-2}$ 49. $\dfrac{3x+8}{x+1}$

EXERCISE 8-7 (Page 316)

1. $\dfrac{3}{(x+1)(x-1)}$ 3. $\dfrac{3+5x}{3-5x}$

5. $\dfrac{(x+2)(x-2)(x+2)(x-2)}{4x^3}$ 7. $\dfrac{x+1}{x-1}$

EXERCISE 8-8 (Page 322)

1. $x = 18$
3. $x = 60$
5. $x = 38\frac{2}{3}$
7. $x = 14$
9. $x = \frac{7}{5}$

11. $x = 2$
13. $x = 10$
15. $x = 4$
17. $x = 2$
19. $y = -\frac{1}{3}$

21. $x = 4$
23. $x = 6$
25. $\frac{15}{25}$
27. $N = 3$
29. $x = -20$

EXERCISE 8-9 (Page 328)

1. 3 hours
3. 7.2 hours (or 7 hours 12 minutes)
5. 1 hour
7. 12 hours
9. $10\frac{2}{3}$ hours (or 10 hours 40 minutes)
11. a. 40 minutes   b. No
13. 21.5 minutes
15. Kathy: $245   LuAnn: $325   Marilyn: $65

REVIEW EXERCISES (CHAPTER 8) (Page 329)

1. $24x^2y^3$
2. $3x^2 - 9x$
3. 2
4. $x^2 - 1$

5. $\frac{1}{x + 1}$
6. $\frac{x + 1}{x + 2}$
7. $\frac{x + 1}{x - 3}$
8. $\frac{3x^2}{2y}$

9. $\frac{x - 1}{2(x - 3)}$
10. $\frac{x^2}{4}$
11. $\frac{x}{2}$
12. 2

13. 1
14. $\frac{(x + 9)}{(x + 5)(x - 5)}$
15. $\frac{-2}{x(x - 1)}$

16. $\frac{3x^2 - 26x + 11}{(x + 2)(x + 1)(x - 3)(x - 1)}$
17. $\frac{2x - 1}{x - 1}$
18. $\frac{(3x + 2)(x + 1)}{x(2x + 1)}$

19. $x = -1$
20. $x = -\frac{1}{4}$
21. $x = 5$
22. $x = 6$

CHAPTER 9

EXERCISE 9-1 (Page 334)

1. $2xy$
3. $(x + 1)$
5. $4y$
7. $2x^2$
9. $12x^3y^2$
11. 4.58
13. 11.83
15. 8.94
17. -10.05
19. 11.00
21. 2.4
23. 3.3
25. 4.4
27. 7.1
29. 10.5

EXERCISE 9-2 (Page 341)

1. $2\sqrt{2}$
3. $6\sqrt{2}$
5. $2\sqrt{5}$
7. $4x\sqrt{2}$
9. $5x^2y^2\sqrt{2xy}$
11. $7xy^2z^3\sqrt{2y}$
13. $6\sqrt{6}$
15. $10x\sqrt{5x}$
17. $2x\sqrt{2x}$
19. $-6x\sqrt{x}$

21. $-4y\sqrt{5x}$
23. $-xz^2\sqrt{yz}$
25. $\frac{3}{4}$
27. $\frac{3}{5}$
29. $\frac{1}{4}$

31. 2
33. $\frac{\sqrt{5}}{2}$
35. $\frac{10\sqrt{2}}{7}$
37. $\frac{3x}{2}$
39. $\frac{2\sqrt{30}}{9}$

41. $2x^2\sqrt{15}$
43. $\frac{x\sqrt{2}}{3}$
45. $\frac{1}{3}$
47. $\frac{x\sqrt{30}}{9}$
49. $\frac{x^2yz\sqrt{xz}}{5}$

EXERCISE 9-3 (Page 345)

1. $5\sqrt{3}$
3. $3\sqrt{2}$
5. $-2\sqrt{5}$
7. $4\sqrt{5} + 3\sqrt{2}$
9. $5\sqrt{3} - \sqrt{5}$
11. $3\sqrt{2}$
13. $\sqrt{5}$
15. $3\sqrt{2}$
17. $\sqrt{6}$
19. $3\sqrt{2}$

21. $6\sqrt{2}$ 23. $4\sqrt{2}$ 25. $-5\sqrt{3}$ 27. $-9\sqrt{2}$ 29. $3\sqrt{5}$

31. $2\sqrt{6}$ 33. 0 35. $\sqrt{11}$ 37. $\sqrt{3}$ 39. $10 - 6\sqrt{3}$

41. $3\sqrt{5}$ 43. $5\sqrt{10}$ 45. $110 - \dfrac{\sqrt{35}}{4}$

EXERCISE 9-4 (Page 350)

1. 3 3. x 5. 4 7. 6

9. $6\sqrt{5}$ 11. $-12\sqrt{3}$ 13. $8\sqrt{15}$ 15. 24

17. $18\sqrt{10}$ 19. $-2\sqrt{6}$ 21. $3\sqrt{2}$ 23. $8\sqrt{3}$

25. $18\sqrt{5}$ 27. $-12\sqrt{2}$ 29. $8\sqrt{5}$ 31. $\sqrt{6}$

33. $30\sqrt{7}$ 35. $\sqrt{10}$ 37. 2 39. $3\sqrt{2}$

41. $\dfrac{\sqrt{3}}{2}$ 43. 1 45. $\dfrac{\sqrt{21}}{6}$ 47. $6x$

49. $2x\sqrt{5y}$ 51. $42x^3y^2$ 53. 3 55. 18

57. $8x$ 59. $2\sqrt{6} - 2\sqrt{15}$ 61. 30 63. $10\sqrt{2} - 20\sqrt{3}$

EXERCISE 9-5 (Page 359)

1. 3 3. 2 5. $\dfrac{5}{2}$ 7. 2

9. $2\sqrt{3}$ 11. $\sqrt{3}$ 13. $5\sqrt{2}$ 15. $x\sqrt{5}$

17. 2 19. $2y\sqrt{x}$ 21. $\dfrac{\sqrt{3}}{3}$ 23. $\dfrac{\sqrt{6}}{3}$

25. $\sqrt{3}$ 27. $\dfrac{2\sqrt{6}}{3}$ 29. $\dfrac{1}{2}$ 31. $\dfrac{\sqrt{2}}{2}$

33. $\dfrac{\sqrt{5}}{5}$ 35. $\dfrac{5\sqrt{2}}{2}$ 37. $\sqrt{3}$ 39. $\dfrac{2\sqrt{2}}{3}$

41. $\dfrac{\sqrt{7}}{4}$ 43. $\dfrac{\sqrt{15}}{6}$ 45. $2\sqrt{5}$ 47. $\sqrt{3}$

49. $4\sqrt{3}$ 51. $\sqrt{5}$ 53. $4\sqrt{10}$ 55. $3\sqrt{10}$

EXERCISE 9-6 (Page 365)

1. $x = 49$ 3. $x = 0$ 5. $x = 6$ 7. No solution

9. $x = 48$ 11. No solution   13. $x = 3$ 15. $x = 3$

17. $x = 5$ 19. $x = 3$ 21. $x = 3$ 23. $x = 85$

25. $x = 7$ 27. $x = 4$

REVIEW EXERCISES (CHAPTER 9) (Page 366)

1. 5.4 2. 11.40 3. 4.8 4. 8.5

5. 14 6. $3x\sqrt{6}$ 7. $4x^2y^3\sqrt{7xy}$ 8. $\dfrac{\sqrt{21}}{7}$

9. $\dfrac{3\sqrt{2}}{2}$ 10. $\dfrac{4\sqrt{10}}{5}$ 11. $2\sqrt{5}$ 12. $13\sqrt{2}$

13. $7\sqrt{3}$ 14. $\sqrt{2}$ 15. $-\sqrt{6}$ 16. $18\sqrt{2}$

17. $84\sqrt{3}$ 18. $\dfrac{\sqrt{6}}{6}$ 19. 20 20. 2

21. $\dfrac{1}{2}$

22. $2\sqrt{6}$

23. $\sqrt{3}$

24. $x = 26$

25. No solution

26. $N = 64$

CHAPTER 10

EXERCISE 10-1 (Page 376)

1., 3., 5., 7., 9.

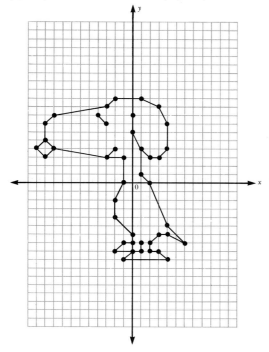

11.

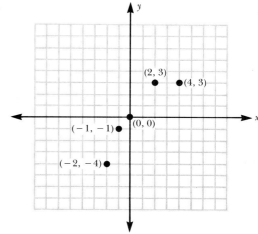

13. $(-4, 6))$

15. $(-9, 0)$

17. $(10, 5)$

19. $(-12, -5)$

21. $(0, -2)$

23. $(11, -3)$

25. Zero

27. a. Second quadrant

b. First quadrant

29.

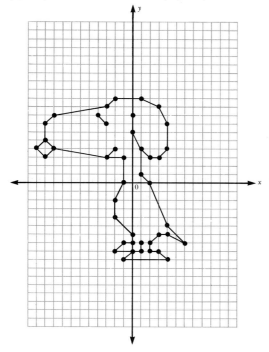

EXERCISE 10-2 (Page 381)

1. Yes   3. No   5. No   7. Yes

9. No   11. $y = 7$ or $(3, 7)$ 13. $y = 2$ or $(4, 2)$

15. $y = -2$ or $(1, -2)$ 17. $y = -8$ or $(-8, -8)$ 19. $y = -2$ or $(7, -2)$

21. $y = -1$ or $(-1, -1)$ 23. $y = 3$ or $(5, 3)$ 25–30. These answers will vary.

EXERCISE 10-3 (Page 390)

1.

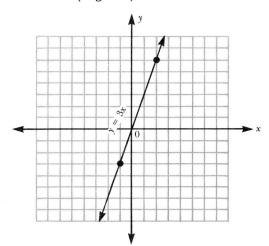

3.

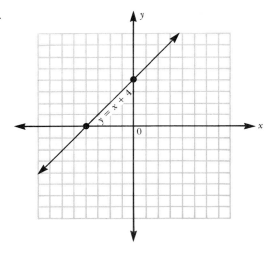

5.

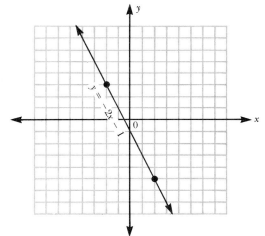

7.

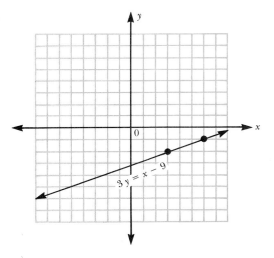

9.

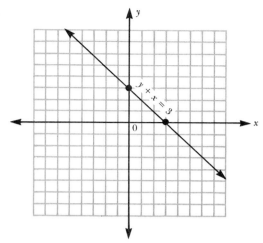

11.

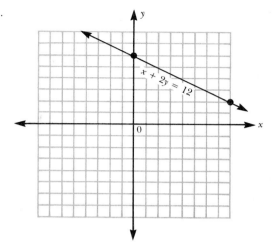

13.

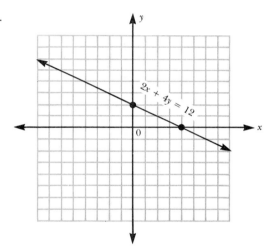

15.

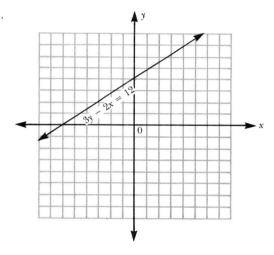

17.

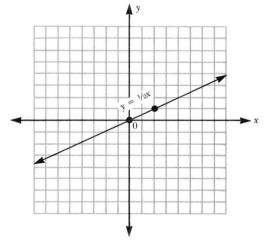

19.

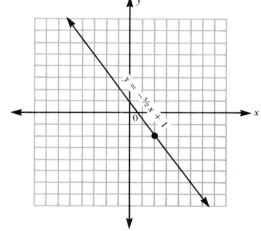

21.

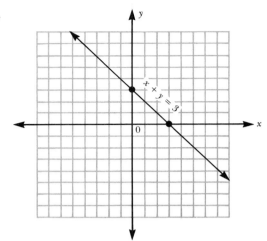

23.

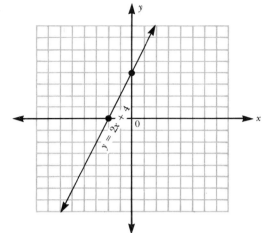

25.

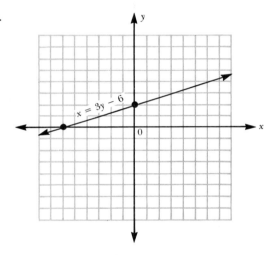

27.

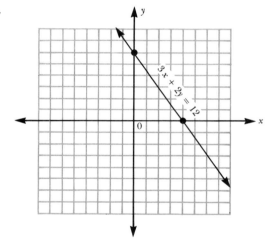

29.

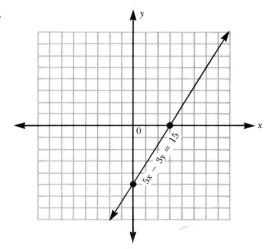

31.

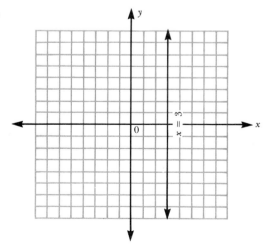

33.

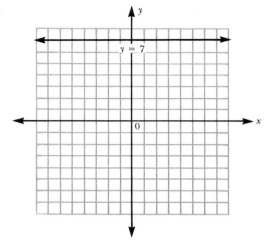

35.

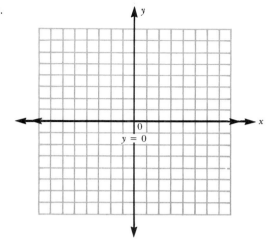

37.

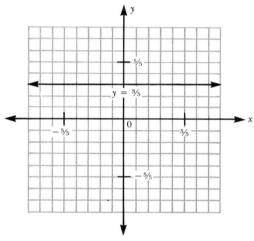

39.

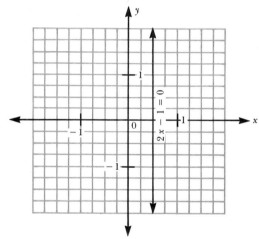

EXERCISE 10-4 (Page 405)

1. Slope $(m) = \dfrac{1}{2}$

3. Slope $(m) = -1$

5. Slope $(m) = \dfrac{4}{-5}$

7. Slope (m) = undefined

9. Slope $(m) = \dfrac{2}{3}$

11.

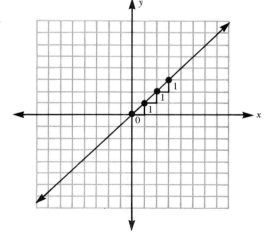

13.

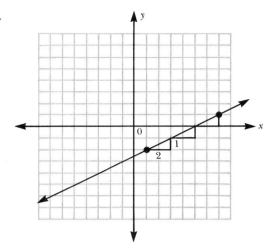

15.

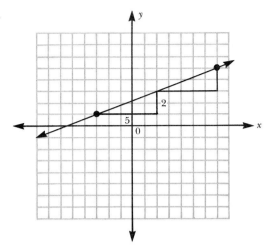

17.

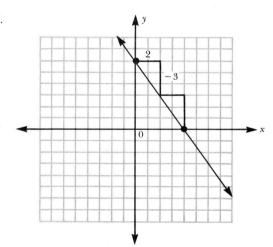

19.

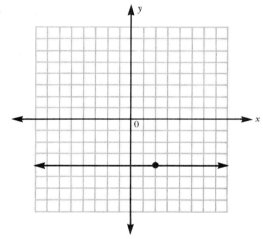

21.

$m = 2$
$b = 1$;
thus, $(0, 1)$

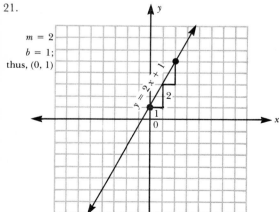

23.

$m = 4$
$b = -3$
$(0, -3)$

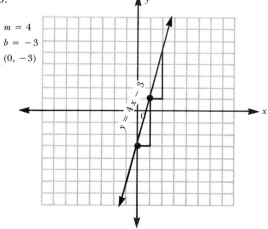

25.

$m = 1$
$b = 0$
$(0, 0)$

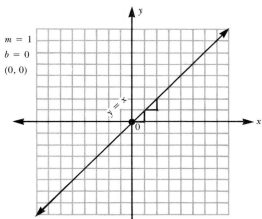

27.

$m = \frac{1}{2}$
$b = 6$
$(0, 6)$

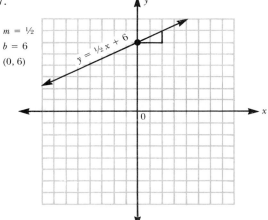

29.

$m = -\frac{2}{3}$
$b = 4$
$(0, 4)$

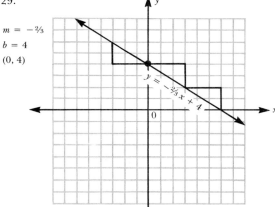

31.

$m = 1$
$b = -3$
$(0, -3)$

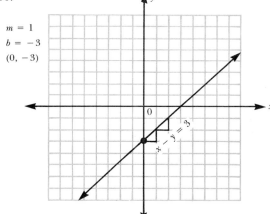

33.

$m = -\frac{2}{3}$
$b = \frac{8}{3}$
$(0, \frac{8}{3})$

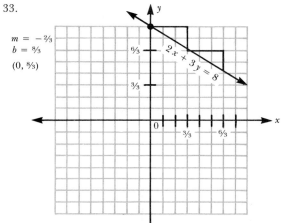

35.

$m = \frac{3}{4}$
$b = -4$
$(0, -4)$

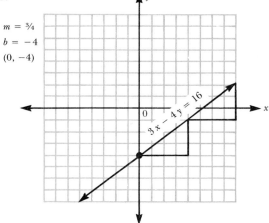

37.

$m = -\frac{2}{3}$
$b = \frac{1}{3}$
$(0, \frac{1}{3})$

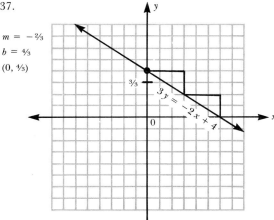

39.

$m = 2$
$b = -6$
$(0, -6)$

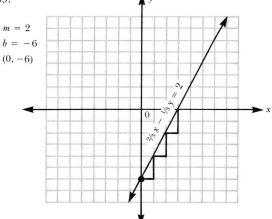

EXERCISE 10-5 (Page 409)

1. $y = 2x - 1$

3. $y = -4x - 7$

5. $y = \dfrac{x}{2} - \dfrac{3}{2}$

7. $y = -\dfrac{5}{4}x + \dfrac{3}{2}$

9. $x = 3$

11. $y = -x + 6$

13. $y = -\dfrac{3}{2}x$

15. $y = x + 2$

17. $y = \dfrac{1}{2}x - 2$

19. $y = \dfrac{1}{2}x + \dfrac{1}{2}$

21. $y = 2x + 1$

23. $y = x - 2$

25. $y = \dfrac{2}{3}x + 1$

EXERCISE 10-6 (Page 418)

1.

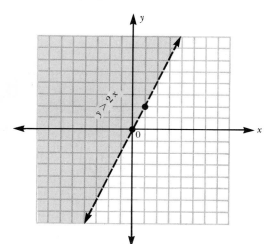

3.

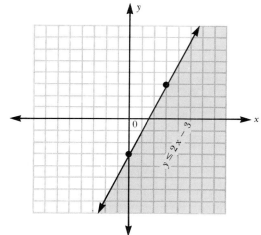

5.

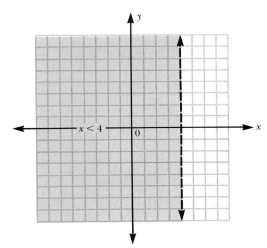

7.

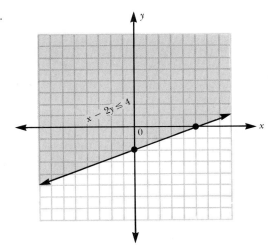

9.

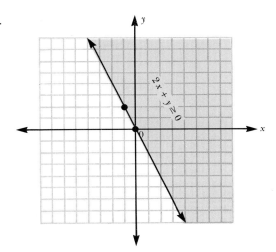

11.

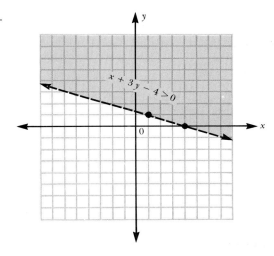

13.

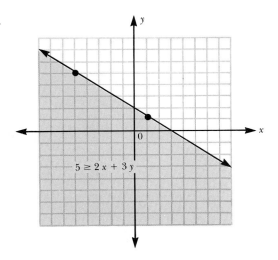

15.

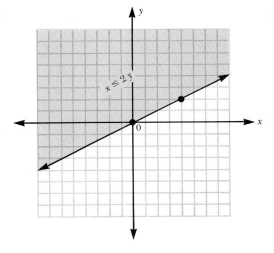

17.

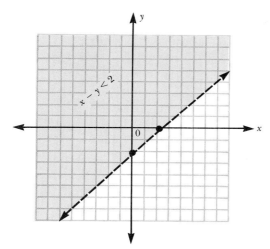

19.

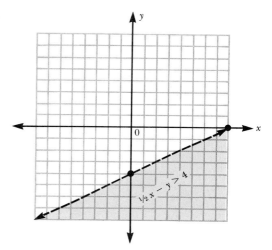

REVIEW EXERCISES (CHAPTER 10) (Page 418)

1.

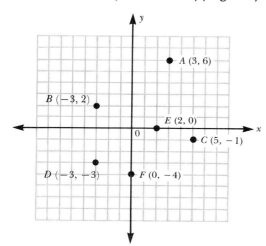

2.

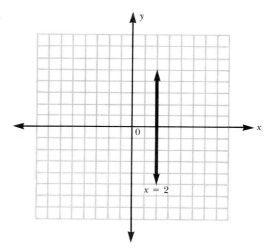

3.

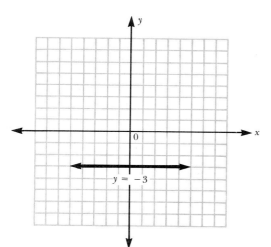

4.

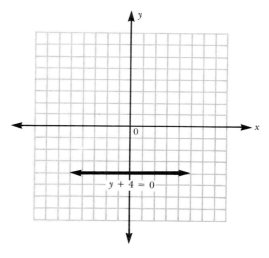

5.

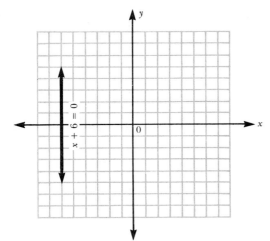

6.

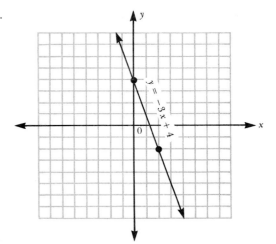

7.

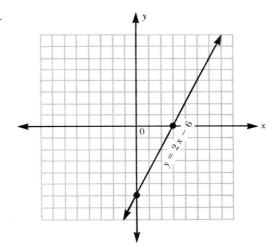

8.

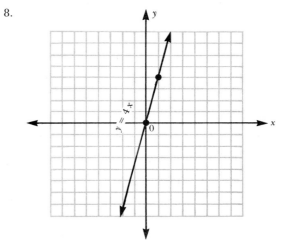

9.

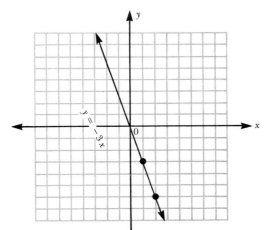

10.

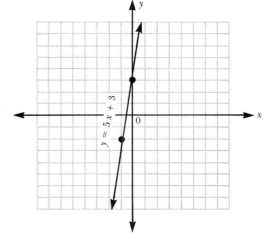

11.

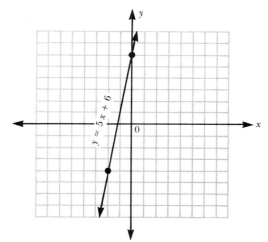

12.

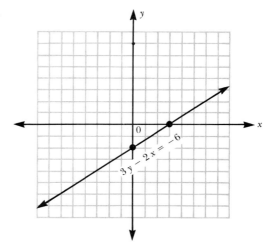

13.

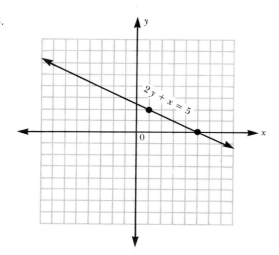

14.

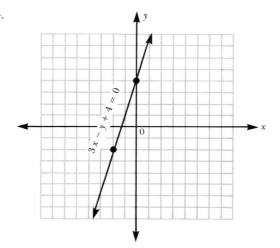

15.

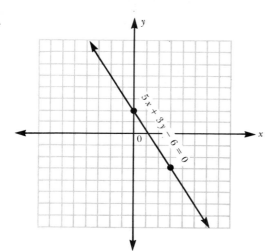

16.

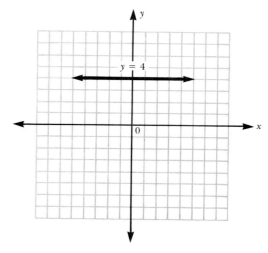

17.

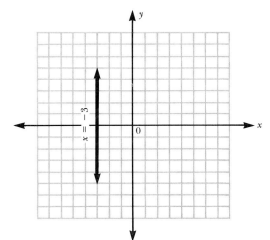

18.

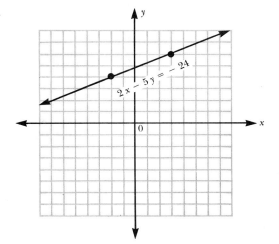

19.

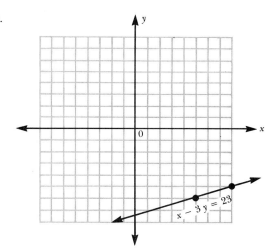

20.

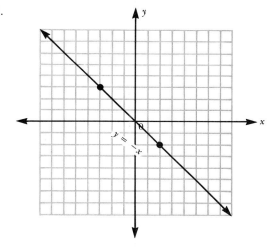

21.

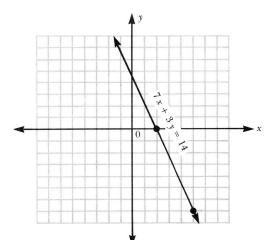

22.

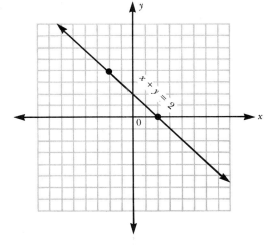

23.

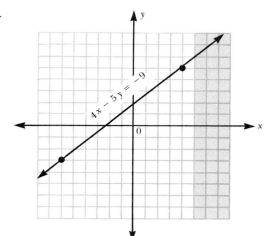

24.

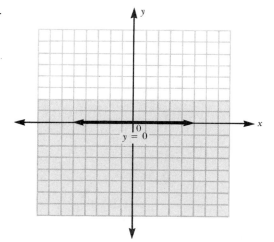

25.

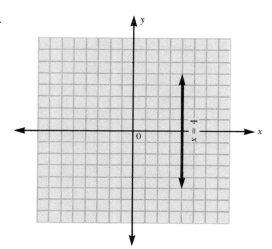

26.

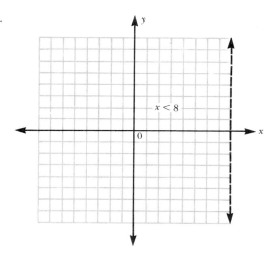

27.

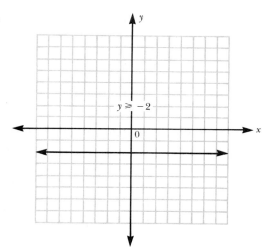

28.

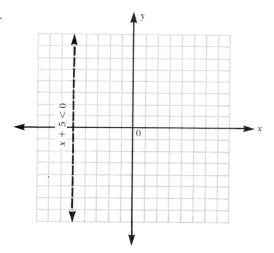

29.

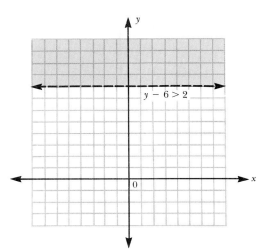

30.

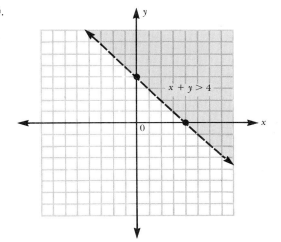

31.

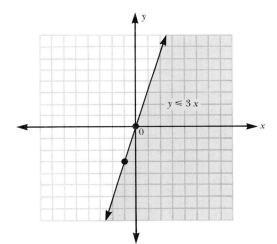

32.

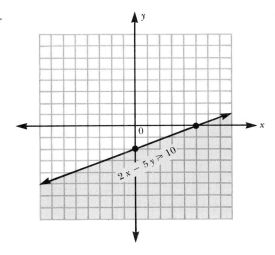

33.

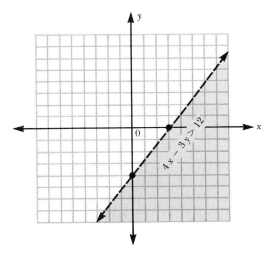

34.

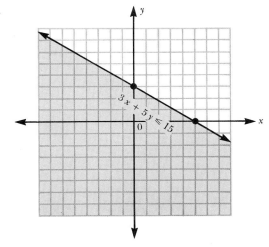

35.

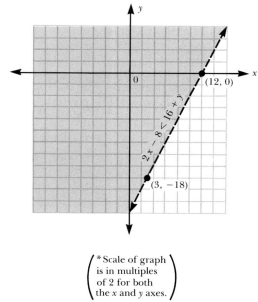

$$\begin{pmatrix} \text{*Scale of graph} \\ \text{is in multiples} \\ \text{of 2 for both} \\ \text{the } x \text{ and } y \text{ axes.} \end{pmatrix}$$

CHAPTER 11

EXERCISE 11-1 (Page 421)

1. No 3. No 5. Yes 7. No 9. No

EXERCISE 11-2 (Page 429)

1.

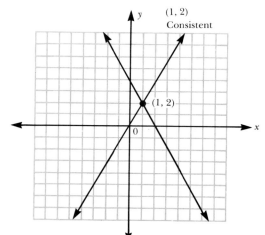

3.

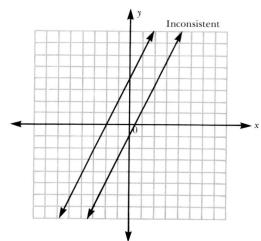

5.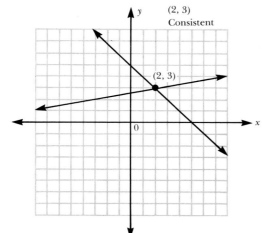

7.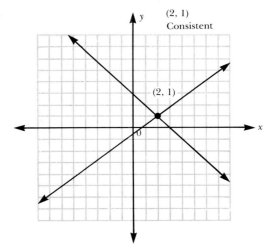

9.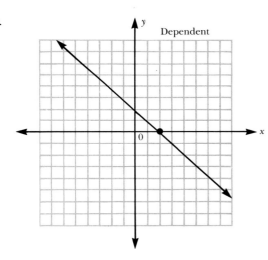

11.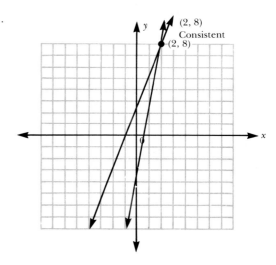

13.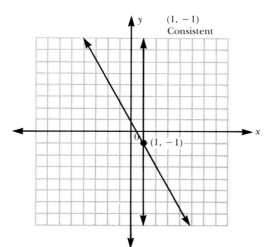

15.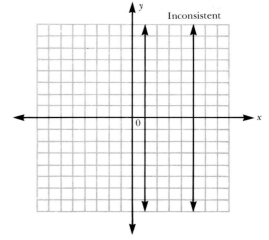

17.

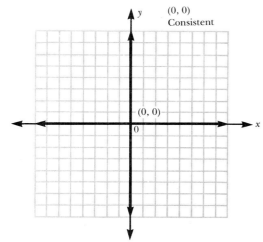

(0, 0)
Consistent

19.

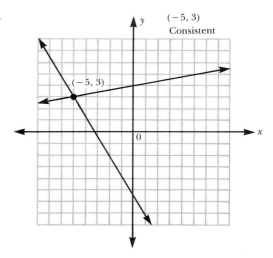

(−5, 3)
Consistent

EXERCISE 11-3 (Page 437)

1. (6, 2)   3. (−36, 15)   5. (5, −3)   7. (1, 0)

9. $\left(\dfrac{2}{3}, -2\right)$ 11. (1, 3)   13. (−6, −7)   15. $\left(\dfrac{1}{2}, \dfrac{1}{2}\right)$

17. Inconsistent; there are no points that are common.   19. $\left(\dfrac{4}{9}, \dfrac{1}{3}\right)$

EXERCISE 11-4 (Page 442)

1. (8, 8)   3. (2, 4)   5. (2, 4)   7. (−6, 0)   9. (−12, −18)

11. (5, 2)   13. (2, −4)   15. (4, 0)   17. (3, −6)   19. $\left(\dfrac{7}{2}, -\dfrac{3}{4}\right)$

EXERCISE 11-5A (Page 446)

1. 48, 25   3. 20, 19   5. 20, 19   7. 20, 93   9. 8, 1

EXERCISE 11-5C (Page 449)

1. 230 nickels
   115 dimes

3. 17 nickels
   148 dimes

5. 27 five dollar chips
   9 ten dollar chips

EXERCISE 11-5D (Page 451)

1. $27\dfrac{1}{2}$ pounds waxed beans

   $22\dfrac{1}{2}$ pounds kidney beans

3. 4 tons of No. 1 cr
   8 tons of No. 2

5. 9 pounds McIntosh
   6 pounds Cortland

EXERCISE 11-5E (Page 453)

1. $800 at 8 percent
   $1700 at 11 percent

3. $7600 at 4 percent

   $5000 at $5\dfrac{1}{2}$ percent

5. $2500

EXERCISE 11-5F (Page 456)

1. 2 miles per hour

3. 50 miles per hour

5. motorboat: $10\dfrac{1}{2}$ miles per hour

   current: $4\dfrac{1}{2}$ miles per hour

REVIEW EXERCISES (CHAPTER 11) (Page 457)

1.
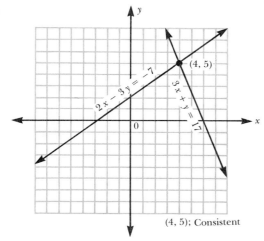
$(4, 5)$; Consistent

2.
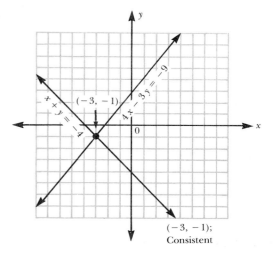
$(-3, -1)$;
Consistent

3.
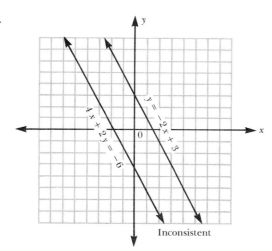
Inconsistent

4.
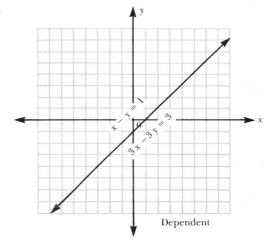
Dependent

5.
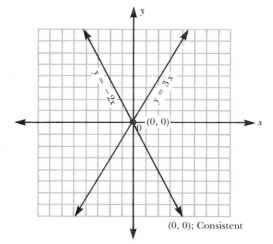
$(0, 0)$; Consistent

6.
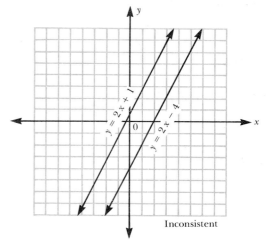
Inconsistent

7.

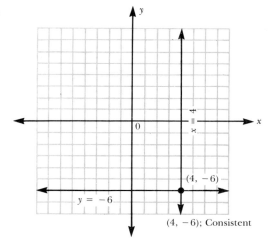

$(4, -6)$; Consistent

8.

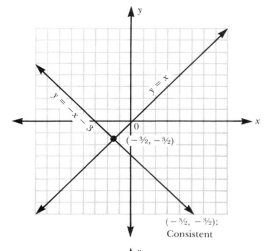

$(-{}^3\!/_2, -{}^3\!/_2)$;
Consistent

9.

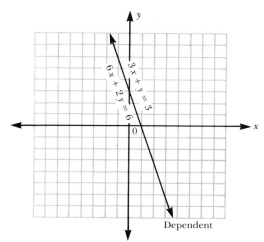

Dependent

10.

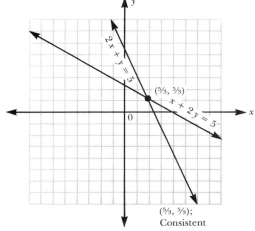

$({}^5\!/_3, {}^5\!/_3)$;
Consistent

11. $(4, 4)$
12. $(-4, 14)$
13. $(-2, 0)$
14. $\left(\dfrac{1}{2}, 5\right)$

15. $(-1, -2)$
16. $\left(\dfrac{5}{3}, 1\right)$
17. $(9, -8)$
18. $\left(-\dfrac{2}{3}, -\dfrac{1}{3}\right)$

19. $\left(\dfrac{5}{11}, -\dfrac{7}{11}\right)$
20. $(-2, -2)$
21. $61, \quad 19$
22. $48, \quad 13$

23. Boat: 18 miles per hour
Current: 6 miles per hour

24. \$5625 at 8 percent
\$4375 at 12 percent

25. $17\dfrac{1}{2}$ pounds of first brand

$12\dfrac{1}{2}$ pounds of second brand

CHAPTER 12

EXERCISE 12-1 (Page 462)

1. $x^2 + 6x - 40 = 0$
3. $x^2 + 4x - 5 = 0$
5. $2x^2 + 16x - 66 = 0$
7. $2x^2 - 6x + 3 = 0$
9. $x^2 - 7x + 2 = 0$
11. $x^2 - 5x - 7 = 0$
13. $x^2 - 4 = 0$
15. $4x - 36 = 0$
17. $x^2 - 13x + 36 = 0$

EXERCISE 12-2 (Page 467)

1. $x = -1$, $x = 7$　　3. $x = -2$, $x = 5$　　5. $x = 5$, $x = -5$　　7. $x = 3$, $x = -3$

9. $x = 0$, $x = 9$　　11. $x = 0$, $x = 4$　　13. $x = 3$, $x = -\dfrac{5}{3}$　　15. $x = -2$, $x = 5$

17. $x = 3$, $x = 3$　　19. $x = -3$, $x = 9$　　21. $x = 6$, $x = -6$　　23. $x = 2$, $x = -\dfrac{5}{4}$

25. $x = 0$, $x = 6$　　27. $x = 0$, $x = 3$　　29. $x = 0$

EXERCISE 12-3 (Page 471)

1. $x = 3$, $x = -3$　　　　3. $x = 3$, $x = -3$　　　　5. $x = 1$, $x = -1$
7. $x = 3$, $x = -3$　　　　9. $x = 2\sqrt{3}$, $x = -2\sqrt{3}$　　11. $x = 2\sqrt{2}$, $x = -2\sqrt{2}$
13. $x = 2\sqrt{7}$, $x = -2\sqrt{7}$　　15. $x = 4\sqrt{2}$, $x = -4\sqrt{2}$

EXERCISE 12-4 (Page 476)

1. $x = -1$, $x = 3$　　3. $x = 1$, $x = -4$　　5. $x = 1$, $x = 4$　　7. $x = -2$, $x = -\dfrac{5}{2}$

9. $x = -1 + \sqrt{10}$, $x = -1 - \sqrt{10}$　　　11. $x = 3 + \sqrt{17}$, $x = 3 - \sqrt{17}$

13. $x = 3 + 2\sqrt{3}$, $x = 3 - 2\sqrt{3}$　　　15. $x = 2 + \sqrt{2}$, $x = 2 - \sqrt{2}$

17. $x = \dfrac{3 + \sqrt{15}}{2}$, $x = \dfrac{3 - \sqrt{15}}{2}$　　　19. $x = -\dfrac{1}{5}$, $x = -1$

EXERCISE 12-5 (Page 481)

1. $x = -3$, $x = 5$　　　　3. $x = -5$, $x = 7$　　　　5. $x = -1$, $x = -\dfrac{2}{3}$

7. $x = -2$, $x = \dfrac{1}{4}$　　　　9. $x = 2$, $x = -2$　　　　11. $x = \sqrt{3}$, $x = -\sqrt{3}$

13. $x = -3 + \sqrt{6}$, $x = -3 - \sqrt{6}$　　　15. $x = -1 + \sqrt{2}$, $x = -1 - \sqrt{2}$

17. $x = \dfrac{13 + \sqrt{137}}{4}$, $x = \dfrac{13 - \sqrt{137}}{4}$　　　19. $x = \dfrac{-1 + \sqrt{33}}{4}$, $x = \dfrac{-1 - \sqrt{33}}{4}$

21. $x = \dfrac{4 + \sqrt{6}}{5}$, $x = \dfrac{4 - \sqrt{6}}{5}$　　23. $x = \dfrac{3 + \sqrt{6}}{2}$, $x = \dfrac{3 - \sqrt{6}}{2}$　　25. $x = \dfrac{7 + \sqrt{41}}{2}$, $x = \dfrac{7 - \sqrt{41}}{2}$

27. $x = 1 + \sqrt{3}$, $x = 1 - \sqrt{3}$　　29. $x = 3$, $x = \dfrac{1}{3}$　　　31. $x = 3.8$, $x = -0.8$

33. $x = 1.6$, $x = -0.6$　　35. $x = 0.4$, $x = -1.7$

SUPPLEMENTARY EXERCISE 12-5 (Page 483)

1. $x = 0$, $x = 6$　　　　　　　3. $x = 3$, $x = -7$

5. $x = \dfrac{-5 + \sqrt{89}}{4}$, $x = \dfrac{-5 - \sqrt{89}}{4}$　　　7. $x = 0$, $x = -3$

9. $x = \dfrac{11 + \sqrt{21}}{10}$, $x = \dfrac{11 - \sqrt{21}}{10}$　　　11. $x = 4$, $x = -\dfrac{7}{2}$

13. $x = \dfrac{3 + \sqrt{7}}{2}$, $x = \dfrac{3 - \sqrt{7}}{2}$　　　15. $x = 4$

EXERCISE 12-6 (Page 490)

1. $y = x^2$

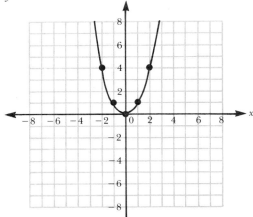

3. $y = 3x^2$

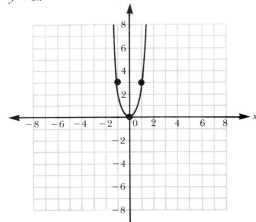

5. $y = x^2 - 1$

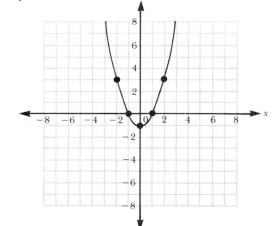

7. $y = x^2 - 2x$

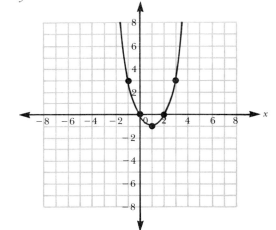

9. $y = 2x^2 - x$

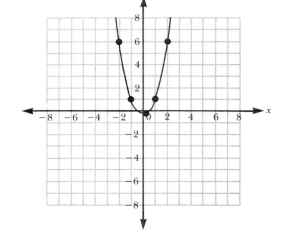

11. $y = x^2 - 2x + 2$

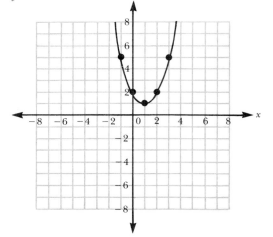

13. $y = x^2 - 6x + 8$

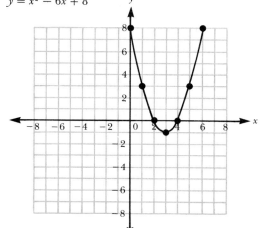

15. $y = 2x^2 + 4x - 5$

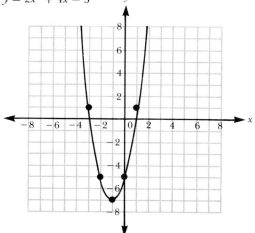

EXERCISE 12-7 (Page 494)

1. $x^2 - 25 = 0$

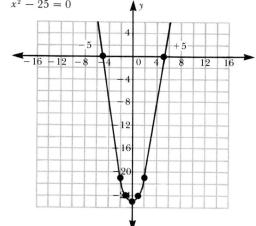

3. $x^2 - 4x = 0$

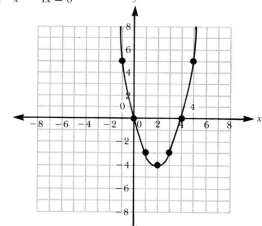

5. $x^2 - 2x - 8 = 0$

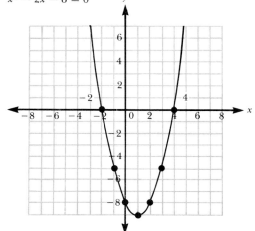

7. $x^2 - x - 1 = 0$

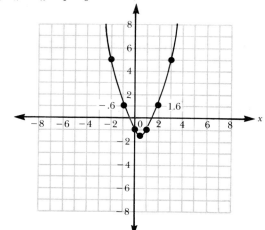

9. $2x^2 + 5x - 3 = 0$

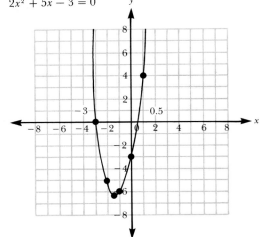

REVIEW EXERCISES (CHAPTER 12) (Page 494)

1. $2x^2 - 5x - 6 = 0$
2. $x^2 + 6x - 4 = 0$
3. $x^2 - 3x + 2 = 0$
4. $x = 0, \quad x = 3$
5. $x = 4, \quad x = -4$
6. $x = -2, \quad x = -4$
7. $x = 2, \quad x = 7$
8. $x = 2, \quad x = -6$
9. $x = -4, \quad x = -4$
10. $x = -3, \quad x = 7$
11. $x = 5, \quad x = -2$
12. $x = 0, \quad x = 0$
13. $x = 4, \quad x = -4$
14. $x = 2\sqrt{2}, \quad x = -2\sqrt{2}$
15. $x = 3\sqrt{2}, \quad x = -3\sqrt{2}$
16. $x = -1, \quad x = -3$
17. $x = 1, \quad x = 2$
18. $x = \dfrac{1}{2}, \quad x = -3$
19. $x = -3 + \sqrt{10}, \quad x = -3 - \sqrt{10}$
20. $x = 2, \quad x = 3$
21. $x = -2 + \sqrt{6}, \quad x = -2 - \sqrt{6}$
22. $x = 0.9, \quad x = -10.9$
23. $y = x^2 - 3x - 4$
24. $y = x^2 - 9$

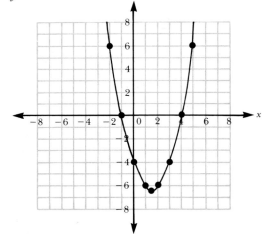

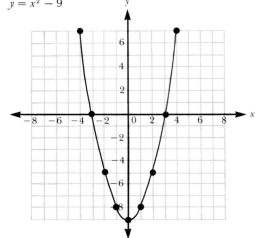

25. $y = x^2 - 3x - 1$

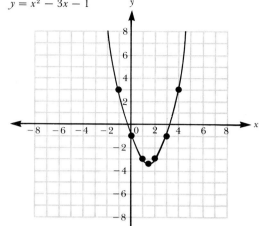

26. $x^2 + x - 6 = 0$

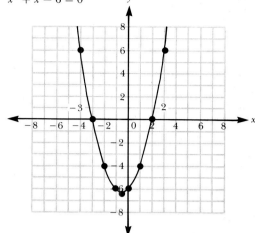

27. $5x^2 - 8x - 1 = 0$

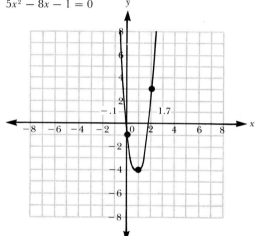

28. $x^2 + 2x - 7 = 0$

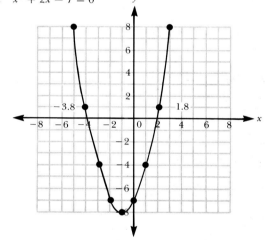

ANSWERS TO APPENDIX

EXERCISE A-1 (Page 500)

1. {m, u, s, k}

3. {Lincoln, Garfield, McKinley, Kennedy}

5. {1, 3, 9}

7. {Albany}

9. {red, orange, yellow, green, blue, indigo, violet}

11. {4}

13. {0}

15. {8}

17. {$x \mid x$ is a multiple of 5 less than or equal to 100}

19. {$x \mid x$ is an ocean}

21. True

23. True   25. False

27. True   29. False

EXERCISE A-2 (Page 503)

1. Finite   3. Infinite

5. Empty   7. Empty

9. Infinite   11. Equivalent

13. Neither   15. Equivalent and equal

EXERCISE A-3 (Page 506)

1. False   3. True   5. True   7. True   9. False
11. True   13. False   15. True   17. True   19. False
21. $\varnothing$, $\{2\}$, $\{8\}$, $\{9\}$, $\{2, 8\}$, $\{2, 9\}$, $\{8, 9\}$, $\{2, 8, 9\}$

EXERCISE A-4 (Page 509)

1. $\{1\}$ 3. $\{3, 7\}$
5. $\{1, 2, 3, 4, 5, 6, 7, 8\}$ 7. $\{3, 5, 6, 7, 9, 10\}$
9. $\{4, 5, 6, 8\}$ 11. $\{1, 2, 3, 5, 6, 7, 9, 10\}$
13. $\{\ \}$ or $\varnothing$ 15. $\{1, 4, 8\}$
17. $\{1, 3, 4, 5, 6, 7, 8, 9, 10\}$ 19. $\{5, 6\}$
21. $P \cap Q = \{3, 7\}$
    $P \cup Q = \{1, 2, 3, 4, 5, 7, 8, 9\}$

Index